AF806052

Lecture Notes in Computer Science

# Lecture Notes in Bioinformatics 14513

The series Lecture Notes in Bioinformatics (LNBI) was established in 2003 as a topical subseries of LNCS devoted to bioinformatics and computational biology.

The series publishes state-of-the-art research results at a high level. As with the LNCS mother series, the mission of the series is to serve the international R & D community by providing an invaluable service, mainly focused on the publication of conference and workshop proceedings and postproceedings.

Martina Vettoretti · Erica Tavazzi ·
Enrico Longato · Giacomo Baruzzo ·
Massimo Bellato
Editors

# Computational Intelligence Methods for Bioinformatics and Biostatistics

## 18th International Meeting, CIBB 2023
## Padua, Italy, September 6–8, 2023
## Revised Selected Papers

Springer

*Editors*
Martina Vettoretti
Università degli Studi di Padova
Padua, Italy

Erica Tavazzi
Università degli Studi di Padova
Padua, Italy

Enrico Longato
Università degli Studi di Padova
Padua, Italy

Giacomo Baruzzo
Università degli Studi di Padova
Padua, Italy

Massimo Bellato
Università degli Studi di Padova
Padua, Italy

ISSN 0302-9743 ISSN 1611-3349 (electronic)
Lecture Notes in Bioinformatics
ISBN 978-3-031-90713-5 ISBN 978-3-031-90714-2 (eBook)
https://doi.org/10.1007/978-3-031-90714-2

LNCS Sublibrary: SL8 – Bioinformatics



This Springer imprint is published by the registered company Springer Nature Switzerland AG
The registered company address is: Gewerbestrasse 11, 6330 Cham, Switzerland

# Preface

This volume contains selected scientific papers from the 18th edition of the International Conference on Computational Intelligence Methods for Bioinformatics and Biostatistics (CIBB 2023), held in Padua, Italy, on September 6–8, 2023.

Since its first edition, CIBB's main goal has been to provide a multidisciplinary forum for researchers interested in the application of computational intelligence, in a broad sense, to open problems in bioinformatics, biostatistics, systems and synthetic biology, and medical informatics. Researchers from the international scientific community interested in advancements and future perspectives in bioinformatics, biostatistics, and medical informatics gathered in Padua from all over the world, contributing to current trends and future opportunities at the interface between computer and life sciences.

The conference program included 72 oral presentations across one main track and seven special sessions on: 1) The many aspects of machine learning explainability in medical imaging; 2) Machine learning in healthcare informatics and medical biology; 3) Prediction uncertainty in machine learning; 4) Advanced statistical and computational methodologies for single-cell omics data; 5) An Italian snapshot on present and future informatics research in bioinformatics; 6) Distributed computing in bioinformatics and computational biology; and 7) Modelling and simulation methods for computational biology and systems medicine.

The conference hosted three eminent scientists in the field, who gave cutting-edge plenary keynote lectures. Blaž Zupan (University of Ljubljana, Slovenia) discussed with us the topic of "Data science for life scientists: should everyone learn programming or is there an alternative?"; Arianna Dagliati (University of Pavia, Italy) introduced us to "Developments in medical informatics: responsible approaches and reliable systems"; and Jessica Barrett (University of Cambridge, UK) presented methods for "Using electronic health records for healthcare research: Opportunities and challenges".

We also hosted three prestigious invited talks, which enriched the conference's special sessions: "Learning gene association networks using single-cell RNA-seq data: a graphical model approach" by Davide Risso (Università degli Studi di Padua, Italy); "Computational Reproducibility in the Life Sciences" by Marco Beccuti (Università degli Studi di Turin, Italy); and "A quick retrospective and prospective survey of simulation in biomedicine and cancer research" by Marco Antoniotti (Università degli Studi di Milano-Bicocca, Italy).

For the first time in CIBB's history, we also opened the floor to student initiatives, with the RSG-Italy Young Symposium hosted by the Italian Regional Student Group of the International Society for Computational Biology (RSG-Italy) and consisting of a roundtable and flash-talk sessions.

This issue of the conference proceedings received 23 submissions, all in the form of full papers, that were accepted after two rounds of peer review carried out by 3 reviewers per contribution, and after a final deliberation by the general and proceedings chairs.

More information about the full program is available on the conference website at https://cibb2023.dei.unipd.it/.

CIBB 2023 was sponsored by Novo Nordisk and Seeweb, and endorsed by the Department of Information Engineering of the University of Padua, the Italian Scientific Society of Biomedical Informatics (SIBIM), and the Veneto Region.

The conference chairs and the proceeding editors warmly thank the conference participants, the authors, the keynote and the invited speakers, the topic chairs Glen Philip Martin, Federica Eduati, and Roberto Gatta, the program committee members, the scientific committee, the local committee, the advisory board, the reviewers, the publicity chair, the CIBB 2021 general chair Davide Chicco, the special sessions and the RSG-Italy Young Symposium organisers, the CIBB steering committee, and anyone involved for their contributions to the success of CIBB 2023.

February 2025

Enrico Longato
Erica Tavazzi
Martina Vettoretti
Giacomo Baruzzo
Massimo Bellato

# Organization

## General Chairs

| | |
|---|---|
| Martina Vettoretti | University of Padua, Italy |
| Erica Tavazzi | University of Padua, Italy |
| Enrico Longato | University of Padua, Italy |

## Topic Chairs

| | |
|---|---|
| Glen Philip Martin | University of Manchester, UK |
| Federica Eduati | TU Eindhoven, The Netherlands |
| Roberto Gatta | Centre Hospitalier Universitaire Vaudois, Switzerland |

## Advisory Board

| | |
|---|---|
| Elia Biganzoli | University of Milan, Italy |
| Davide Chicco | University of Toronto, Canada |
| Barbara Di Camillo | University of Padua, Italy |
| Giovanni Sparacino | University of Padua, Italy |
| Roberto Tagliaferri | University of Salerno, Italy |

## Scientific Committee

| | |
|---|---|
| Keith Abrams | University of Warwick, UK |
| Helena Aidos | University of Lisbon, Portugal |
| Claudio Angione | Teesside University, UK |
| Fátima Sánchez Cabo | Centro Nacional de Investigaciones Cardiovasculares Carlos III, Spain |
| María Fernanda Cabrera-Umpiérrez | Universidad Politécnica de Madrid, Spain |
| Eleni Georga | University of Ioannina, Greece |
| Alberto Giaretta | University of Cambridge, UK |
| Niels Peek | University of Manchester, UK |
| Simone Spagnol | TU Delft, The Netherlands |

## CIBB Steering Committee Delegate

| | |
|---|---|
| Davide Chicco | University of Toronto, Canada |

## CIBB Steering Committee

| | |
|---|---|
| Roberto Tagliaferri | University of Salerno, Italy |
| Davide Chicco | University of Toronto, Canada |
| Francesco Masulli | University of Genoa, Italy |
| Elia Biganzoli | University of Milan, Italy |
| Clelia Di Serio | Università Vita-Salute San Raffaele, Italy |
| Pierre Baldi | University of California Irvine, USA |
| Alexandru Floares | Solutions of Artificial Intelligence Applications Institute, Romania |
| Jon Garibaldi | University of Nottingham, England |
| Nikola Kasabov | Auckland University of Technology, New Zealand |
| Leif Peterson | Rice University, USA |

## Keynote Speakers

| | |
|---|---|
| Blaž Zupan | University of Ljubljana, Slovenia |
| Arianna Dagliati | University of Pavia, Italy |
| Jessica Barrett | University of Cambridge, UK |

## Invited Speakers

| | |
|---|---|
| Davide Risso | University of Padua, Italy |
| Marco Beccuti | University of Turin, Italy |
| Marco Antoniotti | University of Milano-Bicocca, Italy |

## Special Session Organizers

| | |
|---|---|
| Salvatore Calderaro | University of Palermo, Italy |
| Carmelo Militello | National Research Council, Italy |
| Francesco Prinzi | University of Palermo, Italy |
| Filippo Vella | National Research Council, Italy |
| Davide Chicco | University of Toronto, Canada |
| Giuseppe Jurman | Fondazione Bruno Kessler, Italy |

| | |
|---|---|
| Giuseppe Agapito | Magna Græcia University of Catanzaro, Italy |
| Joao Ribeiro Pinto | Bosch, Portugal |
| Abbas Alameer | Kuwait University, Kuwait |
| Paolo A. Netti | Italian Institute of Technology, Italy |
| Filippo Causa | University of Naples Federico II, Italy |
| David Dannhauser | University of Naples Federico II, Italy |
| Annamaria Carissimo | National Research Council, Italy |
| Dario Righelli | University of Padua, Italy |
| Vincenzo Bonnici | University of Parma, Italy |
| Giacomo Baruzzo | University of Padua, Italy |
| Simone Pernice | University of Turin, Italy |
| Davide Chicco | University of Toronto, Canada |
| Umberto Ferraro Petrillo | Sapienza University of Rome, Italy . |
| Giuseppe Cattaneo | University of Salerno, Italy |
| Raffaele Giancarlo | Università di Palermo, Italy |
| Lorenzo Di Rocco | Sapienza University of Rome, Italy |
| Giulia Grani | Sapienza University of Rome, Italy |
| Chiara Damiani | University of Milano-Bicocca, Italy |
| Bruno Giovanni Galuzzi | University of Milano-Bicocca, Italy . |
| Vincenzo Bonnici | University of Parma, Italy |
| Marco Beccuti | University of Turin, Italy |
| Rosalba Giugno | University of Verona, Italy |
| Manuel Tognon | University of Verona, Italy |
| Simone Avesani | University of Verona, Italy |
| Eva Viesi | University of Verona, Italy |

## Publicity Chair

| | |
|---|---|
| Marta Lovino | University of Modena and Reggio Emilia, Italy |

## Local Committee

| | |
|---|---|
| Mikele Milia | University of Padua, Italy |
| Giacomo Baruzzo | University of Padua, Italy |
| Massimo Bellato | University of Padua, Italy |
| Giulia Cesaro | University of Padua, Italy |
| Isotta Trescato | University of Padua, Italy |
| Michele Atzeni | University of Padua, Italy |

| | |
|---|---|
| Francesca Longhin | University of Padua, Italy |
| Sara Poletto | University of Padua, Italy |
| Matteo Baldan | University of Padua, Italy |
| Alessandro Bernardi | University of Padua, Italy |
| Alessandro Guazzo | University of Padua, Italy |
| Ilaria Patuzzi | University of Padua, Italy |
| Gaia Tussardi | University of Padua, Italy |

## Program Committee

| | |
|---|---|
| Giuseppe Agapito | Magna Græcia University of Catanzaro, Italy |
| Helena Aidos | University of Lisbon, Portugal |
| Abbas Alameer | Kuwait University, Kuwait |
| Luca Alessandri | University of Turin, Italy |
| Claudia Angelini | National Research Council, Italy |
| Claudio Angione | Teesside University, UK |
| Antonino Aparo | University of Verona, Italy |
| Michele Atzeni | University of Padua, Italy |
| Riccardo Aucello | University of Turin, Italy |
| Sansanee Auephanwiriyakul | Chiang Mai University, Thailand . |
| Simone Avesani | University of Verona, Italy |
| Daniele Baccega | University of Turin, Italy |
| Matteo Baldan | University of Padua, Italy |
| Ileana Baldi | University of Padua, Italy |
| Krzysztof Bartoszek | Linköping University, Sweden |
| Giacomo Baruzzo | University of Padua, Italy |
| Petra Baumann | Medical University of Graz, Austria |
| Marco Beccuti | University of Turin, Italy |
| Massimo Bellato | University of Padua, Italy |
| Pietro Belloni | University of Padua, Italy |
| Anna Bernasconi | Politecnico di Milano, Italy |
| Gilles Bernot | Université Côte d'Azur, France |
| Giovanni Birolo | University of Turin, Italy |
| Vincenzo Bonnici | University of Parma, Italy |
| Pietro Bosoni | University of Pavia, Italy |
| Davide Bressan | University of Trento, Italy |
| Maria Fernanda | Universidad Politécnica de Madrid, Spain |
| Cabrera-Umpierrez | Universidad Politécnicade Madrid, Spain |
| Salvatore Calderaro | University of Palermo, Italy |
| Nunzio Camerlingo | Pfizer, Inc., USA |

| | |
|---|---|
| Andrea Campagner | University of Milano-Bicocca, Italy |
| Giacomo Cappon | University of Padua, Italy |
| Annamaria Carissimo | National Research Council, Italy |
| Giuseppe Cattaneo | University of Salerno, Italy |
| Filippo Causa | University of Naples Federico II, Italy |
| Paolo Cazzaniga | University of Bergamo, Italy |
| Matteo Cesari | Medical University of Innsbruck, Austria |
| Giulia Cesaro | University of Padua, Italy |
| Yuwen Chen | University of Cambridge, UK |
| Davide Chicco | University of Toronto, Canada |
| Giovanni Cicceri | University of Palermo, Italy |
| Matteo Comin | University of Padua, Italy |
| Rosanna Comoretto | University of Turin, Italy |
| Corrado Coppola | Sapienza University of Rome, Italy |
| Anna Livia Croella | Sapienza University of Rome, Italy |
| Fabio Cumbo | Cleveland Clinic, USA |
| Luisa Cutillo | University of Leeds, UK |
| Chiara Damiani | University of Milano-Bicocca, Italy |
| David Dannhauser | University of Naples Federico II, Italy |
| Pierluigi Francesco De Paola | National Research Council, Italy . |
| Barbara Di Camillo | University of Padua, Italy |
| Giorgio Maria Di Nunzio | University of Padua, Italy . |
| Lorenzo Di Rocco | Sapienza University of Rome, Italy |
| Guglielmo Faggioli | University of Padua, Italy |
| Piero Fariselli | University of Turin, Italy |
| Carlo Ferrari | University of Padua, Italy |
| Umberto Ferraro Petrillo | Sapienza University of Rome, Italy . |
| Christoph M. Friedrich | University of Applied Sciences and Arts Dortmund, Germany |
| Alessio Funari | Sapienza University of Rome, Italy |
| Bruno Giovanni Galuzzi | University of Milano-Bicocca, Italy . |
| Gennaro Gambardella | Telethon Institute of Genetics and Medicine, Italy |
| Roberto Gatta | University of Brescia, Italy |
| Francesco Gentile | University of Catanzaro, Italy |
| Eleni Georga | University of Ioannina, Greece |
| Filippo Geraci | National Research Council, Italy |
| Raffaele Giancarlo | University of Palermo, Italy |
| Alberto Giaretta | University of Cambridge, UK |

| | |
|---|---|
| Rosalba Giugno | University of Verona, Italy |
| Yair Goldberg | Technion - Israel Institute of Technology, Israel |
| Giorgio Grani | Sapienza University of Rome, Italy |
| Alessandra Grossi | University of Milano-Bicocca, Italy |
| Alessandro Guazzo | University of Padua, Italy |
| Henrik Imberg | University of Gothenburg, Sweden |
| Giuseppe Jurman | Fondazione Bruno Kessler, Italy |
| Corrado Lanera | University of Padua, Italy |
| Youngro Lee | Seoul National University, South Korea |
| Nicola Licheri | University of Turin, Italy |
| Giosue Lo Bosco | University of Palermo, Italy |
| Enrico Longato | University of Padua, Italy |
| Francesca Longhin | University of Padua, Italy |
| Marta Lovino | University of Modena and Reggio Emilia, Italy |
| Zeinab Mahmoudi | Novo Nordisk, Denmark |
| Stefano Marchesin | University of Padua, Italy |
| Glen Martin | University of Manchester, UK |
| Andreia Martins | University of Lisbon, Portugal |
| Francesca Marturano | Harvard University, USA |
| Carlotta Maschiocchi | Fondazione Policlinico Universitario Gemelli, Italy |
| Giancarlo Mauri | University of Milano-Bicocca, Italy |
| Claudia Mengoni | University of Verona, Italy |
| Mikele Milia | University of Padua, Italy |
| Carmelo Militello | National Research Council, Italy |
| Alexander Monzon | University of Padua, Italy |
| Margherita Mutarelli | National Research Council, Italy |
| Paolo Antonio Netti | Italian Institute of Technology, Italy |
| Shahryar Noei | Fondazione Bruno Kessler, Italy |
| Lisa Novello | Fondazione Bruno Kessler, Italy |
| Stefania Orini | University of Brescia and IRCCS Istituto Centro San Giovanni di Dio Fatebenefratelli, Italy |
| Alessia Paglialonga | National Research Council, Italy |
| Francesco Palini | Sapienza University of Rome, Italy |
| Niels Peek | University of Manchester, UK |
| Leonardo Pellegrina | University of Padua, Italy |
| Marzio Pennisi | University of Eastern Piedmont, Italy |
| Simone Pernice | University of Turin, Italy |
| Wellington Pinheiro Dos Santos | Universidade Federal de Pernambuco, Brazil . |
| Cinzia Pizzi | University of Padua, Italy |
| Marco Podda | University of Pisa, Italy |

| | |
|---|---|
| Francesco Prendin | University of Padua, Italy |
| Francesco Prinzi | University of Palermo, Italy |
| João Ribeiro Pinto | INESC TEC and University of Porto, Portugal |
| Dario Righelli | University of Padua, Italy |
| Riccardo Rizzo | National Research Council, Italy |
| Aurora Saibene | University of Milano-Bicocca, Italy |
| Tiziana Sanavia | University of Turin, Italy |
| Ilie Sarpe | University of Padua, Italy |
| Fabio Scarpa | University of Padua, Italy |
| Rakesh Shiradkar | Emory University, USA |
| Pasquale Sibilio | Sapienza University of Rome, Italy |
| Gianmaria Silvello | University of Padua, Italy |
| Dario Simionato | University of Padua, Italy |
| Diogo Soares | University of Lisbon, Portugal |
| Andrea Sottosanti | University of Padua, Italy |
| Simone Spagnol | Iuav University of Venice, Italy |
| Antonino Staiano | University of Naples Parthenope, Italy |
| Roberto Tagliaferri | University of Salerno, Italy |
| Erica Tavazzi | University of Padua, Italy |
| Michele Tebaldi | University of Verona, Italy |
| Irene Terrone | University of Turin, Italy |
| Jacopo Tessadori | Fondazione Bruno Kessler, Italy |
| Manuel Tognon | University of Verona, Italy |
| Mirko Treccani | University of Verona, Italy |
| Isotta Trescato | University of Padua, Italy |
| Filippo Utro | IBM, USA |
| Filippo Vella | National Research Council, Italy |
| Laura Veschetti | University of Verona, Italy |
| Martina Vettoretti | University of Padua, Italy |
| Eva Viesi | University of Verona, Italy |
| Veronica Vinciotti | University of Trento, Italy |

## Reviewers

| | |
|---|---|
| Giuseppe Agapito | Magna Græcia University of Catanzaro, Italy |
| Michele Atzeni | University of Padua, Italy |
| Eleonora Auletta | University of Lisbon, Portugal |
| Simone Avesani | University of Verona, Italy |
| Matteo Baldan | University of Padua, Italy |
| Mattia Ballo | University of Padua, Italy |
| Vincenzo Bonnici | University of Parma, Italy |

| | |
|---|---|
| Pietro Bosoni | University of Pavia, Italy |
| Salvatore Calderaro | University of Palermo, Italy |
| Nunzio Camerlingo | Pfizer, Inc., USA |
| Andrea Campagner | University of Milano-Bicocca, Italy |
| Filippo Causa | University of Naples Federico II, Italy |
| Paolo Cazzaniga | University of Bergamo, Italy |
| Marina Ceccon | University of Padua, Italy |
| Giulia Cesaro | University of Padua, Italy |
| Davide Chicco | University of Milano-Bicocca, Italy |
| Giovanni Cicceri | University of Palermo, Italy |
| Giovanna Cilluffo | University of Palermo, Italy |
| Chiara Cimolato | University of Padua, Italy |
| Vasco Coelho | University of Milano-Bicocca, Italy |
| Debora Dallera | University of Pavia, Italy |
| Chiara Damiani | University of Milano-Bicocca, Italy |
| David Dannhauser | University of Napoli Federico II, Italy |
| Davide De Marchi | University of Pavia, Italy |
| Davide Dei Cas | University of Padua, Italy |
| Bruno Giovanni Galuzzi | University of Milano-Bicocca, Italy |
| Giulia Grani | Sapienza University of Rome, Italy |
| Alessandro Guazzo | University of Padua, Italy |
| Qilin Hong | University of Wisconsin-Madison, USA |
| Elena Idi | University of Padua, Italy |
| Wei Liu | Fujian Agriculture and Forestry University, China |
| Enrico Longato | University of Padua, Italy |
| Francesca Longhin | University of Padua, Italy |
| Stefano Maestri | University of Camerino, Italy |
| Elena Marinello | University of Padua, Italy |
| Glen Martin | University of Manchester, UK |
| Francesca Marturano | Harvard University, USA |
| Nicola Melillo | University of Pavia, Italy |
| Mikele Milia | University of Padua, Italy |
| Giovanna Nicora | University of Pavia, Italy |
| Marco Nobile | Ca' Foscari University of Venice, Italy |
| Shahryar Noei | Fondazione Bruno Kessler, Italy |
| Daniele Maria Papetti | University of Milano-Bicocca, Italy |
| Leonardo Pellegrina | University of Padua, Italy |
| Simone Pernice | University of Turin, Italy |
| Michela Quadrini | University of Camerino, Italy |
| Aurora Saibene | University of Milano-Bicocca, Italy |
| Gabriele Sales | University of Padua, Italy |

| | |
|---|---|
| Roman Shaposhnikov | University of Milano-Bicocca, Italy |
| Simone Spagnol | Iuav University of Venice, Italy |
| Chaimaa Tarzi | Teesside University, UK |
| Erica Tavazzi | University of Padua, Italy |
| Mirko Treccani | University of Verona, Italy |
| Anne Tromelin | Centre INRAE Bourgogne-Franche-Comté, France |
| Filippo Vella | National Research Council of Italy, Italy |
| Suraj Verma | Teesside University, UK |
| Eva Viesi | University of Verona, Italy |

## CIBB Founders

| | |
|---|---|
| Roberto Tagliaferri | University of Salerno, Italy |
| Francesco Masulli | University of Genoa, Italy |
| Antonina | University of Pisa, Italy |

## CIBB Past Editions

CIBB 2021, Virtual
CIBB 2019, Bergamo, Italy
CIBB 2018, Caparica, Portugal
CIBB 2017, Cagliari, Italy
CIBB 2016, Stirling, UK
CIBB 2015, Naples, Italy
CIBB 2014, Cambridge, UK
CIBB 2013 (co-organized with PRIB 2013), Nice, France
CIBB 2012, Houston, USA
CIBB 2011, Gargnano, Italy
CIBB 2010, Palermo, Italy
CIBB 2009, Genoa, Italy
CIBB 2008, Vietri sul Mare, Italy
CIBB 2007 (within WILF 2007), Camogli, Italy
CIBB 2006 (within FLINS 2006), Genoa, Italy
CIBB 2005 (within WILF 2005), Crema, Italy
CIBB 2004 (within WIRN 2004), Perugia, Italy

# Contents

# A Network Approach to Aquatic Food Web Dynamics

Davide Torre[1](✉), Giuseppe F. Italiano[1,2], and Blerina Sinaimeri[1,2]

[1] Luiss Guido Carli, Viale Romania, 32, 00197 Rome, RM, Italy
{dtorre,gitaliano,bsinaimeri}@luiss.it
[2] Erable Team, Inria Lyon, Lyon, France

**Abstract.** Network-based methods have gained significant popularity in ecological research over recent decades. However, network analysis seems to have been restricted mostly to some specific interactions, such as plant-pollinators or host-parasites, with food webs receiving less attention.

Food webs represent feeding interactions among species in a habitat revealing how species depend on each another for survival. As such, they are essential for understanding ecosystems and predicting their evolution. Among food webs, aquatic habitats are considered one of the most vulnerable ecosystems, primarily due to factors such as climate change and the introduction of non-native species. Understanding the robustness of these networks to species removal and identifying pivotal nodes is essential to mitigate the risk of key species extinction and preserve ecosystem stability.

**Keywords:** Food webs · Robustness · Core and periphery structure · Motif representation

## 1 Introduction

Ecological networks provide valuable insights into the complex interactions among species of an ecosystem (see, *e.g.*, [12,14,23]). Although over the last few decades the use of traditional network-based methods has gained popularity in ecological research, network analysis seems to have been restricted mostly to some specific interactions, such as plant-pollinators [13,27,28] or host-parasites [35,37]. In this scenario, further network-based research could contribute to a deeper understanding of ecological networks and of their interactions. To address this issue, in this work we investigate the structural properties of (aquatic) food webs. Food webs depict the flow of energy and matter between different organisms in an ecosystem, and they reveal how species depend on each another for survival. This structure can provide important information about the

The work of Giuseppe F. Italiano and Blerina Sinaimeri was partially supported by MUR of Italy, under PRIN Project n. 2022TS4Y3N - EXPAND: scalable algorithms for EXPloratory Analyses of heterogeneous and dynamic Networked Data.


M. Vettoretti et al. (Eds.): CIBB 2023, LNBI 14513, pp. 1–15, 2025.
https://doi.org/10.1007/978-3-031-90714-2_1

potential impacts of species loss or invasions. Understanding food webs seems thus essential for better understanding the intricate relationships and interactions that occur within the entire ecosystem.

Our study considered 173 aquatic (*e.g.*, lakes, marine, shallow waters) food webs. We first focused on identifying structural patterns within these networks, such as a core and a peripheral structure. A better understanding of those patterns can help us shed more light on the organization of an ecological system and on its resilience to perturbations [10,11,28]. Next, we study their robustness with respect to node deletions, revealing which species are holding the network together and are thus crucial for maintaining its underlying structure. Finally, we analyze the motif representation (i.e., patterns of interconnections occurring in complex networks at numbers significantly higher than those in random networks [29]) of these food web to uncover their mesoscopic structure.

To the best of our knowledge, this study represents the largest network analysis ever done on an aquatic dataset, with as many as 173 aquatic food webs. The code as well as the dataset employed for this study are available at https://github.com/davidetorre92/AquaNet.

## 2 Data

For this study, we collected a dataset of 173 aquatic food webs, 18 of these were taken from `igraphdata` [9] library, 12 were manually extracted from [7] and 2 directly from the published papers [20,25]. The remaining 141 were downloaded from the Ecopath With Ecosim repository [8][1]. All the food webs are available in XML `graphml` format on https://github.com/davidetorre92/AquaNet/ and a summary of the most representative metrics are provided in Table 1 in the Supplementary Material.

To model food webs, we focus on the presence or absence of trophic interactions, thereby creating an unweighted directed network. Nodes within the network represent compartments, which refers either to a community of species with similar feeding relations or non-living organic deposits (e.g., detritus, particulate organic carbon, dissolved organic carbon). We refer to the nodes representing a living community of species as *living compartments* and the organic deposits as *non-living compartments*. The directed links in the network represent the transfer of biomass among compartments. More specifically an edge from compartment $c$ to compartment $c'$ indicates that $c'$ preys upon $c$. We also account for the phenomenon of intraspecific predation (*i.e.*, cannibalism), by incorporating self-loops in the network. Self-loops are links that connect a node back to itself, indicating cases where a species preys upon its own members. Throughout, we use the notation $S_G$ and $L_G$ to denote the number of nodes and links in the network $G$, respectively. When $G$ is clear from the context, we will use $S$ and $L$ instead of $S_G$ and $L_G$.

[1] To download the dataset from this repository, we used the API available at https://ecobase.ecopath.org/.

Table 1 summarizes the key statistical measures of these aquatic food webs and Fig. 1 illustrates the geographic locations from which these webs originate.

**Table 1.** Summary of Food Web Metrics

| Metric | Mean | Std Dev | Min | Max |
|---|---|---|---|---|
| Number of Compartments ($S$) | 41.77 | 23.12 | 18 | 182 |
| Number of Trophic Interactions ($L$) | 424.94 | 448.04 | 55 | 2614 |
| Connectance ($C = L/S^2$) | $2.27 \times 10^{-1}$ | $7.39 \times 10^{-2}$ | $4.02 \times 10^{-2}$ | $4.08 \times 10^{-1}$ |
| Percentage of Non-living Compart. | 3.93% | 2.01% | 0.55% | 14.3% |

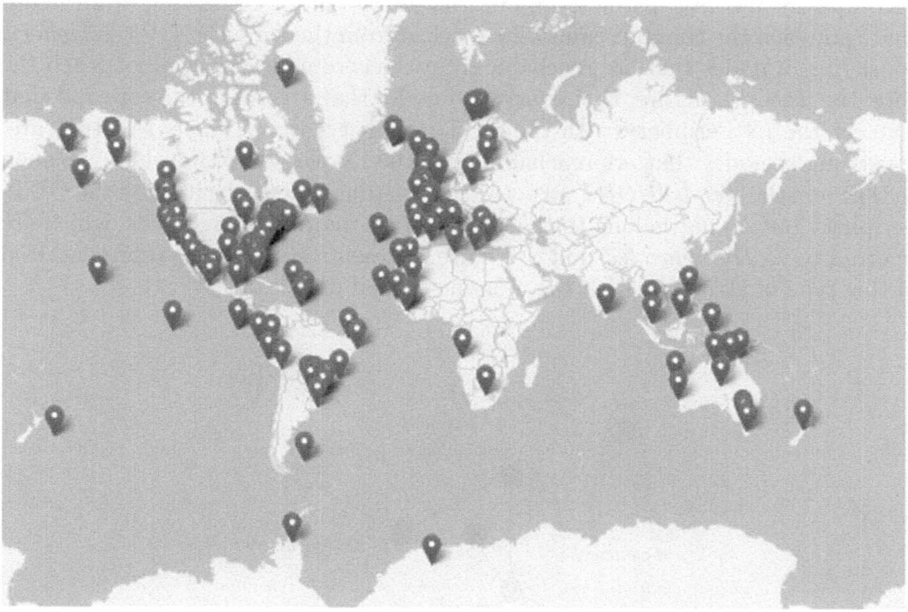

**Fig. 1.** Global map depicting the approximate locations of 173 food webs. Markers found in a continental zone indicate a food web that relates to one of the coasts of that region.

## 3 Core-Periphery Structure

The core-periphery structure is a basic pattern in a network, marked by two distinct components: a "core" with closely linked nodes, and a "periphery" where nodes are loosely connected, both to the core and to each other [18]. In ecological networks, the core-periphery structure is linked to the concept of nestedness (see, *e.g.*, [2,3,28]). This arrangement plays an important role in the network adaptability, enabling them to cope with environmental variations and internal disturbances. To identify this structure in aquatic food webs, we classified *core* nodes as those in the largest strongly connected component of the graph, and all other nodes as the *periphery*. A strongly connected component is a subset of nodes where each node can reach and is reachable from any other along a directed path. We focus on strongly connected components as they represent subset of nodes that are connected by directed cycles. The existence of directed cycles is related to the stability of the network as it allows for alternative pathways of energy flow and resource exchange, enhancing the system's resilience to disturbances. Consequently, the presence of strongly connected pairs within a food web, indicating their involvement in directed cycles, assumes a crucial role in determining ecosystem stability.

Following a standard terminology (see, e.g., [5]), the periphery can be further decomposed into five distinct subsets: the *IN-periphery*, consisting of all nodes that can reach the core but cannot be reached from the core; the *OUT-periphery*, consisting of nodes that are reachable from the core nodes but cannot reach the core; the *Tubes* structure, that consists of nodes that are not in the core and that connect the IN-Periphery with the OUT-periphery; the *Tendrils-IN* structure, consisting of nodes that are reachable from the IN-periphery, but do not belong to the core; the *Tendrils-OUT* structure, consisting of nodes that reach the OUT-periphery but do not belong to the core; and all the remaining nodes, which are referred to as *Disconnected* nodes. Figure 2 presents a schematic representation of this type of structure with the aforementioned definitions.

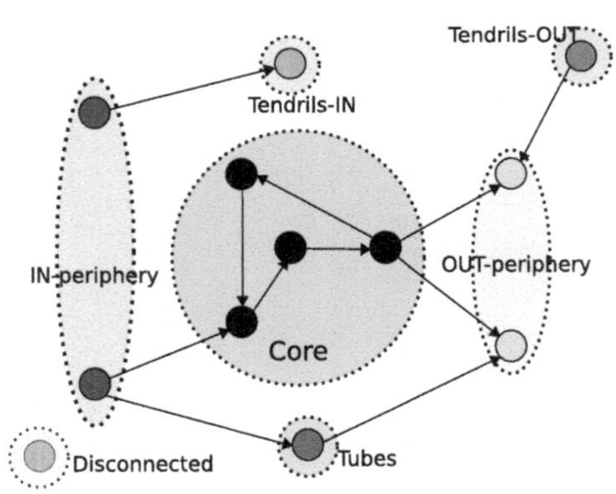


**Fig. 2.** A sketch of the core-periphery structure. Core nodes belong to the giant strongly connected components, while other nodes belong to the periphery.

We computed the core-periphery structure for all the webs in our dataset. In Fig. 3, we present for each food web the proportion of the nodes in the core and each peripheral structure. Our analysis shows that the networks in our dataset share a very similar structure: a large core in which compartments are

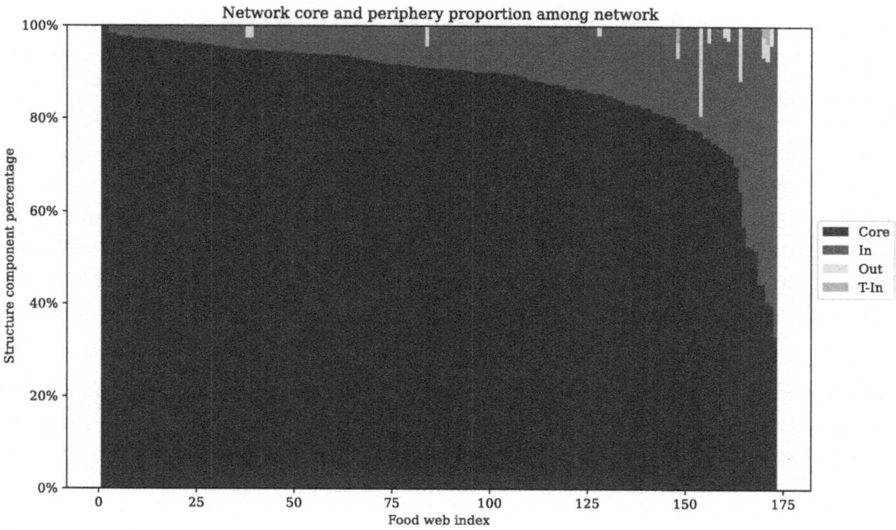


**Fig. 3.** Proportion of core periphery relative to the food web size. The x-axis indicates the index of the food web, while the y-axes indicates the proportion of the nodes belonging to the the relative structure component. The food webs in the dataset are sorted according to their core size proportion.

highly interconnected and a smaller IN-periphery that carries biomass to the core nodes. The nodes in the IN-periphery correspond mainly to producers and are crucial because they are the foundation of the energy pyramid, supporting all other trophic levels above them. On the other side, nodes in the OUT-periphery represent consumers that feed directly on the core of the food web. They can also include decomposers, which play a critical role in breaking down organic matter and recycling nutrients back to the producers.

The other peripherical structures: Tendrils-IN, Tendrils-OUT and Tubes contain species that neither reach nor are reached by the core of the network and thus do not seem to have a strong overall influence on the network. In our datasets only 3 food webs contain Tendrils-IN structure and none of the them contains Tubes and Tendrils-OUT. The three food webs that contain Tendrils-IN along with the species that form these structure, are as follows: in the Lower Chesapeake Bay, rotifers feeding on phytoplankton; in the Tasek Bera swamp of Malaysia, swallows feeding on various organisms of the IN-periphery; and in the Salt meadow, New Zealand, redpolls, feeding on seeds. Typically, these species can be thought as predators that occasionally feed within the system but primarily operate externally (e.g., a bird that mostly feeds in one ecosystem but nests in another).

Furthermore, it is interesting to see how the network structure and the structural position of species within the network adapt to different changes of the environment that may occur over time [28]. Within our datasets, we have 22 networks for which the available information spans different contexts: for some networks, data is collected across various years, and for others, we have data related to both the wet and dry seasons. Our analysis showed that the core-periphery structure of these datasets remained consistent over time, with core species staying in the core and peripheral species remaining on the periphery across different years and seasons. The persistence of species positions suggests that these ecosystems possess a high level of ecological resilience. Despite seasonal fluctuations and potential disturbances in particular associated with wet and dry periods, the core species maintain their central roles, ensuring the stability and functionality of the ecosystem.

**Horizontal and Vertical Diversity of Core and Periphery.** Biodiversity is another crucial factor affecting the functionality and stability of food webs. The biodiversity of food webs is composed of horizontal (*i.e.*, within trophic levels) and vertical diversity (*i.e.*, the number of trophic levels) [15,38] and they are believed to have opposing effects on stability. Vertical diversity tends to reduce stability in simple food chains by lengthening recovery times, suggesting why natural food webs have few trophic levels [26,33]. Whereas, horizontal generality, which usually considers the measure of generality and vulnerability of the species, is believed to influence positively the stability of the network [34]. Indeed, generalists, with their broad diets, can provide robustness to a food web by filling ecological gaps when certain species are removed or environmental conditions change.

We define now formally these concepts. Consider a food web represented by its adjacency matrix $a$, that is $a_{ij} = 1$ if the is a directed edge from $i$ to $j$, and $a_{ij} = 0$ otherwise. Given a species $i$, the amount $\sum_{j=1}^{S} a_{ji}$, is an indicator of the generality of $i$ and measures the breadth of its diet, that is, the set of relationships that a consumer has with its prey. The vulnerability of $i$, is the amount $\sum_{j=1}^{S} a_{ij}$, and measures the number of consumers that prey upon this species. Note that in graph-theoretical terms, the generality of a node corresponds to its in-degree, while the vulnerability corresponds to its out-degree. To be able to compare these quantities across different networks, they are usually normalized by the average number of links per species $L/S$. Formally, to measure the generality and vulnerability of a species $i$ in the food web $G$ we use the following definitions provided in [39]: $Gen_i = \frac{S}{L} \sum_{j=1}^{S} a_{ji}$ and $Vul_i = \frac{S}{L} \sum_{j=1}^{S} a_{ij}$.

A species $i$ is a generalist if $Gen_i > 1$ and non-generalist otherwise, and vulnerable if $Vul_i > 1$ and non-vulnerable otherwise. The trophic level $TL_i$ of the species $i$ is defined recursively as in [40]:

$$TL_i = 1 + \frac{1}{\sum_j a_{ji}} \sum_j a_{ji} TL_j \tag{1}$$

Notice that, the matrix $a$ can be a weighted, with $a_{ji}$ denoting the fraction that species $j$ constitutes within the diet of species $i$. The base case sets the trophic level to 1 for primary producers, which have no incoming edges, and also assigns by convention a level of 1 to non-living compartments [8, 40].

To better understand the food web robustness we explore the vertical and horizontal diversity of the core and periphery of food networks. In examining the horizontal diversity, we focus on how generalist and vulnerable living species are distributed across the cores and peripheries of each food web. Figure 4a illustrates that in the core of food webs, the proportions of generalist and vulnerable species are quite similar, with generalists making up an average of 42.4% and vulnerable species 39.8%. Conversely, Fig. 4b reveals that, with few exceptions, food web peripheries predominantly consist of vulnerable nodes. Specifically, on average, peripheries contain 1.8% generalist species and 48.1% vulnerable nodes. This is in accordance with what is observed in other ecological networks where generalist nodes of the network predominantly inhabit the core of ecological networks, emphasising their crucial role in ecosystem stability [1, 30].

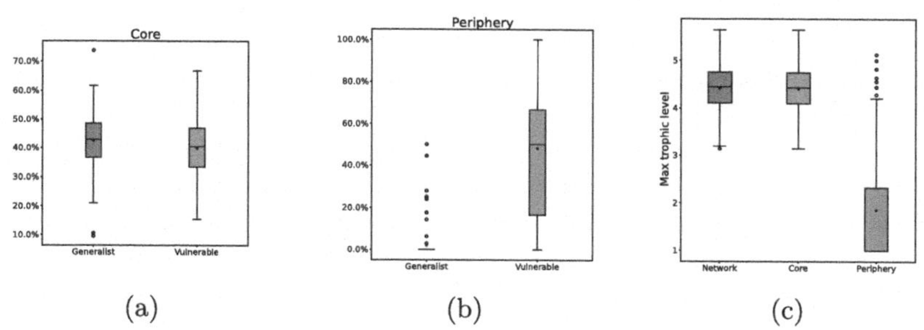


**Fig. 4.** Proportion of generalist and vulnerable species in: (c) the core and b) the periphery of food webs. (c) Distribution of maximum trophic level inside food webs. The symbol "+" denotes the mean value.

Regarding vertical diversity in each food web, we calculated the maximum trophic level across the entire food web, as well as within its core and periphery. Figure 4c presents the distribution of these maximum trophic levels. The figure distinctly shows that species with the highest trophic levels are predominantly found in the core, whereas those with lower trophic levels belong to the periphery.

## 4 Critical Nodes and Robustness

In this section we evaluate the impact of nodes in the stability of a food web. Since in general this problem is hard (see, e.g., [31]), this evaluation is done empirically, by assessing how much the removal of a node (corresponding to the disappearance of a species in the ecological network) disrupts the underlying structure of the network. This provides also a notion of robustness of the network: if removing few nodes has no noticeable impact, then the network structure is robust. On the other hand, if removing few nodes quickly affects the underlying structure of the network, then the network is more fragile. Currently, the most used strategies in the literature to find the sequence of the most critical species are based on: (1) the most connected species; (2) randomly chosen species; and (3) the least connected species [16,17]. Our approach differs from that seen in the past because we look for which species have a disruptive effect on the entire network, rather than picking a sequence based upon the *local* features of a node. To identify the presence of such species, we searched for that sequence of nodes that results in the highest minimisation of the *reachable pair fraction* upon their removal. This quantity is defined as [4]:

$$r(G) := R(G)/S^2 \tag{2}$$

where $R(G)$ are the reachable pairs of the graph $G$, *i.e.*, the number of pairs $(u, v)$ such that there exists a directed path from $u$ to $v$, and $S$ is the number of nodes of the graph $G$. This quantity is normalized after dividing by the number of all possible pairs of nodes.

**Notion of Robustness.** In the context of mutualistic ecological networks, Sheykhali et al. [35] provided a notion of robustness by measuring the area under the curve of the relative size of the largest connected component vs. the fraction of nodes that have been removed in the web. Inspired by this idea, we established a notion of robustness by measuring the area under the $r(G)$ vs. the fraction of removed nodes. Consider a sequence of nodes in a network, denoted as $CSeq_G = \{s_1, \ldots, s_i, \ldots, s_{|S_G|}\}$. Thus, we define

$$\rho_G = \frac{1}{|S_G|} \sum_i r(G^{(i)}) \tag{3}$$

where $G^{(i)}$ is the graph obtained after removing from $G$ the edges incident to the vertices $\{s_1, \ldots, s_i\}$. We set $G^0 = G$. Notice that, in this context, we are looking for a sequence of critical nodes of living compartments. Therefore, $S_G$ in Eq. 3 is the number of living compartments of the graph $G$ and the sequence of nodes $CSeq_G$ is a set of nodes containing all living compartments of the graph $G$.

**Description of the Algorithm for Computing $\rho_G$.** To determine the robustness $\rho_G$ of a food web $G$, we need to find the sequence of most critical nodes. More specifically, given a food web $G = (S_G, L_G)$, the goal is to find a sequence of nodes from $S_G$, $CSeq_G$, whose removal minimizes $\rho(G)$. We proceed in a greedy manner. Starting from an empty sequence of nodes $CSeq_G$, for each node $s$ of $G$, we measure $r(G^{(1)})$ where $G^{(1)}$ is obtained from $G$ by removing the links that are incident to $s$. We add to $CSeq_G$, the node that minimizes the quantity $r(G^{(1)})$. In case of ties, i.e., if there is more than one node minimizing $r(G^{(1)})$, we choose the node with the lowest trophic index (Eq. 1). This choice is motivated by the fact that, intuitively, we expect that a node with lower trophic index will be more critical for the overall network robustness. We continue this process until there are no more links incident to the nodes representing living compartments. Once the sequence is set, we compute $\rho_G$ accordingly using Eq. 3.

We computed the robustness of all the food webs in our dataset. Table 2 presents the networks arranged in descending order of robustness. Due to space limitations, we display only the top three networks and the bottom three networks in the table. The complete ranking can be found in the Supplementary Material (Table 3). Furthermore, for each network, we provide the first three nodes found in the sequence, corresponding to the top three most critical nodes according to our algorithm.

**Table 2.** The food webs are sorted in non-increasing order of robustness, with the highest values corresponding to the top rows. The first column corresponds to the rank in the ordering. The third column contains the sequence of the most critical nodes.

| Rank | Food web | $[s_1^G, s_2^G, s_3^G]$ |
|---|---|---|
| 1 | Sand beach, South Africa | [Gastrosaccus, Callianassa, Donax; bacteria; Cumacea] |
| 2 | Sítios Novos reservoir (2011) | [Camarão, Cladocera, Copepoda] |
| 3 | Terminos Lagoon (1980) | [Meiofauna, Polychaetes, Zooplankton] |
| ... | ... | ... |
| 171 | Chesapeake Bay Mesohaline | [bacteria in sediment poc, zooplankton, blue crab] |
| 172 | St. Marks River (Florida) | [Benthic bact, Bacterio plankton, Micro protozoa] |
| 173 | Salt meadow, New Zealand | [amphipods, collembola, mites] |

The measure of robustness of a food web appears to be positively correlated (Pearson correlation = 0.60) with the proportion of nodes belonging to the core, (and similarly, negatively correlated with the proportion of nodes belonging to the periphery). This measure is not directly correlated with other simple food web metrics, such as the number of species, the number of links, or the maximum trophic level. This is interesting as it suggests that this robustness captures an aspect of food web structure not directly reflected by these traditional measures. Moreover, the critical nodes identified by our algorithm for computing the food web robustness also appear to be biologically significant for its stability. In the Sand beach, South Africa food web shrimp-like species and clams significantly contribute to the biota, as demonstrated by their rich concentration, serving as vital food sources for many fish species [6]. In the St. Mark River food web, benthic microbial communities are crucial for sustaining connectivity by driving organic matter decomposition, nutrient regeneration, and influencing dissolved oxygen levels [22]. Similarly on Salt meadow, New Zealand food web, small arthropods like amphipods, collembola and mites which occupy both aquatic and terrestrial habitats, have a significant role in the ecosystem, as either detritivores or herbivores [24,32].

Finally, in robust food webs, the top critical nodes exhibit minimal variability in trophic levels and predominantly belong to higher trophic levels. Conversely, in the least robust food webs, critical nodes display greater variability, and correspond to lower trophic levels. Figure 2 in the Supplementary Material shows the normalized trophic level of the top 5 (i.e. the trophic level of a species mapped in the range [0, 1] relative to that food web) for each food web, sorted by robustness. The 5 critical nodes in the most robust webs are in the highest trophic level, while the least robust are in the lowest trophic level and have the greatest variability.

## 5 Representation of Three Nodes Motifs

We further investigated the structural pattern of the aquatic food web by considering three nodes food webs motifs, i.e., connected subgraphs containing interactions of three nodes [29,36]. More specifically, we quantified the over- and under-representation of 13 distinct motifs when neglecting cannibalism. The complete set of 3 nodes motifs is shown in Fig. 5. Motifs denoted with $S_*$ feature single-link relations, while $D_*$ feature double link relations.

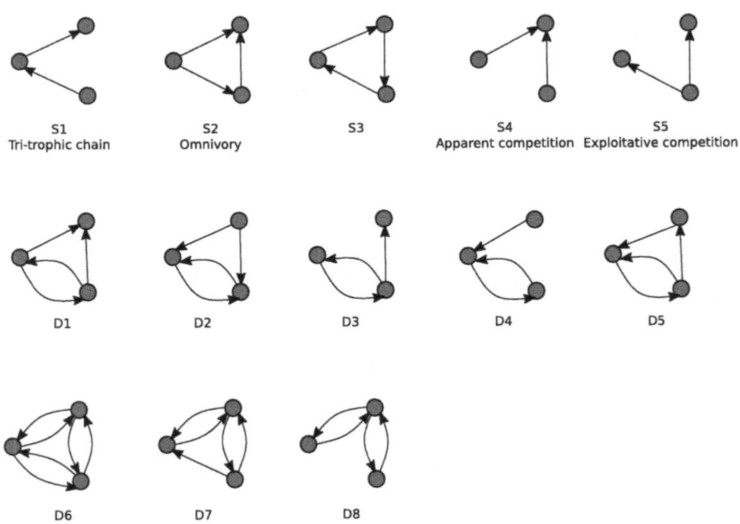


**Fig. 5.** Connected three nodes motif.

$S_1$ represent a tri-trophic chain in which a top predator eats a single consumer, which in turn eats a single resource, $S_2$ omnivory, where the species at the top of the chain feeds on both the middle and bottom species; $S_3$ represent a cycle in which it is not possible to define a hierarchy between species; $S_4$ is the apparent competition where two resources are fed by the same predator; and $S_5$ is the exploitative competition where a resource shared by two consumers [19]. To quantify the over- and under-representation of such motifs, we employed a method defined in [36], which consists of evaluating the quantity:

$$z_i = \frac{N_{i,real} - \langle N_i \rangle}{\sigma_i} \tag{4}$$

where $N_{i,real}$ is the number of the $i$-th motif in a given network, $\langle N_i \rangle$ and $\sigma_i$ are the average number of the $i$ motif and its standard deviation measured on an ensemble of 50 random networks. These networks were generated from the original graph via the swap method described in [21,36], which preserves the in-degree and out-degree of each node. To generate each random network, we performed $100 \times L_G$ swaps. Given a food web $G$, say that a motif $i$ is under-represented in $G$ if $z_i < 0$ and over-represented if $z_i > 0$. The complete table of representation of motifs for all food webs is available in the Supplementary Material (Table 4).

To compare the representation of the motifs in the food webs of our dataset, we evaluated the profile of the z-score of each motif in each food web [29,36]:

$$\mathcal{P}_i = z_i / \sqrt{(\sum_i z_i^2)} \tag{5}$$

The distribution of the profile of the z-score is shown in Fig. 6a.

We followed the analysis of Stouffer et al. [36] and investigated the relationship between the representation of the $S_2$ motifs (omnivory), and $S_4$ and $S_5$ motifs (competition).

Thus, we consider two groups of food webs: those in which $S_2$ is under-represented and those in which it is over-represented. In Fig. 6b we plot the profile distribution of the $z$-score based in this categorisation. We observe that, in accordance to the results of [36], aquatic food webs display a strong relationship between the representation of motif $S_2$, $S_4$ and $S_5$. More specifically, when in a food web the motif $S_2$ is over-represented then the motifs $S_4$ and $S_5$ are under-represented, whereas an under-representation of motif $S_2$ corresponds to an over-representation of motifs $S_4$ and $S_5$. As suggested in [36], this pattern of under-representation of motif $S_2$ and over-representation of motif $S_4, S_5$ could be explained with the nested-hierarchy model.

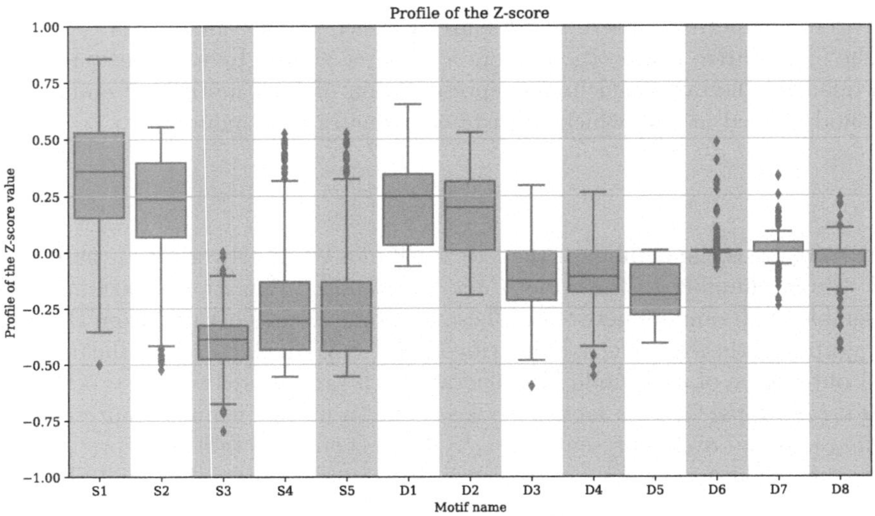


(a)

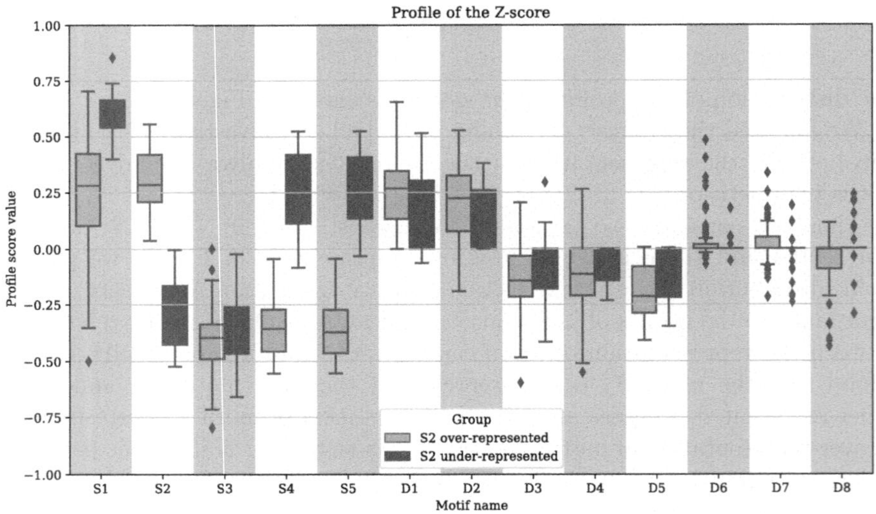


(b)

**Fig. 6.** (a) Z-score profiles for each motif across all food webs. (b) Z-score profiles for motifs in food webs with under-representation (blue boxes) and over-representation (red boxes) of omnivory (motif $S_2$). (Color figure online)

## 6 Discussion and Conclusions

In this paper we have studied the structural characteristics of aquatic food webs. We collected a dataset of 173 aquatic food webs, which are made available on https://github.com/davidetorre92/AquaNet/. Our analysis revealed a common pattern across networks: a dense core of generalist species encircled by distinct peripheries. We examine the core-periphery vertical and horizontal biodiversity showing that the core is mainly composed by generalist species. Additionally, species with the highest trophic levels are predominantly found in the core, whereas those with lower trophic levels belong to the periphery.

Next, we evaluate species removal effects on food webs using a network robustness metric, which turned out to be helpful in identifying the species that are vital to ecosystem stability. The biological significance of the critical species identified by our algorithm appears to be consistent with existing knowledge in the literature. Furthermore, the robustness measure is not directly correlated with other simple food web metrics, such as the number of species, the number of links, or the maximum trophic level. This is interesting, as it suggests that this measure captures an aspect of the food web structure which is not directly reflected by more traditional measures. In this direction it is interesting to explore more the relationship between network complexity and stability in food webs. Additionally, it would be intriguing to study how the removal of critical species influences the ecosystem's biodiversity beyond its impact on network robustness.

Lastly, we examine motifs in food webs that occur more frequently than in random networks-and their significance for network stability and robustness. Our results indicate that specific motifs are consistently prevalent across different food networks, and we find motif pairs with interconnected occurrences. It would be interesting to better explore how do these motifs contribute to the overall functionality and stability of ecosystems. To answer this question we probably need to use data related to the strength of interaction as well as species abundance. Finally, as the food webs considered in this paper vary in network resolution, it would be worth to explore its influence on motif representation patterns.

**Acknowledgements.** Davide Torre is a PhD student enrolled in the National PhD in Artificial Intelligence for Health and Life Sciences, XXXVII cycle (Università Campus Bio-Medico di Roma).

## References

1. Anderson, T.K., Sukhdeo, M.: Host centrality in food web networks determines parasite diversity. PLoS ONE **6**(10), e26798 (2011). https://doi.org/10.1371/journal.pone.0026798
2. Atmar, W., Patterson, B.D.: The measure of order and disorder in the distribution of species in fragmented habitat. Oecologia **96**, 373–382 (1993)
3. Bascompte, J., Jordano, P., Melián, C.J., Olesen, J.M.: The nested assembly of plantanimal mutualistic networks. Proc. Natl. Acad. Sci. **100**(16), 93839387 (2003). https://doi.org/10.1073/pnas.1633576100

4. Boldi, P., Rosa, M., Vigna, S.: Robustness of social networks: comparative results based on distance distributions. In: Datta, A., Shulman, S., Zheng, B., Lin, S.-D., Sun, A., Lim, E.-P. (eds.) SocInfo 2011. LNCS, vol. 6984, pp. 8–21. Springer, Heidelberg (2011). https://doi.org/10.1007/978-3-642-24704-0_7
5. Broder, A.Z., et al.: Graph structure in the web. Comput. Networks **33**(1–6), 309–320 (2000). https://doi.org/10.1016/S1389-1286(00)00083-9
6. Brown, A.: Food relationships on the intertidal sandy beaches of the cape peninsula. South Afr. J. Sci. **60**(2), 35–41 (1964). https://doi.org/10.10520/AJA00382353_1153
7. Cohen, J.E., Briand, F., Newman, C.M.: Data on 113 community food webs, pp. 203–279. Springer, Heidelberg (1990). https://doi.org/10.1007/978-3-642-83784-5_4
8. Colléter, M., et al.: EcoBase: a repository solution to gather and communicate information from EwE models. Ph.D. thesis, Fisheries Centre, University of British Columbia, Canada (2013)
9. Csardi, G., Nepusz, T.: The igraph software package for complex network research. Int. J. Complex Syst. **1695** (2006). https://igraph.org
10. Csermely, P., London, A., Wu, L.Y., Uzzi, B.: Structure and dynamics of core/periphery networks. J. Complex Netw. **1**(2), 93–123 (2013). https://doi.org/10.1093/comnet/cnt016
11. Csete, M., Doyle, J.: Bow ties, metabolism and disease. Trends Biotechnol. **22**(9), 446–450 (2004). https://doi.org/10.1016/j.tibtech.2004.07.007
12. Dale, M.R.T.: Applying Graph Theory in Ecological Research. Cambridge University Press, Cambridge (2017). ISBN: 9781316105450
13. Dáttilo, W., et al.: Unravelling darwin's entangled bank: architecture and robustness of mutualistic networks with multiple interaction types. Proc. Roy. Soc. B Biol. Sci. **283**(1843), 20161564 (2016)
14. Delmas, E., et al.: Analysing ecological networks of species interactions. Biol. Rev. **94**(1), 16–36 (2018)
15. Duffy, J.E., Cardinale, B.J., France, K.E., McIntyre, P.B., Thébault, E., Loreau, M.: The functional role of biodiversity in ecosystems: incorporating trophic complexity. Ecol. Lett. **10**(6), 522–538 (2007). https://doi.org/10.1111/j.1461-0248.2007.01037.x
16. Dunne, J.A., Williams, R.J., Martinez, N.D.: Network structure and biodiversity loss in food webs: robustness increases with connectance. Ecol. Lett. **5**(4), 558–567 (2002)
17. Dunne, J.A., Williams, R.J., Martinez, N.D.: Network structure and robustness of marine food webs. Mar. Ecol. Prog. Ser. **273**, 291–302 (2004)
18. Gallagher, R.J., Young, J.G., Welles, B.F.: A clarified typology of core-periphery structure in networks. Sci. Adv. **7**(12) (2021). https://doi.org/10.1126/sciadv.abc9800
19. Giling, D.P., et al.: Plant diversity alters the representation of motifs in food webs. Nat. Commun. **10**(1), 1226 (2019)
20. Huxham, M., Beaney, S., Raffaelli, D.: Do parasites reduce the chances of triangulation in a real food web? Oikos **76**(2), 284–300 (1996)
21. Itzkovitz, S., Milo, R., Kashtan, N., Newman, M., Alon, U.: Reply to comment on subgraphs in random networks. Phys. Rev. E **70**(5), 058102 (2004)
22. Ömer K. Coskun, Özen, V., Wankel, S.D., Orsi, W.D.: Quantifying population-specific growth in benthic bacterial communities under low oxygen using H218O. ISME J. **13**(6), 1546–1559 (2019). https://doi.org/10.1038/s41396-019-0373-4

23. Landi, P., Minoarivelo, H.O., Brännström, Å., Hui, C., Dieckmann, U.: Complexity and stability of ecological networks: a review of the theory. Popul. Ecol. **60**(4), 319–345 (2018). https://doi.org/10.1007/s10144-018-0628-3
24. Luxton, M.: The marine littoral mites of the New Zealand region. J. R. Soc. N. Z. **20**(4), 367418 (1990). https://doi.org/10.1080/03036758.1990.10426719
25. Martinez, N.D.: Artifacts or attributes? Effects of resolution on the little rock lake food web. Ecol. Monogr. **61**(4), 367–392 (1991). https://doi.org/10.2307/2937047
26. McHugh, P.A., McIntosh, A.R., Jellyman, P.G.: Dual influences of ecosystem size and disturbance on food chain length in streams. Ecol. Lett. **13**(7), 881890 (2010). https://doi.org/10.1111/j.1461-0248.2010.01484.x
27. Memmott, J., Waser, N.M., Price, M.V.: Tolerance of pollination networks to species extinctions. Proc. Roy. Soc. London Ser. B Biol. Sci. **271**(1557), 2605–2611 (2004)
28. Miele, V., Ramos-Jiliberto, R., Vázquez, D.P.: Core–periphery dynamics in a plant–pollinator network. J. Anim. Ecol. **89**(7), 1670–1677 (2020). https://doi.org/10.1111/1365-2656.13217
29. Milo, R., Shen-Orr, S., Itzkovitz, S., Kashtan, N., Chklovskii, D., Alon, U.: Network motifs: Simple building blocks of complex networks. Science **298**(5594), 824–827 (2002). https://doi.org/10.1126/science.298.5594.824
30. Palacio, R.D., Valderrama-Ardila, C., Kattan, G.H.: Generalist species have a central role in a highly diverse plantfrugivore network. Biotropica **48**(3), 349–355 (2016)
31. Paudel, N., Georgiadis, L., Italiano, G.F.: Computing critical nodes in directed graphs. ACM J. Exp. Algorithmics **23** (2018). https://doi.org/10.1145/3228332
32. Paviour-Smith, K.: The biotic community of a salt meadow in New Zealand. Trans. Proc. Roy. Soc. New Zealand **83**, 525–554 (1956)
33. Sabo, J.L., Finlay, J.C., Kennedy, T., Post, D.M.: The role of discharge variation in scaling of drainage area and food chain length in rivers. Science **330**(6006), 965967 (2010). https://doi.org/10.1126/science.1196005
34. Schoener, T.W.: Food webs from the small to the large: the Robert H. Macarthur award lecture. Ecology **70**(6), 15591589 (1989). https://doi.org/10.2307/1938088
35. Sheykhali, S., et al.: Robustness to extinction and plasticity derived from mutualistic bipartite ecological networks. Sci. Rep. **10**(1) (2020). https://doi.org/10.1038/s41598-020-66131-5
36. Stouffer, D.B., Camacho, J., Jiang, W., Amaral, L.: Evidence for the existence of a robust pattern of prey selection in food webs. Proc. Roy. Soc. B Biol. Sci. **274**(1621), 1931–1940 (2007). https://doi.org/10.1098/rspb.2007.0571
37. Thébault, E., Fontaine, C.: Stability of ecological communities and the architecture of mutualistic and trophic networks. Science **329**(5993), 853–856 (2010)
38. Wang, S., Brose, U.: Biodiversity and ecosystem functioning in food webs: the vertical diversity hypothesis. Ecol. Lett. **21**(1), 920 (2017). https://doi.org/10.1111/ele.12865
39. Williams, R.J., Martinez, N.D.: Simple rules yield complex food webs. Nature **404**(6774), 180–183 (2000)
40. Williams, R.J., Martinez, N.D.: Limits to trophic levels and omnivory in complex food webs: theory and data. Am. Nat. **163**(3), 458–468 (2004)

# Leveraging Diffuser Data Augmentation to Enhance ViT-Based Performance on Dermatoscopic Melanoma Images Classification

Tiziana Currieri[1], Giovanni Cicceri[1](✉), Sergio Cannata[2], Giansalvo Cirrincione[3], Marta Lovino[4], Carmelo Militello[5], Francesco Prinzi[1], Eros Pasero[2], and Salvatore Vitabile[1]

[1] Department of Biomedicine, Neuroscience and Advanced Diagnostics (BiND), University of Palermo, 90127 Palermo, Italy
{tiziana.currieri,giovanni.cicceri,francesco.prinzi, salvatore.vitabile}@unipa.it

[2] Department of Electronics and Telecommunications, Politecnico di Torino, 10129 Turin, Italy
{sergio.cannata,eros.pasero}@polito.it

[3] Département Electronique - Electrotechnique - Automatique (EEA), University of Picardie Jules Verne, 80000 Amiens, France
giansalvo.cirrincione@u-picardie.fr

[4] Department of Engineering Enzo Ferrari, University of Modena and Reggio Emilia, 41125 Modena, Italy
marta.lovino@unimore.it

[5] Institute for High-Performance Computing and Networking (ICAR-CNR), Italian National Research Council, 90146 Palermo, Italy
carmelo.militello@cnr.it

**Abstract.** The worldwide prevalence of skin cancer, particularly malignant melanoma, has greatly increased. As a consequence, optimal disease management for early diagnosis and intervention is required for improved patient prognosis. To facilitate this, new technologies such as Machine Learning and Deep Learning have been employed to analyze dermatoscopic images. Some applications involve abnormal tissue structure identification and lesion classification. In this study, we introduce a Denoising Diffusion Probabilistic Models (DDPM) approach for data augmentation (DA) to enhance the predictive classification capability of Vision Transformer (ViT) for dermatoscopic image analysis. The ViT was used as deep architecture to evaluate the contribution of diffuser-based and classical DA techniques. The classifier was trained to distinguish between Melanoma, Nevus and Seborrheic Keratosis patches. Experimental trials performed on a publicly available melanoma dataset showed enhanced performance of Diffuser-based DA over classical DA. Furthermore, this study presents, analyzes, and discusses the findings, emphasizing the practicality and efficiency of the proposed method.


M. Vettoretti et al. (Eds.): CIBB 2023, LNBI 14513, pp. 16–27, 2025.
https://doi.org/10.1007/978-3-031-90714-2_2

**Keywords:** Skin Lesion Classification · Data Augmentation · Deep Learning · Dermatoscopic Image · Diffuser · Melanoma · ViT-based Architecture

## 1 Introduction

Skin cancer is recognized by the World Health Organization (WHO) as the most frequently occurring oncological condition globally, with more than a million new cases reported each year, accounting for nearly one-third of all cancer diagnoses [15]. Research from WHO highlights ultraviolet (UV) radiation as the primary contributing factor, recommending measures such as minimizing sun exposure and routinely monitoring the skin for unusual lesions. Melanoma, a type of skin cancer that originates in pigment-producing melanocytes, has shown a marked increase in prevalence over the past three decades [23]. Prompt diagnosis and intervention are critical for improving patient outcomes and curbing the progression of this disease. In this regard, Computer-Aided Diagnosis (CAD) systems have proven to be valuable tools for aiding clinicians in the identification of skin cancer [20]. These systems leverage Machine Learning (ML) and DL techniques to evaluate dermatoscopic images, including those of melanoma, by detecting atypical tissue patterns and classifying lesions as malignant or benign [16]. Specifically, many works have focused on using ML models to classify cutaneous melanoma illness [19,27]. Deep Learning (DL) approaches, including convolutional neural networks (CNNs) and Vision Transformers (ViTs), have been widely applied to the detection and classification of melanoma and other skin cancers, yielding highly promising outcomes [2,18,21,29]. Indeed, CNN-based approaches [14,30] achieve performance around 0.84–0.86 in accuracy, and 0.84–0.88 in sensitivity and specificity. Other methods based on residual networks remain in the same range of results [12,28].

This study explores the potential of synthetic images, generated through a novel Diffuser Data Augmentation (DA) method, to improve the performance of a ViT-based classifier on dermatoscopic images. Diffusers represent a novel technique to increase the number of images to enhance the generalization capacity of the classifier. Unlike conventional DA approaches, Diffusers allow the distribution of data samples to be modeled and replicated. To this aim, we wanted to highlight the advantages offered by the use of diffuser-based DA compared with the classical one. We selected melanoma detection as a case study, composed of dermatoscopic images, to compare the two DA methods. Leveraging a publicly available melanoma dataset [25], the experimental results confirm improved accuracy of diffuser-based DA compared to the classical one. Despite the impressive performance of existing models in detecting and classifying skin diseases, these findings suggest that incorporating synthetic data can further optimize model training.

The main contributions of this work are summarized as follows:

- an advanced Diffusers-based DA procedure, improving the generalization capabilities of the ViT-based classifier compared to classical DA;

– encouraging experimental trials, with classification performance exceeding the state-of-the-art results on the same dataset.

### 1.1 Related Works

Techniques for generating and augmenting data have been widely explored to address challenges in image classification, particularly in the field of medical imaging. Traditional approaches, such as geometric transformations, brightness modifications, and noise addition, are commonly used to expand dataset variability and improve model robustness [6]. Beyond these basic methods, advanced generative models, including Generative Adversarial Networks (GANs) and Variational Autoencoders (VAEs), have proven effective in producing synthetic images that closely mimic real data distributions [9,17]. Recently, flow-based models have also gained attention for their ability to generate high-quality images by explicitly modelling the likelihood of the data [7]. Diffusion models represent another innovative approach, demonstrating substantial potential in generating high-fidelity images through iterative noise removal processes [13]. In this study, we employ Diffusion Models as a generative technique for creating synthetic images aimed at melanoma classification. Unlike traditional transformations or GAN-based approaches, Diffusion Models excel in capturing the underlying data distribution with high fidelity, resulting in realistic and diverse synthetic samples. While this work primarily compares the proposed approach with classical data augmentation methods, we acknowledge the importance of benchmarking against other state-of-the-art generative techniques, such as GANs and flow-based models, as part of future efforts to contextualize the contributions of Diffusion Models within the broader field.

## 2 Materials and Methods

### 2.1 Dataset Description

The dataset analyzed comprises 2000 dermoscopic images of skin lesions, provided in JPEG format and publicly accessible through the ISIC Archive [25]. This dataset was utilized as part of the ISIC 2017 Challenge [3], with the primary goal of lesion classification. The images are categorized into three classes: *Melanoma* (MM), *Nevus* (NCN), and *Seborrheic Keratosis* (SK). Specifically, MM is a malignant tumor originating from melanocytes, the cells responsible for producing melanin, and is considered the most aggressive form of skin cancer due to its high metastatic potential [22]. NCN, commonly known as a mole, is a benign proliferation of melanocytes, often appearing as small, pigmented spots on the skin, and generally poses no health risk [10]. SK is a non-cancerous skin condition characterized by the proliferation of keratinocytes, resulting in raised, wart-like lesions that are typically benign [1]. The training set comprises 374 cases of MM, 254 cases of SK, and 1372 cases of NCN. The validation set consists of 150 images, while the test set includes 600 images, featuring a variety of dimensions, orientations, and lighting conditions (see Table 1). The competition committee provided this division.

**Table 1.** ISIC dataset summary.

| Class | Training Set | Validation Set | Test Set |
|---|---|---|---|
| Melanoma (MM) | 374 | 30 | 118 |
| Nevus (NCN) | 1372 | 78 | 392 |
| Seborrheic Keratosis (SK) | 254 | 42 | 90 |
| Total | 2000 | 150 | 600 |

## 2.2 Vision Transformer

ViT revolutionizes image processing by leveraging the self-attention mechanism traditionally used in Natural Language Processing (NLP) tasks [8,26]. Unlike CNNs, which rely on convolutional operations, the ViT divides an input image into a sequence of non-overlapping patches, treating these patches as "tokens" similar to words in a sentence. Each patch is flattened into a vector, and a learnable positional embedding is added to each token to preserve spatial information. This ensures that the model retains awareness of the relative positioning of each patch within the original image. The sequence of patch embeddings is subsequently processed through a series of Transformer encoder layers, with each layer executing two primary operations: a *Multi-Head Attention* mechanism and a position-wise *Feed-Forward Neural Network* (FFNN). At the heart of this architecture is the Multi-Head Attention mechanism [4,26], which operates by computing three key components for each patch embedding: the *queries*, which represent what the model is looking for in the image; the *keys*, which identify the features of interest; and the *values*, which provide the actual information to be aggregated. The mechanism computes a compatibility score between the query and each key (via dot product), scales the score, and normalizes it through a softmax function to assign attention weights. These weights are then used to combine the values, effectively allowing the model to focus on the most relevant regions of the image while still considering the global context. By using multiple attention heads, the model learns independent and complementary representations of the image, significantly enhancing its ability to capture both local and global patterns. Residual connections and layer normalization are incorporated after each Multi-Head Attention and FFNN block to stabilize training, prevent vanishing gradients, and ensure an efficient flow of information throughout the architecture. One of the primary strengths of this architecture lies in its ability to process image patches in parallel, enabling the extraction of detailed local features while simultaneously constructing a coherent global understanding of the image. This makes the ViT suitable also for image analysis, where it has been successfully applied to tasks such as tumour detection, organ segmentation, and disease classification [8,24]. As depicted in Fig. 1, our implementation begins by dividing the input image into patches, vectorizing them, and enriching each patch with a positional encoding. These encoded patches are then passed through a series of Transformer encoder layers, each consisting of a Multi-Head Attention block and an FFNN. The final output from the Transformer encoder is pooled into a classification token, which is processed by a Multilayer Perceptron (MLP) head to generate the predicted class label.

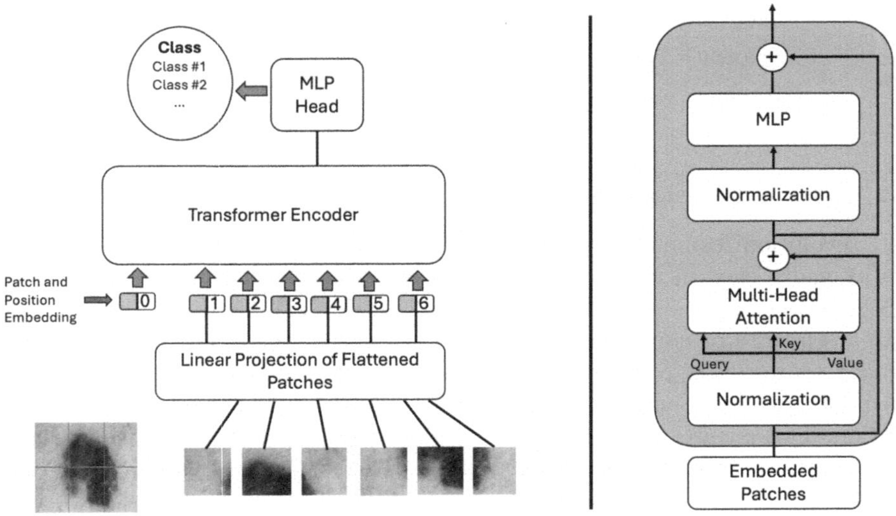


**Fig. 1.** Overview of the architecture used in this work, highlighting the key components of the Vision Transformer (ViT). The three arrows between the normalization layer and the multi-head attention block represent the sequential flow of the three main components of the attention mechanism: *query*, *key*, and *value* matrices.

## 2.3 Classical Data Augmentation

Classical-based DA transformations were implemented to compare the diffuser-based DA. DA methods were employed to enhance the ability to train the model in a wider range of scenarios, thus enhancing its ability to generalize and diagnose in real clinical situations. In particular, increased brightness plays a crucial role in expanding the diversity of illumination conditions. This transformation improves the model's robustness to light variations while helping to simulate the nuances that might emerge in heterogeneous clinical environments. The dynamic contrast balance realized through the Contrast Limited Adaptive Histogram Equalisation (CLAHE) technique results in a tangible improvement in detail discrimination [11]. Geometrical transformation is introduced to improve the model's capacity to recognize injuries regardless of orientations and geometrical deformations. The application of gamma corrections and colour adjustments, such as altering saturation and hue, contributes to the further enrichment of the colour information in the images. The introduction of Gaussian noise and augmentation of saturation help to mimic real-life image acquisition conditions, making the model more resistant to random fluctuations and more adaptable to heterogeneous clinical contexts. In addition, the use of contrast enhancement further increased the robustness and realism of the augmented dataset.

Applying the aforementioned techniques, starting from an initial set of 374 melanoma images, we generated 2992 new images (374 for 8 transformations). The transformations applied are detailed in Table 2. Subsequently, a random elimination process was employed to arrive at a final count of 998 images, ensuring an equitable balance between melanoma and nevus samples in the training set. The same procedure was performed for the SK class, by applying the above-mentioned transformations and random elimination. At the end of this process, a balanced dataset was generated. Figure 2 shows some classical transformations for image enhancement applied to melanoma images.

**Table 2.** Details about the classical DA techniques applied in the study.

| Augmentation Technique | Description |
|---|---|
| Contrast Adjustment | Rescales image intensity using the 2nd and 98th percentiles. |
| Saturation Increases | Enhances the saturation channel in HSV color space. |
| Contrast Limited Adaptive Histogram Equalization (CLAHE) | Improves contrast by applying CLAHE on the L channel in LAB color space. |
| Diagonal Flip | Applies diagonal rotations within a range of $\pm 60^{\circ}$. |
| Hue and Saturation Adjustment | Modifies hue by a random factor and scales saturation by a random multiplier. |
| Brightness Adjustment | Alters brightness using a linear scale and offset. |
| Gamma Correction | Adjusts image brightness and contrast using gamma power-law transformation. |
| Gaussian Noise Addition | Introduces Gaussian noise with a mean of 0 and sigma of 25 to the image. |

### 2.4 Diffusers-Based Data Augmentation

Diffusers have rapidly become popular architectures for image generation [5], especially when used in text-to-image applications. Diffusers are an improved version of GANs, VAEs, and flow-based models. Diffusers are trained by corrupting an image at each time step until its data distribution becomes equivalent to an isotropic Gaussian - and reverse - through Markov chains. This is implemented through a UNet enhanced with skip connections (e.g. residual blocks) and self-attention, allowing it to handle different conditioning information for image generation. The model is conditioned on the current time step, which is embedded and added channel-wise to the input of the network through every level of the UNet. The work described in this study presents the deployment of a Denoising Diffusion Probabilistic Model (DDPM) to generate artificial MM and SK images. In Fig. 3, the Diffuser architecture used for the generation of synthetic images is shown.

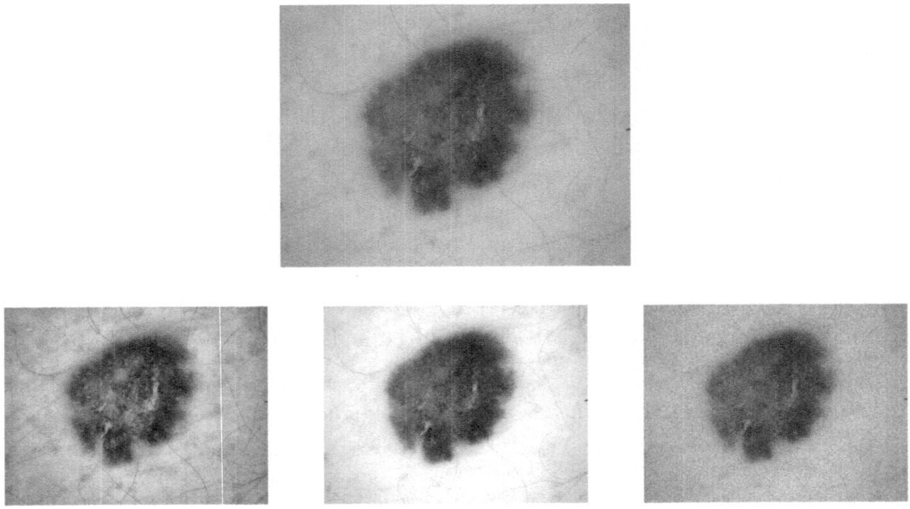

**Fig. 2.** In the top row is the original melanoma image. In the bottom row: (left) CLAHE application; (center) contrast enhancement application; (right) Gaussian noise application.

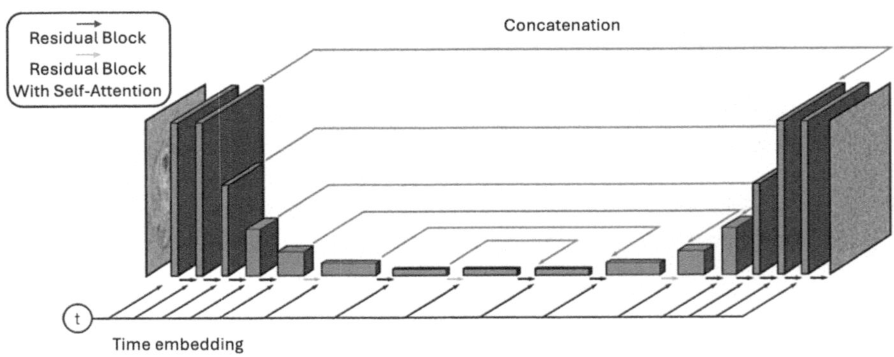


**Fig. 3.** The Diffuser architecture used for the generation of synthetic images.

### 2.5 Implementation Details

The ViT-Large model was fine-tuned using 32 × 32 patches and hyperparameters detailed in [2], aiming to develop a high-performing model in distinguishing between different skin lesion categories. To achieve this, diffuser-based and classical DA were performed and the model was trained using 998 samples for the MM class and 1118 for the SK class, matching the cardinality of the Nevus category, the largest class in the dataset. Table 3 summarizes the samples per class. In addition, an example of synthetic melanoma images is displayed in Fig. 4. Finally, the augmented dataset was utilized to train and evaluate the fine-tuned model described earlier, leading to improved sensitivity in detecting melanoma cases and an overall enhancement in accuracy.

**Table 3.** ISIC dataset after DA procedure.

| Class | Training Set | Validation Set | Test Set |
|---|---|---|---|
| Melanoma (MM) | 1372 (374+998) | 30 | 118 |
| Nevus (NCN) | 1372 | 78 | 392 |
| Seborrheic Keratosis (SK) | 1372 (254+1118) | 42 | 90 |

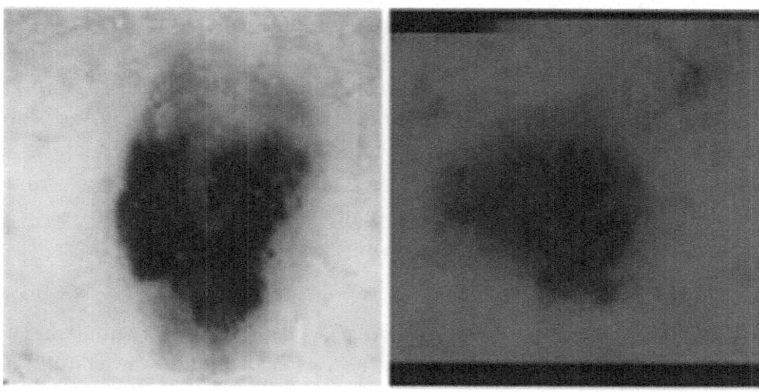

**Fig. 4.** Examples of a real melanoma image (left) and a synthetic one (right) generated via the Diffuser-based DA.

## 3 Results

Experimental tests have shown DA improvement. The effectiveness of the two methodologies (i.e. without and with DA) is assessed and compared using the confusion matrices. Figure 5 - reporting the confusion matrices computed on the test set - shows a decrease of false negatives, *i)* from 17 to 10 considering ViT-based without DA vs. Diffusers DA) and *ii)* from 13 to 10 considering ViT-based with classical DA vs. Diffusers DA.

Table 4 presents a comparative analysis of the classification performance achieved by the ViT-based classifier without DA, with classical DA, and with Diffusers-based DA. The *Nevus* category exhibits no relevant variations on the confusion matrix or any other metrics, except for a minor decline in specificity. *Seborrheic keratosis* experienced slight increases in accuracy and specificity, at the expense of reduced sensitivity and AUROC scores. *Melanoma*, on the other

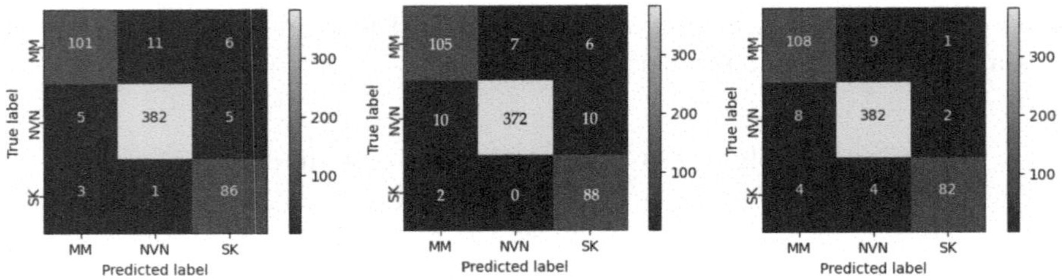


**Fig. 5.** Confusion matrices obtained by the ViT-based classifier (left) without DA, (center) with classical DA; (right) with Diffuser-based DA.

hand, suffered a marginal specificity decrease, while showing increased scores on AUROC, accuracy, and sensitivity. In addition, by observing the results obtained with classic DA, although there were improvements in specificity and sensitivity of the *Nevus* class, and the AUROC metric of the *Seborrheic keratosis* class, on the *Melanoma* class there are no marked advances. Overall, it is possible to assert that DA improves classification performance: classical DA doesn't have an evident behaviour among all classes, while Diffuser-based DA shows a slight improvement in all three classes (as shown in the last row of the same table). The results highlight the advantages of utilizing Diffusers to address the initial class imbalance by augmenting the training set with synthetic data.

**Table 4.** Results of the comparison between ViT-based model without and with DA using classical transformations and the Diffuser. Performance was computed in the one-*vs.*-rest strategy.

| Class | Evaluation metric | ViT-based | ViT-based + Classic DA | Vit-based + Diffusers DA |
|---|---|---|---|---|
| Melanoma (MM) | $\text{accuracy}_{MM}$ | 0.958 | 0.958 | **0.963** |
| | $\text{sensitivity}_{MM}$ | 0.856 | 0.889 | **0.915** |
| | $\text{specificity}_{MM}$ | **0.983** | 0.975 | 0.975 |
| | $\text{AUROC}_{MM}$ | 0.919 | 0.932 | **0.945** |
| Nevus (NVN) | $\text{accuracy}_{NVN}$ | **0.963** | 0.955 | 0.962 |
| | $\text{sensitivity}_{NVN}$ | **0.974** | 0.948 | **0.974** |
| | $\text{specificity}_{NVN}$ | 0.942 | **0.966** | 0.938 |
| | $\text{AUROC}_{NVN}$ | **0.958** | 0.957 | 0.956 |
| Seborrheic Keratosis (SK) | $\text{accuracy}_{SK}$ | 0.975 | 0.970 | **0.982** |
| | $\text{sensitivity}_{SK}$ | 0.955 | **0.977** | 0.911 |
| | $\text{specificity}_{SK}$ | 0.978 | 0.968 | **0.994** |
| | $\text{AUROC}_{SK}$ | 0.967 | **0.973** | 0.953 |
| MM+NVN+SK | $\text{accuracy}_{overall}$ | 0.948 | 0.941 | **0.953** |

## 4 Conclusions

The availability of predictive models, able to effectively and efficiently support physicians' decision-making, represents a factor enabling early diagnosis and personalized medicine. ViTs can model long-range spatial relationships in images: for this reason, the classification of large dermatoscopic images is an optimal case study. As with all DL techniques, ViTs require a large number of samples to provide a complete view of the reality being modeled as possible. Diffusers represent the state-of-the-art in DA, making it possible to augment the dataset by generating synthetic images. Unlike conventional DA approaches, Diffusers allow us to model and replicate a distribution close to the original samples. Through synthetic images it is possible to improve the generalization capacity of the classifiers. To this aim, we wanted to focus on the advantage offered by Diffusers on melanoma detection and to compare two DA methods, by evaluating their impact on the dermatoscopic image classification. This study implemented a ViT-based classifier, able to discriminate between MM, NVN, and SK, using the classical and Diffuser-based DA procedures to increase the number of training images. This work showed the advantage of DA: in the evaluations conducted using the one-*vs.*-rest strategy, all metrics recorded values exceeding 90% (with the sole exception being the sensitivity related to melanoma). Notably, the employment of synthetic images for melanoma class reduced by 41% the false negatives (ViT-based without DA vs. Diffusers DA) and by 23% (classical DA vs. Diffusers DA). Future research will focus on three primary objectives. The first concerns evaluating potential biases introduced by the excessive augmentation of images used in training DL models. Such future work will rigorously assess the distribution of the augmented data to ensure its appropriateness and representativeness. This will involve developing methodologies to minimize data-induced bias while maintaining model robustness and accuracy. The second is to interpret the ViT-based model by analyzing its decision-making mechanisms by applying Explainable AI (XAI) techniques. The third involves undertaking a comprehensive evaluation of Diffusion Models compared to other advanced DA techniques, such as GANs and flow-based models. Finally, while this study primarily examines diffusion models alongside classical DA methods, future efforts will expand this scope to offer a more detailed assessment of their relative strengths and limitations. To this aim, further tests will be conducted to statistically quantify the contribution of diffusion-based data augmentation (*e.g.* using bootstrapping or cross-validation).

## References

1. Braun, R.P., et al.: Dermoscopy of pigmented seborrheic keratosis: a morphological study. Arch. Dermatol. **138**(12), 1556–1560 (2002)
2. Cirrincione, G., et al.: Transformer-based approach to melanoma detection. Sensors **23**(12), 5677 (2023)
3. Codella, N.C.F., et al.: Skin lesion analysis toward melanoma detection: a challenge at the 2017 international symposium on biomedical imaging (ISBI), hosted by the international skin imaging collaboration (ISIC). In: 2018 IEEE 15th International

Symposium on Biomedical Imaging (ISBI 2018), pp. 168–172 (2018). https://doi.org/10.1109/ISBI.2018.8363547
4. Cordonnier, J.B., Loukas, A., Jaggi, M.: Multi-head attention: collaborate instead of concatenate. arXiv preprint arXiv:2006.16362 (2020)
5. Corvi, R., Cozzolino, D., Zingarini, G., Poggi, G., Nagano, K., Verdoliva, L.: On the detection of synthetic images generated by diffusion models. In: ICASSP 2023-2023 IEEE International Conference on Acoustics, Speech and Signal Processing (ICASSP), pp. 1–5. IEEE (2023)
6. Cubuk, E.D., Zoph, B., Shlens, J., Le, Q.V.: Randaugment: practical automated data augmentation with a reduced search space. In: Proceedings of the IEEE/CVF Conference on Computer Vision and Pattern Recognition Workshops, pp. 702–703 (2020)
7. Dinh, L., Krueger, D., Bengio, Y.: Nice: non-linear independent components estimation. arXiv preprint arXiv:1410.8516 (2014)
8. Dosovitskiy, A.: An image is worth 16x16 words: transformers for image recognition at scale. arXiv preprint arXiv:2010.11929 (2020)
9. Goodfellow, I., et al.: Generative adversarial nets. In: Advances in Neural Information Processing Systems, vol. 27 (2014)
10. Happle, R.: What is a nevus: a proposed definition of a common medical term. Dermatology **191**(1), 1–5 (1995)
11. Hayati, M., et al.: Impact of CLAHE-based image enhancement for diabetic retinopathy classification through deep learning. Procedia Comput. Sci. **216**, 57–66 (2023). https://doi.org/10.1016/j.procs.2022.12.111. https://www.sciencedirect.com/science/article/pii/S1877050922021895. 7th International Conference on Computer Science and Computational Intelligence 2022
12. He, K., Zhang, X., Ren, S., Sun, J.: Deep residual learning for image recognition (2015)
13. Ho, J., Jain, A., Abbeel, P.: Denoising diffusion probabilistic models. Adv. Neural. Inf. Process. Syst. **33**, 6840–6851 (2020)
14. Hu, J., Shen, L., Sun, G.: Squeeze-and-excitation networks. In: Proceedings of the IEEE Conference on Computer Vision and Pattern Recognition, pp. 7132–7141 (2018)
15. Hu, W., Fang, L., Ni, R., Zhang, H., Pan, G.: Changing trends in the disease burden of non-melanoma skin cancer globally from 1990 to 2019 and its predicted level in 25 years. BMC Cancer **22**(1), 836 (2022)
16. Jutzi, T.B., et al.: Artificial intelligence in skin cancer diagnostics: the patients' perspective. Front. Med. **7**, 233 (2020)
17. Kingma, D.P.: Auto-encoding variational bayes. arXiv preprint arXiv:1312.6114 (2013)
18. Krishna, G.S., Supriya, K., Sorgile, M., et al.: Lesionaid: vision transformers-based skin lesion generation and classification. arXiv preprint arXiv:2302.01104 (2023)
19. Kumar, N.V., Kumar, P.V., Pramodh, K., Karuna, Y.: Classification of skin diseases using image processing and SVM. In: 2019 International Conference on Vision Towards Emerging Trends in Communication and Networking (ViTECoN), pp. 1–5. IEEE (2019)
20. Lucieri, A., Bajwa, M.N., Braun, S.A., Malik, M.I., Dengel, A., Ahmed, S.: Exaid: a multimodal explanation framework for computer-aided diagnosis of skin lesions. Comput. Methods Programs Biomed. **215**, 106620 (2022)
21. Malo, D.C., Rahman, M.M., Mahbub, J., Khan, M.M.: Skin cancer detection using convolutional neural network. In: 2022 IEEE 12th Annual Computing and Communication Workshop and Conference (CCWC), pp. 0169–0176. IEEE (2022)

22. Miller, A.J., Mihm, M.C., Jr.: Melanoma. N. Engl. J. Med. **355**(1), 51–65 (2006)
23. Naik, P.P.: Cutaneous malignant melanoma: a review of early diagnosis and management. World J. Oncol. **12**(1), 7 (2021)
24. Parvaiz, A., Khalid, M.A., Zafar, R., Ameer, H., Ali, M., Fraz, M.M.: Vision transformers in medical computer vision—a contemplative retrospection. Eng. Appl. Artif. Intell. **122**, 106126 (2023)
25. The International Skin Imaging Collaboration: The International Skin Imaging Collaboration. https://www.isic-archive.com//. Accessed 11 May 2023
26. Vaswani, A.: Attention is all you need. In: Advances in Neural Information Processing Systems (2017)
27. Verma, A.K., Pal, S., Kumar, S.: Classification of skin disease using ensemble data mining techniques. Asian Pac. J. Cancer Prev. APJCP **20**(6), 1887 (2019)
28. Wang, F., et al.: Residual attention network for image classification (2017)
29. Xie, J., Wu, Z., Zhu, R., Zhu, H.: Melanoma detection based on swin transformer and simam. In: 2021 IEEE 5th Information Technology, Networking, Electronic and Automation Control Conference (ITNEC), vol. 5, pp. 1517–1521. IEEE (2021)
30. Zhang, B., Jin, S., Xia, Y., Huang, Y., Xiong, Z.: Attention mechanism enhanced kernel prediction networks for denoising of burst images. arXiv preprint arXiv:1910.08313 (2019)

# Thyroid Nodule Diagnosis Using a New Supervised Autoencoder Neural Network with $\ell_{1,\infty}$ Projection

Grégoire D'Andrea[2,3], Cyprien Gille[1], Thierry Pourcher[3], and Michel Barlaud[1](✉)

[1] Laboratoire d'Informatique, Signaux et Systèmes de Sophia Antipolis (I3S), Centre National de la Recherche Scientifique (CNRS), Université Côte d'Azur, Sophia Antipolis, Nice, France
barlaud@i3s.unice.fr

[2] Otorhinolaryngology and Head and Neck Surgery Department, Institut Universitaire de la Face et du Cou, GHS Nice University Hospital - Antoine Lacassagne Centre, Université Côte d'Azur, 06103 Nice, France

[3] Transporters in Imaging and Radiotherapy in Oncology (TIRO) Laboratory, Direction de la Recherche Fondamentale (DRF), Institut des sciences du vivant Fréderic Joliot, Commissariat à l'Energie Atomique et aux Énergies Alternatives (CEA), Université Côte d'Azur, Nice, France

**Abstract.** Thyroid nodules are more and more widespread, with the frequency of thyroid cancers increasing accordingly. In routine practice, the clinician's goal is to determine if a nodule is malignant or suspicious, and thus requires surgery. However, nearly one-third of thyroid nodules fall into the 'indeterminate' classification segment, often leading to the use of surgery as a diagnostic tool, when it could have been avoided with a better preoperative diagnosis. Molecular tests exist to address this issue, but they are markedly onerous.

In this paper, we propose an accurate, low-cost method to predict the malignancy of thyroid nodules using a categorical thyroid nodule medical dataset. The prediction is achieved through the use of a new supervised autoencoder neural network. Results show that the supervised autoencoder outperforms classical classification methods on all evaluation metrics. A proof-of-concept Graphical User Interface was developed to demonstrate the medical usefulness of the latent space of the trained network.

Aside from the prediction itself, the supervised autoencoder neural network provides three medically interesting benefits: a prediction confidence score, consistent feature ranking, and an interpretative latent space. All of those benefits are valuable from the clinician's point of view, as they allow for an efficient, low-cost, and accurate diagnosis.

**Keywords:** Thyroid nodules diagnosis · feature selection · Low-cost computation · Supervised Autoencoder and projections


M. Vettoretti et al. (Eds.): CIBB 2023, LNBI 14513, pp. 28–42, 2025.
https://doi.org/10.1007/978-3-031-90714-2_3

# 1 Study Objectives

## 1.1 Medical State of the Art

Thyroid nodules (TNs) are widespread in the human population, and their frequency is expected to continue to increase [4,6,13]. The last few decades have been marked by a real 'epidemic' of thyroid cancers, which account for 5–10% of all TNs [4,13,19,21,34]. In routine practice, the aim is to identify benign nodules that may benefit from surveillance, and malignant or suspicious nodules for which surgery may be required. However, despite optimal diagnostic management, approximately 25–30% of all TNs fall into the gray area of the Bethesda System for Reporting Thyroid Cytopathology (TBSRTC) cytological classification, where the diagnosis of benignity or malignancy remains uncertain. For these so-called cytologically indeterminate TNs (ITNs), a surgery is often performed for a diagnostic and sometimes therapeutic purpose [4,13,34]. However, the majority of those ITNs are benign on final pathological examination [4,6,34]. As a consequence, patients with benign ITNs have undergone surgery that could be avoided with a more accurate preoperative diagnosis.

To address this diagnostic issue, molecular tests have been developed to refine the risk of malignancy of ITNs [13,28,34]. Based on the detection of genomic or transcriptomic abnormalities, they are designed to select TNs that could benefit from surveillance or surgery. The Thyroseq v3 and the AFIRMA Genomic Sequencing Classifier are the main marketed molecular tests [13,28,34]. Despite their performance which permits a reduction of the number of unnecessary surgical procedures, their significant cost is a barrier to their widespread use worldwide where they are frequently unavailable [23,28]. In fact, although the recommendations of the American and European Thyroid Associations suggest considering the use of a molecular test for ITNs, they do not clearly recommend their systematic and indiscriminate use in routine practice [22,24].

Thus, there is an urgent need to find an accurate and low-cost method to improve the diagnosis of the nature of ITNs, in order to reduce the number of unnecessary surgical procedures, and to reduce the individual and medico-economic burden of these lesions [23,34].

## 1.2 Medical Dataset

In this paper, we base our efforts on a database compiled over two years by Dr. G. D'Andréa, comprised of mostly categorical clinical and cytological data [8]. Our goal is to reach accurate predictions that are also useful from the physician's point of view.

To improve the assessment of ITNs, efforts have been made in the last decades to develop algorithms and nomograms with the intent of refining the risk of malignancy of ITNs, and to help in the management of patients. In fact, classical nodule subclassifications, such as the TI-RADS classification, only use a few features of the nodule [10,29]. Moreover, more advanced efforts were realized using only echography or cytology data, or tried to integrate all sources of nodule information, to the availability of a none generalizable model because of a lacking modus operandi [8,9,30,37]. The only successful methods exhibit prohibitive costs. In this paper, we took an integrative approach and used three sources of nodule information: clinical, echography and cytology data (See Table 1). The new in-house dataset used in this study contains 1335 nodules from 1288 patients. They are described through 159 mostly categorical features (clinical, from an echography, from a cytology, from surgery or from an extemporaneous biopsy...) [8]. Among these, we pick 12 biologically significant features, detailed in Table 1.

**Table 1.** Description of the dataset: 1288 patients, 1335 nodules (473 cancers), 12 features.

| Feature | Value | Categorical values |
|---|---|---|
| Echography Nodule size | 20.8 mm +/- 11.3 | See Table 2 |
| TIRADS | Echography | See Remark 1 |
| Age | 52.9 +/- 14.5 | See Table 2 |
| Sex (nodules) | 291 males and, 1044 females | Binary |
| Thyroidopathy | 1149 | Binary |
| Hypothyroidism | 31 | Binary |
| Hyperthyroidism | 55 | Binary |
| HASHIMOTO | 100 | Binary |
| Cyto-BETHESDA | AUS/FLUS (253), sFN/FN/HN (693),sM (389) | Categorical (3,4,5) |
| Cellularity | Cytology | Categorical(1,2,3,4) |
| Colloid | Cytology | Categorical(0,1,2,3) |
| Cytology sub-classification | Cytology | Categorical(1,2,3,4) |

*Remark 1.* The TIRADS is a 6-tier ultrasonographic classification system of TNs which categorizes them by attributing an increasing risk of malignancy (TIRADS 1: 0%; $TIRADS2 :< 3\%; TIRADS3 :$ 2–5%; $TIRADS4A :$ 5–10%; $TIRADS4B :$ 10–80%; $TIRADS5 :> 80\%$). Despite no longer being used routinely, the TIRADS classification allows for more precise risk malignancy stratification within intermediate-risk nodules as compared to the actual EU-TIRADS system. In the manuscript, the TI-RADS stages 4A, 4B, and 5 were mapped respectively to categories 4, 5 and 6 of the TI-RADS feature.

The Bethesda System for Reporting Thyroid Cytopathology is a 6-tier classification system which categorizes thyroid nodules based on cytologic features. In this classification, each category is associated with an estimated risk of malignancy as follows: I - Non-diagnostic/unsatisfactory, 5–10%; $II - Benign$ 0 – 3%; $III-$ Atypia of undetermined significance (AUS), 6–30%; $IV-$ Follicular neoplasm or suspicious for a follicular neoplasm, 10–40 $V-$ Suspicious for malignancy, 50–75%; $VI - Malignant > 90\%$. Due to the significant variability in the risk of malignancy for TNs classified as Bethesda III-V, they are commonly referred to as ITNs. In this study, only ITNs were included. The 'cellularity' classification is based on the abundance of cells identified in the cytological smear. It is divided into 4 categories: 1, abundant; 2, increased; 3, low; 4, average. The 'colloid' classification is based on the presence of a colloidal substance identified in the cytological smear. It is divided into 4 categories: 0, absent; 1, poor; 2, present; 3, abundant. The cytopathological sub-classification was interpreted and completed on the basis of the data collected, following the recommendations of Valderrabano et al. [32]. It is divided into 4 categories only applicable to Bethesda III and IV categories: 1, nuclear atypia; 2, architectural atypia; 3, oncocytic characteristics; 4, other atypia. Although they do not strictly speaking form part of the Bethesda classification, these data were found to be of interest (or already associated with specific malignancy rates like the cytologic subclassification) and retained in order to assess their usefulness.

*Remark 2.* Transformation of the numerical features into categorical features using clustering.

All of these features are categorical, with two to six possible values, except for the age and the echography size.

For the two numerical features age and the echography size, the classical method to obtain categories is to use supervised clustering algorithm [17] or unsupervised clustering (k-means). However, the clustering method provides no guarantee that the bins will convey sensible medical information. Thus, we prefer to use categorical bins provided by the physician, as described in Table 2.

**Table 2.** Transformation of the numerical features into categorical features.

| Age | Category | Echography size | Category |
|---|---|---|---|
| <30 | 1 | < 11 | 1 |
| [30 – 60] | 2 | [11 – 21[ | 2 |
| >60 | 3 | [21 – 41[ | 3 |
| — | — | ≥41 | 4 |

*Remark 3.* Missing values and imputation

In our dataset, only 144(10.7%) nodules had a missing value for at least one feature. We addressed this classical problem by imputation of missing values using *scikit-learn.org/1.5/modules/impute.html.*

## 2 Supervised Autoencoder Neural Network

Artificial Intelligence - beyond the classical methods mentioned previously - has proved to be very potent in medical applications in the last few years. Deep neural networks are used in state-of-the-art practices to perform both classification tasks (for example, on echography or cytology-based imagery) [32]. Overall, they have proven their effectiveness in bioinformatics for classification and feature selection [18,27,31]. Although there is a large literature on deep learning methods using categorical data [12], to the best of our knowledge there is no paper reporting on the use of clinical categorical data to train neural networks.

### 2.1 A New Supervised Autoencoder Architecture

Classical stacked autoencoders [35] were used recently in metabolomic studies. Their biggest strength is their ability to learn a representation of the data, typically in a space of lower dimension than the input space. As such, they are often used for dimensionality reduction and have applications in the medical field as data denoisers and relevant feature selectors.

The architecture of an autoencoder can be described in two parts: an encoder, that maps the high-dimensional input to a lower-dimension space called the latent space; and a decoder, that tries to reconstruct the original input using only the encoded data. The reconstruction loss is then used to train the network [11,14]. A widely used type of autoencoders is the Variational Autoencoder. This variant adds the assumption that the encoded data follows a prior, a gaussian distribution, and thus combines the reconstruction loss with a distance function (between the gaussian prior and the actual learned distribution) [16].

However, the Gaussian prior and its recently proposed alternatives (e.g. Gaussian mixtures) encourage the latent space to fit those distributions, and are chosen for their computational simplicity rather than their compatibility with the latent structure of the data.

In this paper, we use a Supervised AutoEncoder (SAE) neural network (NN), introduced in 2021 by Barlaud and Guyard in [2], where we relax the parametric distribution assumption. In our SAE-NN, we use labels available in the data (in our case, the malignancy of a nodule) and a classification loss to encourage the latent space to fit the actual distribution and clustering of the data. Thus, two effects are at play during training: the classification loss trains the encoder to be a good classifier between malignant and benign nodules, and the reconstruction loss trains the encoder to learn a good representation of the data to facilitate the task of the decoder (which is also trained during this process). See [2,5,26] and Fig. 1 for more details.

The SAE and its training and evaluation routines were programmed using the PyTorch package.

### 2.2 Supervised Autoencoder Neural Network Algorithm

Let $X$ be the raw data matrix $(n \times d)$, with $Y$ the labels vector $(n \times 1)$. Let $Z$ be the encoded latent matrix $(n \times 2)$. The matrix $\widehat{X}$ $(n \times d)$ is the reconstructed data. $W$ is the matrix of the weights of the linear fully connected autoencoder neural network.

The goal is to compute the network weights $W$ minimizing the total loss which includes both the classification loss and the reconstruction loss. To perform

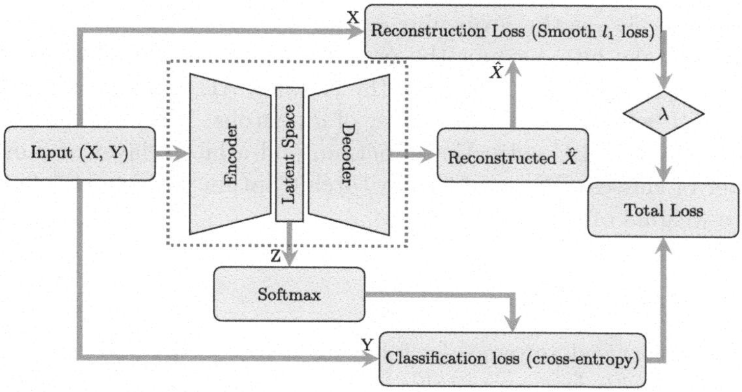


**Fig. 1.** Supervised Autoencoder Neural Network Architecture.

feature selection, we propose to minimize the following constrained approach:

$$Loss(W) = \mathcal{H}(Z, Y) + \lambda\psi(\widehat{X} - X) \text{ s.t. } P_{\eta}^{1,\infty} \tag{1}$$

We use the Cross Entropy (CE) Loss for the classification loss $\mathcal{H}$. We use the robust Smooth $\ell_1$ (Huber) Loss [14], more robust than the mean square error (MSE) as the reconstruction loss $\psi$ and $P_{\eta}^{1,\infty}$ is the projection onto the ball $\mathcal{B}_{1,\infty}^{\eta}$. $\eta$ is the radius of the ball, and $\lambda$ is a weighting parameter between the classification loss and the reconstruction loss.

### 2.3 Feature Selection Using the $\ell_{1,1}$ or the $\ell_{1,\infty}$ Projection

Let us recall that the classical $\ell_2$ norm constraint does not induce any sparsity, and the $\ell_1$ constraint induces unstructured sparsity.

Thus, in this paper, we use the $\ell_{1,\infty}$ projection [25] or the fast bilevel $\ell_{1,\infty}$ projection [3]. Let recall the $\ell_{1,\infty}$ projection. Let $Y \in \mathbb{R}^{n\times m}$ be a real matrix of dimensions $n \times m$, the elements of $Y$ are denoted by $Y_{i,j}$, $i = 1, \ldots, n$, $j = 1, \ldots, m$. The $\ell_{1,\infty}$ norm of $Y$ is

$$\|Y\|_{1,\infty} := \sum_{j=1}^{m} \max_{i=1,\ldots,n} |Y_{i,j}|. \tag{2}$$

Given a radius $\eta \geq 0$, the goal is to project $Y$ onto the $\ell_{1,\infty}$ norm ball of radius $\eta$, denoted by

$$\mathcal{B}_{1,\infty}^{\eta} := \left\{ X \in \mathbb{R}^{n\times m} \ : \ \|X\|_{1,\infty} \leq \eta \right\}. \tag{3}$$

The projection $P_{\mathcal{B}_{1,\infty}^{\eta}}$ onto $\mathcal{B}_{1,\infty}^{\eta}$, also noted $P_{\eta}^{1,\infty}$ is given by:

$$P_{\eta}^{1,\infty}(Y) \ = \ P_{\mathcal{B}_{1,\infty}^{\eta}}(Y) \ = \ arg \min_{X\in\mathcal{B}_{1,\infty}^{\eta}} \frac{1}{2}\|X - Y\|_{\mathrm{F}}^{2} \tag{4}$$

where $\| \cdot \|_F = \| \cdot \|_{2,2}$ is the Frobenius norm.

In order to take into account the projection, we use a double descent algorithm with $\ell_{1,\infty}$ projection [2] to train the Netbio SAE, a linear fully connected network (LFC), which has an input layer of $d$ neurons, 1 hidden layer of 96 neurons followed by a ReLU activation function, and a latent layer of dimension 2 (the number of classes). We used the PyTorch Captum package [20] to compute the feature weights of the trained SAE.

### 2.4 Statistical Evaluation

We train the supervised autoencoder following a standard 4-fold cross validation process https://scikit-learn.org/stable/modules/cross_validation.html.

Note that the test set should still be held out for final evaluation, but the validation set is no longer needed when doing CV. In the basic approach, called 4-fold CV, the training set is split into 4 smaller sets. We train the model on three quarters of the dataset, evaluating it on the remaining quarter, and repeating this process so that every quarter gets used for evaluation. We then report the mean metrics over the 4 steps of cross-validation as a function of parameter $\eta$.

### 2.5 Feature Selection Using $\ell_{1,\infty}$ Projection

We estimate the optimal parameter $\eta$ of the $\ell_{1,\infty}$ projection from the data. The parameter $\eta$ controls the sparsity, and thus the number of selected features. The parameter was optimized using cross-validation with multiple $\eta$.

Using the SHAP method provided in the Captum python package, we compute feature importance for the supervised autoencoder.

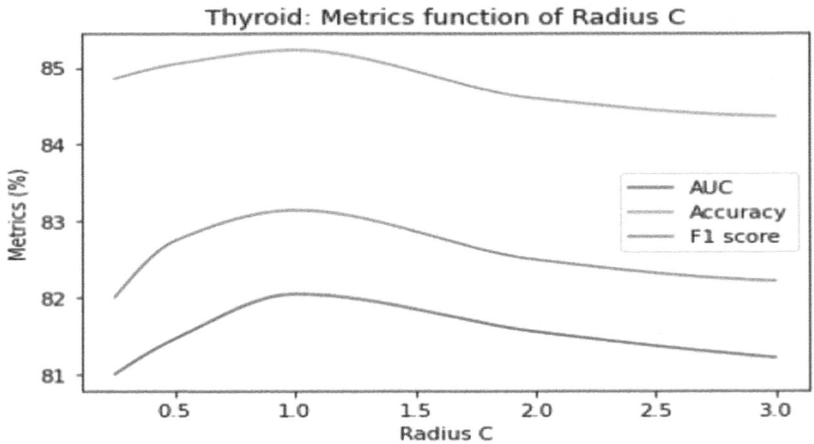


**Fig. 2.** Metrics (Accuracy, AUC and F1 score) as a function of the radius parameter $\eta$

Figure 2 shows the curve of the accuracy versus the radius ($\eta$). We can see that there exist a $\ell_{1,\infty}$ projection with radius $\eta = 1$ for which all metrics are optimum.

**Table 3.** Best Metrics for the SAE.

| Metrics | Accuracy | AUC | F1 | Precision | Recall |
|---|---|---|---|---|---|
| % | $85.22 \pm 1.77$ | $82.04 \pm 1.89$ | $83.14 \pm 1.89$ | $85.09 \pm 2.06$ | $82.04 \pm 1.89$ |

Table 3 reports the best metrics ($\eta = 1$).

Table 4 shows the feature rankings, and displays the second advantage of using the supervised autoencoder over classical methods: Features are selected more strongly. Starting from the fifth most important feature, the importance already become relatively negligible. Features are also ranked very consistently across runs; both of these qualities help with interpretability, which is crucial for clinical use.

**Table 4.** Feature ranking by the autoencoder (Mean over 4 seeds and 4 folds).

| Feature | Mean | Std |
|---|---|---|
| TI-RADS | 0.97 | 0.05 |
| Cyto-BETHESDA | 0.73 | 0.05 |
| Cyto-sub-classif | 0.44 | 0.02 |
| Echo Size | 0.12 | 0.05 |
| Colloid | 0.03 | 0.02 |
| Age | 0.02 | 0.004 |
| Hyperthyroidism | 0.007 | 0.002 |
| HASHIMOTO | 0.003 | 0.004 |
| Sex | 0.0003 | 0.005 |

## 2.6 Comparison with Classical Classification Methods

We compare our method with Partial Least Square (PLS) [36], LASSO [33] and RANDOM FORESTS (RF) [1] using PyTorch scikit-learn classification tool with standard parameters.

Table 5 show that the SAE obtains superior performance metrics than most of the classical methods: The autoencoder outperforms all the classical methods in terms of accuracy, AUC and F1 score by at least 2%. Note the larger STD of the random Forest.

Note that, as reported in Table 6, the top 3 ranked features of the SAE and PLS are similar while the 2 top ranked features of LASSO and RF are inverted. Note that the feature importance in Random Forest is computed using a metric called Gini importance. Moreover, all classical methods except RF give a sizeable importance to the patient's age, whereas the SAE does not.

**Table 5.** Metrics Comparison with classical classification methods.

| Method | PLS | LASSO | RF | SAE |
|---|---|---|---|---|
| Accuracy | 83.24 ± 1.62 | 83.11 ± 1.73 | 83.30 ± 3.48 | **85.19** ± 1.77 |
| AUC | 83.6 ± 1.62 | 81.21 ± 1.73 | 81.20 ± 3.48 | **84.02** ± 1.77 |
| Precision | 83.38 ± 1.161 | 81.21 ± 1.73 | 81.20 ± 3.4 | **84.02** ± 1.77 |
| Recall | 81.53 ± 1.56 | 81.21 ± 1.51 | 81.20 ± 3.4 | **84.02** ± 1.77 |
| F1 score | 82.25 ± 1.5 | 82.15 ± 1.51 | 82.36 ± 3.4 | **82.99** ± 1.77 |

**Table 6.** Feature Comparison with classical methods.

| Feature | PLS | LASSO | RF | SAE |
|---|---|---|---|---|
| TI-RADS | 1 | 0.85 | 0.65 | **0.97** |
| Cyto-BETHESDA | 0.88 | 1 | 1 | **0.73** |
| Cytology-subclassification | 0.34 | 0.33 | 0.42 | **0.44** |
| Age | 0.20 | 0.17 | Echo Size: 0.18 | **Colloid: 0.03** |
| Echo Size | 0.14 | 0.14 | Colloid: 0.05 | **0.12** |

## 3 Prognosis in the Latent Space of the Supervised Autoencoder with a Confidence Score

Above better performance and better feature importance ranking, the other main advantage of the autoencoder is the projection of the data in the latent space by the encoder: as our latent space is of dimension 2 (two target classes), it can very easily be represented, and give accurate and visually explained predictions. As we can see in Fig. 3, the two classes are distinctly clustered in the latent space. Note that among classical methods, only the PLSDA offers any projection capabilities, and even then, does not cluster data in any meaningful way [5]. The softmax used on the latent space (see Fig. 1) gives us a confidence score ($CS(i)$) on the predicted label for each patient $i$.

$$CS(i) = softmax(Z) = \frac{exp(Z_i)}{\sum_{j=1}^{k} exp(Z_j)} \ \forall i = 1, 2 \quad (5)$$

This benefit is only offered by the supervised autoencoder: none of the classical methods provide a sense of how confident the prediction was.

To take full advantage of this, we developed a graphical user interface (GUI) that allows for quick visualization of the impact of each feature on the prediction (Fig. 3). Using the location of the nodule in the latent space, coupled with the confidence score, the physician can get a grasp on how confident the network is in its prediction, and why. The code to reproduce these results is available on Github[1].

[1] https://github.com/CyprienGille/Diagnostic-SAE-GUI.

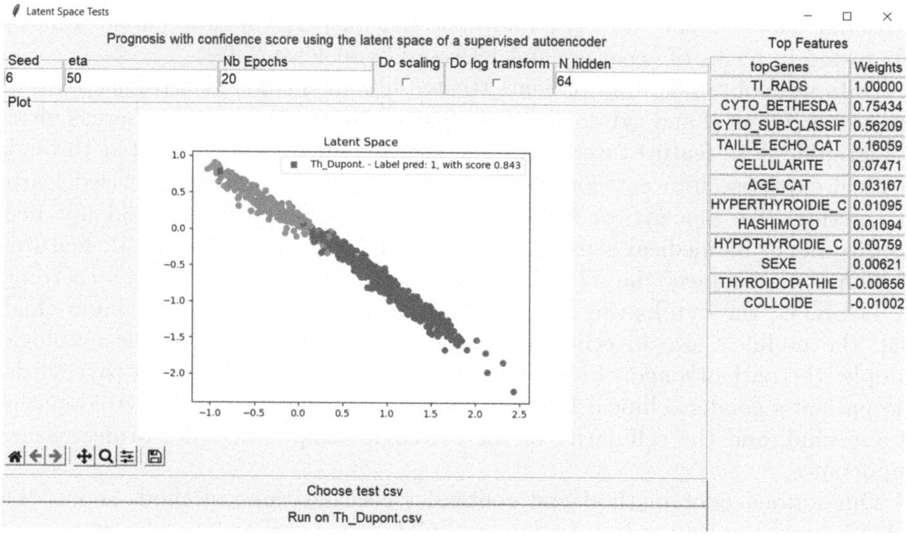


**Fig. 3.** Graphical User Interface for feature testing.

In this example, we can see that the test patient has been classified in class 1 with a confidence score $CS(i) = 0.843$. Note that the order of the less important features differs from the one shown in Table 4 because fewer epochs were used.

# 4 Discussion

Because of their high prevalence in the worldwide population, thyroid nodules remain a major public health problem. In particular, indeterminate thyroid nodules raise the problem of diagnosing their malignancy. Consequently, their medical and surgical management is inevitably burdened with a significant individual and medico-economic impact [4, 13, 23, 34].

Most of the diagnostic methods of the malignancy of ITNs are based on the measurement of omics data [13,28]. Among them, the Thyroseq v3 and the AFIRMA Genomic Sequencing Classifier are the most developed current molecular tests with high performances. The only effective methods, namely the main commercially available molecular tests, are associated with prohibitively high costs, as previously mentioned.

In this paper we proposed a low-cost and accurate thyroid nodule diagnostic classification method based on a combination of clinical, ultrasound, and cytological data using a supervised autoencoder method. Our model displayed high performances reaching 85.22%, 82.99% and 84.02% in terms of accuracy, AUC and F1 score, respectively. We also evaluated the metrics of three classical machine learning methods on the same dataset: a partial least squares-discriminant analysis model, a lasso regression analysis model, and a random forest model. Of particular interest, the supervised autoencoder outperformed all classical methods tested in terms of accuracy and F1 score by at least 2.4%.

Data were obtained from patients treated in four tertiary referral hospitals of two towns. Clinical and cytological data features were categorical, whereas ultrasonographic data features were continuous real-valued data, converted through medical expertise into categorical data rather than by a non-supervised learning method like kmeans, or supervised like CAIM [17] which would not necessarily make biomedical sense. Of the initial features, we picked 12 features of higher importance: the TI-RADS score, the cytological category according to TBSRTC, the cytological subclassification according to Valderrabano et al. [34], the nodule's size in echography, the presence of colloid in the cytologic sample, the patient's age, clinical hyperthyroidism, the Hashimoto's thyroiditis, the patient's gender, clinical hypothyroidism, the existence of a thyroidopathy of any kind, and the cellularity of the cytologic sample, in order of decreasing importance.

Our autoencoder method and contender classifications methods report TI-RADS score, the Bethesda category, and Cyto-subclassif as 3 top features.

Features like the TI-RADS score, the Bethesda category are well known to be associated with the risk of malignancy of one TN, and remain the spearheads of the current diagnostic strategy [9,30]. The cytologic subclassification according to Valderrabano et al. is an emerging classification, which has already proven its ability to refine the risk of malignancy among ITNs [34].

There is wide evidence that thyroid autoimmune disease (Hashimoto's thyroiditis being the most frequent) and thyroid cancer frequently coexist, even if the underlying mechanisms remain unclear.

Beyond its performances, the use of a supervised autoencoder allowed us to create a graphical representation of the malignity prediction of one TN, as a projection of the data into a latent space. It has only two dimensions, making it easy for clinicians to represent, visualize, and interpret the prediction. Our model also gives a confidence score on the predicted label, which allows estimations of the degree of reliability of the prediction. We have created it in such a way that it is not only intuitive to use, but also very simple and fast, allowing it to be used in routine practice. In addition, it is now freely available on Github under the MIT License, which avoids any extra cost related to the use of these tools, thereby seeking to reduce the medico-economic impact of the management of ITNs.

The main methodological arguments that support the performance of our supervised autoencoder are the validation results. Indeed, our method was validated by a random cross-validation method on the whole dataset, which guarantees its reliability. Moreover, the cohort was composed of a large number of TNs, with many features included in the initial database, despite its retrospective design which is the main limitation of our work. Note that there was very little missing data considering the selected features. Those factors allowed us to create a robust and accurate neural network. Other pitfalls that may limit the robustness of our work reside in the absence of proofreading of the US images and cytological slides by another expert. Finally, the choice to include patients with lesions suspected of malignancy may be debatable. Nevertheless, the wide variability in the risk of malignancy in this cytological category found in the literature raises questions, and we have therefore sought to improve the risk stratification of malignancy within this cytological category [6,28]. In addition, the inclusion of patients with suspected malignant nodules was mainly used to train our mathematical model to identify ITNs with aspects suggestive of papillary thyroid carcinoma. As the follicular variant of the papillary thyroid carcinoma is the most frequent variant found in the cytologically indeterminate categories, our strategy was able to better diagnose them.

To further improve accuracy, in a future work we propose to add metabolomic data which is a very promising approach, particularly adapted to routine clinical practice [7,15].

## 5 Conclusion

In this paper, we showed that using a supervised autoencoder on categorical medical data produced a prognosis on the malignancy of thyroid nodules that was perfectly suited for a physician's use. Beyond high statistical metrics and accuracy, the autoencoder offered features that were not available with classical classification methods, and that are greatly relevant from the medical practitioner's point of view: strong feature selection, output of a confidence score, and a visually meaningful 2-dimensional latent space.

**Data Availibility Statement.** The datasets generated and/or analyzed during the current study are available in the Diagnostic-SAE-GUI repository, at https://github.com/CyprienGille/Diagnostic-SAE-GUI/tree/main/data.

**Ethics Approval and Consent to Participate.** The authors confirm that all methods were carried out in accordance with the relevant guidelines and regulations.
The authors confirm that informed consent was obtained from all subjects.
The authors confirm that this retrospective study has been approved by the institutional ethics committees of the University Hospital of Nice and the University Hospital of Toulouse.
In addition, this study met the requirements of the French reference method 04 of the "*commission nationale de l'informatique et des libertés*" and was declared under the numbers 2204177v0, and F20210608133203. (Health Data Hub, https://www.health-data-hub.fr/projets/amelioration-du-diagnostic-et-de-la-prise-en-charge-chirurgicale-des-nodules-thyroidiens).

**Consent for Publication.** Not applicable.

**Funding.** Not applicable.

**Competing Interests.** The authors declare that they have no competing interests.

**Author Contributions.** GD provides the medical dataset and comments medical results, MB write the machine learning section, CG and MB designed the Pytorch code and the experiment. All authors read and approved the final manuscript.

## References


1. Altmann, A., Toloşi, L., Sander, O., Lengauer, T.: Permutation importance: a corrected feature importance measure. Bioinformatics **26**(10), 1340–1347 (2010)
2. Barlaud, M., Guyard, F.: Learning a sparse generative non-parametric supervised autoencoder. In: ICASSP 2021 - 2021 IEEE International Conference on Acoustics, Speech and Signal Processing (ICASSP), pp. 3315–3319 (2021). https://doi.org/10.1109/ICASSP39728.2021.9414410. https://ieeexplore.ieee.org/document/9414410
3. Barlaud, M., Perez, G., Marmorat, J.P.: Linear time bi-level l1, $\infty$ projection; application to feature selection and sparsification of auto-encoders neural networks. arXiv:2407.16293v1 [cs.LG] (2024)
4. Bongiovanni, M., Bellevicine, C., Troncone, G., Sykiotis, G.P.: Approach to cytological indeterminate thyroid nodules. Gland Surg. **8**, S98–S104 (2019). https://doi.org/10.21037/gs.2018.12.06. https://www.ncbi.nlm.nih.gov/pmc/articles/PMC6694028/
5. Chardin, D., Gille, C., Pourcher, T., Humbert, O., Barlaud, M.: Learning a confidence score and the latent space of a new supervised autoencoder for diagnosis and prognosis in clinical metabolomic studies. BMC Bioinform. **23**(1), 1–16 (2022). https://doi.org/10.1186/s12859-022-04900-x. https://bmcbioinformatics.biomedcentral.com/articles/10.1186/s12859-022-04900-x
6. Cibas, E.S., Ali, S.Z.: The 2017 bethesda system for reporting thyroid cytopathology. Thyroid Official J. Am. Thyroid Assoc. **27**(11), 1341–1346 (2017). https://doi.org/10.1089/thy.2017.0500
7. D'Andréa, G., et al.: Pilot study on the use of untargeted metabolomic fingerprinting of liquid-cytology fluids as a diagnostic tool of malignancy for thyroid nodules. Metabolites **13**(7), 782 (2023). https://doi.org/10.3390/metabo13070782
8. d'Andréa, G.C.J.: Apport de l'intelligence artificielle dans la prise en charge des nodules thyroïdiens indéterminés, p. 85 (2021). https://dumas.ccsd.cnrs.fr/dumas-04244459
9. Deng, X.H., Tang, L.N., Liu, S.Q., Li, X.L., He, Y.P., Xu, H.X.: A proposal to stratify the intermediate-risk thyroid nodules according to the AACE/ACE/AME guidelines with ultrasound features. Sci. Rep. **7**(1), 17901 (2017). https://doi.org/10.1038/s41598-017-18207-y
10. Gharib, H., et al.: AACE/AME/ETA Task Force on Thyroid Nodules: American association of clinical endocrinologists, associazione medici endocrinologi, and european thyroid association medical guidelines for clinical practice for the diagnosis and management of thyroid nodules: executive summary of recommendations. J. Endocrinol. Invest. **33**(5), 51–56 (2010)
11. Goodfellow, I., Bengio, Y., Courville, A.: Deep Learning. MIT Press (2016)
12. Hancock, J.T., Khoshgoftaar, T.M.: Survey on categorical data for neural networks. J. Big Data **7**(1), 1–41 (2020). https://doi.org/10.1186/s40537-020-00305-w. https://journalofbigdata.springeropen.com/articles/10.1186/s40537-020-00305-w
13. Haugen, B.R., et al.: 2015 American thyroid association management guidelines for adult patients with thyroid nodules and differentiated thyroid cancer: the American

thyroid association guidelines task force on thyroid nodules and differentiated thyroid cancer. Thyroid Official J. Am. Thyroid Assoc. **26**(1), 1–133 (2016). https://doi.org/10.1089/thy.2015.0020
14. Huber, P.J.: Robust Statistics. Wiley (1981)
15. Kell, D.B.: Metabolomics and systems biology: making sense of the soup. Curr. Opin. Microbiol. **7**(3), 296–307 (2004). https://doi.org/10.1016/j.mib.2004.04.012
16. Kingma, D.P., Welling, M.: Auto-encoding variational bayes (2022). https://doi.org/10.48550/arXiv.1312.6114. http://arxiv.org/abs/1312.6114
17. Kurgan, L., Cios, K.: CAIM discretization algorithm. IEEE Trans. Knowl. Data Eng. **16** (2004)
18. Leclercq, M., et al.: Large-scale automatic feature selection for biomarker discovery in high-dimensional OMICs data. Front. Genet. **10**, 452 (2019). https://doi.org/10.3389/fgene.2019.00452
19. Li, M., Dal Maso, L., Vaccarella, S.: Global trends in thyroid cancer incidence and the impact of overdiagnosis. Lancet Diab. Endocrinol. **8**(6), 468–470 (2020). https://doi.org/10.1016/S2213-8587(20)30115-7
20. Lundberg, S.M., Lee, S.I.: A unified approach to interpreting model predictions. In: Proceedings of the 31st International Conference on Neural Information Processing Systems, NIPS 2017, pp. 4768–4777. Curran Associates Inc. (2017)
21. Mazzaferri, E.L.: Management of a solitary thyroid nodule. N. Engl. J. Med. **328**(8), 553–559 (1993). https://doi.org/10.1056/NEJM199302253280807
22. Morris, L.: Molecular profiling of thyroid nodules-are these findings meaningful, or merely measurable?: a review. JAMA Otolaryngol. Head Neck Surg. **146**(9), 845–850 (2020). https://doi.org/10.1001/jamaoto.2020.1851
23. Nicholson, K.J., Roberts, M.S., McCoy, K.L., Carty, S.E., Yip, L.: Molecular testing versus diagnostic lobectomy in bethesda III/IV thyroid nodules: a cost-effectiveness analysis. Thyroid Official J. Am. Thyroid Assoc. **29**(9), 1237–1243 (2019). https://doi.org/10.1089/thy.2018.0779
24. Paschke, R., Cantara, S., Crescenzi, A., Jarzab, B., Musholt, T.J., Sobrinho Simoes, M.: European thyroid association guidelines regarding thyroid nodule molecular fine-needle aspiration cytology diagnostics. Eur. Thyroid J. **6**(3), 115–129 (2017). https://doi.org/10.1159/000468519
25. Perez, G., Condat, L., Barlaud, M.: Near-linear time projection onto the l1,∞ ball application to sparse autoencoders. In: IEEE International Conference on Tools with Artificial Intelligence, Washington, USA (2024)
26. Pognonec, P., Gustovic, A., Djabari, Z., Pourcher, T., Barlaud, M.: Mitotic index determination on live cells from label-free acquired quantitative phase images using a supervised autoencoder. IEEE/ACM Trans. Comput. Biol. Bioinf. **18**(6), 2828–2834 (2021). https://doi.org/10.1109/TCBB.2021.3115876
27. Ravi, D., et al.: Deep learning for health informatics. IEEE J. Biomed. Health Inform. **21**(1), 4–21 (2017). https://doi.org/10.1109/JBHI.2016.2636665
28. Rossi, E.D., Pantanowitz, L., Faquin, W.C.: The role of molecular testing for the indeterminate thyroid FNA. Genes **10**(10), 736 (2019). https://doi.org/10.3390/genes10100736
29. Russ, G., Royer, B., Bigorgne, C., Rouxel, A., Bienvenu-Perrard, M., Leenhardt, L.: Prospective evaluation of thyroid imaging reporting and data system on 4550 nodules with and without elastography. Eur. J. Endocrinol. **168**(5), 649–655 (2013). https://doi.org/10.1530/EJE-12-0936. https://academic.oup.com/ejendo/article/168/5/649/6659664

30. Sands, N.B., et al.: McGill thyroid nodule score (MTNS): "rating the risk," a novel predictive scheme for cancer risk determination. J. Otolaryngol. Head Neck Surg. = Le Journal D'oto-Rhino-Laryngologie Et De Chirurgie Cervico-Faciale **40 Suppl. 1**, S1–13 (2011)
31. Song, Y., et al.: Deep learning enables accurate diagnosis of novel coronavirus (COVID-19) with CT images. IEEE/ACM Trans. Comput. Biol. Bioinf. **18**(6), 2775–2780 (2021). https://doi.org/10.1109/TCBB.2021.3065361
32. Sparano, C., et al.: Choosing the best algorithm among five thyroid nodule ultrasound scores: from performance to cytology sparing—a single-center retrospective study in a large cohort. Eur. Radiol. **31**(8), 5689–5698 (2021). https://doi.org/10.1007/s00330-021-07703-5
33. Tibshirani, R.: Regression shrinkage and selection via the lasso. J. Roy. Stat. Soc. Ser. B (Methodol.) 267–288 (1996)
34. Valderrabano, P., et al.: Cancer risk associated with nuclear atypia in cytologically indeterminate thyroid nodules: a systematic review and meta-analysis. Thyroid **28**(2), 210–219 (2018). https://doi.org/10.1089/thy.2017.0419. https://www.ncbi.nlm.nih.gov/pmc/articles/PMC7869885/
35. Vincent, P., Larochelle, H., Lajoie, I., Bengio, Y., Manzagol, P.A.: Stacked denoising autoencoders: learning useful representations in a deep network with a local denoising criterion. J. Mach. Learn. Res. **11**, 3371–3408 (2010)
36. Wold, S., Sjostrom, M., Eriksson, L.: Pls-regression: a basic tool of chemometrics. Elsevier **58**(2), 109–130 (2001)
37. Yousefi, E., Sura, G.H., Somma, J.: The gray zone of thyroid nodules: Using a nomogram to provide malignancy risk assessment and guide patient management. Cancer Med. **10**(8), 2723–2731 (2021). https://doi.org/10.1002/cam4.3866

# Can Smoothing Methods Recognize the Patterns of the Hazard Function in Complex Clinical Scenarios?
## A Simulation Study Using Discrete-Time Survival Models

Giuseppe Marano[1(✉)], Giacomo Biganzoli[1], Ester Luconi[1], Elia Mario Biganzoli[1,2], and Patrizia Boracchi[1,2]

[1] Department of Biomedical and Clinical Sciences, University of Milan, Milan, Italy
{giuseppe.marano,Giacomo.biganzoli,ester.luconi,elia.biganzoli, patrizia.boracchi}@unimi.it

[2] Data Science and Research Center (DSRC), University of Milan, Milan, Italy

**Abstract.** Simulation studies were performed to assess the performances in realistic estimation problems of B-splines (BS) and restricted cubic splines (RCS) within flexible models for survival data. Several theoretical distributions, designed to match the estimates from clinical cancer studies, were adopted for data generation, including a proportional hazards (PH) model and a non-PH one. The simulation plan included sample sizes equal to 500 and 2000, with 10% censored times within the follow-up period. The investigated performances were: estimation error (through MSE), agreement of the estimates with the theoretical target function (through $R^2$), and the ability to identify the correct number of peaks of the target hazard, along with the ability to detect peak times and peak heights within clinically relevant ranges. We found reduced performances in the pattern identification task, even for the simplest distributional models, for which estimation error was optimal. Overall, BS slightly outperformed RCS in pattern identification, while estimation error was equivalent. These results will be compared to further simulations concerning a more extensive set of smoothing methods.

**Keywords:** smoothed estimation · discrete-time survival models · regression modeling strategies · non-proportional hazards

## 1 Introduction

In survival analysis, accurate estimation of the hazard function is of uttermost importance for providing useful information and assessing specific research hypotheses in real studies. Simes and Zelen [1] have discussed the relevance of this function for the interpretation of time-to-event processes within the medical field. Furthermore, accurate estimation of the hazard is a prerequisite for obtaining accurate estimates of survival probabilities, and, in the multivariable modeling framework, it is essential when estimation of the functional form of


M. Vettoretti et al. (Eds.): CIBB 2023, LNBI 14513, pp. 43–57, 2025.
https://doi.org/10.1007/978-3-031-90714-2_4

complex time-dependent covariate effects is required [2]. In simulation studies, the theoretical models for this target function usually fall into either parametric distributions (e.g.: Weibull, log-logistic) or purely mathematical functions. Although essential for evaluation purposes, the functions in the former group exhibit simple behavior, which limits their ability to represent complex scenarios. The latter group is adopted to specify more complex patterns. For example, the following expression was used in [3] to specify a two-peaked hazard function: $h(t) = 1.4 \sin(3\pi t) + 2.1$. However, these functions could not be flexible enough to be representative of the "real" ones. This holds true, for example, in studies about non-severe diseases with a long follow-up, where the shape of the hazard is expected to be non-monotonic, possibly characterized by a multi-peaked behavior [4–6].

Following the above considerations, we set out to evaluate the ability of smoothing estimation strategies to recognize the major patterns of realistic hazard functions. To such end, two scenarios were considered in this study, in which a biostatistician is asked to perform an exploratory analysis on the shape of the hazard of a determined clinical event, either for a homogeneous population or a heterogeneous one. The term 'homogeneous' indicates a population characterized by individuals very similar in terms of covariate values. Thus, regression models are not needed to account for the relationship between covariates and hazard. Instead, 'heterogeneous' refers to a population where subjects with distinct covariates show a different event occurrence over time, thus requiring a regression approach to account for the relationship between covariates and hazard. Several original distributional models for a single time-to-event were used for data generation, which resemble the estimates of the hazard obtained in studies on breast cancer patients [4,7], thus providing a valid prototype set for performing simulations akin to "real clinical behavior". Among the set of available estimation methods, we resorted to discrete-time survival models [8], which are well suited for estimating the hazard function, and enable the use of standard techniques of analysis already implemented in scientific software, also including smoothing methods for regression models (GLM family), such as regression splines and penalized estimation techniques.

The aims of the present study are twofold: the first one consists of evaluating the estimation error and the accuracy of several flexible polynomials, namely restricted cubic splines (RCS) and basis splines (BS) [9]. To this end, standard evaluation criteria were adopted based on the mean square error (MSE) and the bias/variance decomposition. The second aim was to evaluate the ability of the above splines to identify the distinctive patterns of the target hazard. More specifically, in the present study, the interest lies in the timings of the local maxima, or, in other words, on peak times: this choice was motivated by the fact that peak times characterize those time periods with a higher incidence of events and are, thus, useful for clinical decision making. However, there are no widely accepted evaluation criteria for assessing this feature, and the task is complicated by the possibility that the estimates could exhibit spurious peaks or spurious flattened behavior (due, respectively, to a slight over-estimation or under-estimation of the degrees of freedom of the splines). For this reason, the simulation plan was extended to additionally evaluate the ability to detect the peaks in the "true" hazard function.

## 2 Data and Methods

### 2.1 Methodological Notes

In survival analysis, the targets for analysis are specific functions that represent the key characteristics of time-to-event processes, such as the probability of remaining "free" of the event over time (survival function) and the rate of occurrence of the event over time (hazard function). In the continuous framework, the hazard function is defined as the limit of the probability that the event-time T falls within a interval $[t, t + \Delta t)$ conditionally to the fact that the event has not occurred before t:

$$h(t) = \lim_{\Delta t \to 0} \frac{\Pr(t \leq T < t + \Delta t \,|\, T \geq t)}{\Delta t}. \tag{1}$$

It is strictly linked to the survival function: $S(t) = Pr(T > t)$; through the relationship:

$$S(t) = \exp\left(-\int_0^t h(u)\, du\right). \tag{2}$$

For a discrete time-to-event variable, let's indicate with $I_1, \ldots, I_K$ a partition of $K$ adjacent time intervals: $I_1 = [0, t_1), \ldots, [t_{k-2}, t_{k-1}), [t_{k-1}, +\infty)$. In a similar fashion to the continuous case, the hazard function is defined as the probability that the event-time $T$ lies in an interval $I_k$, $(k = 1, \ldots, K)$ conditionally to the fact that the event has not occurred before:

$$h(I_k) = P\{t_{k-1} \leq T < t_k \,|\, T \geq t_{k-1}\}, \quad k = 1, \ldots, K, t_0 = 0, t_K = \infty; \tag{3}$$

and the relationship with the discrete survival function is:

$$S(I_k) = \prod_{h=1}^{k} (1 - h(I_k)). \tag{4}$$

Beyond the differences between theoretical models, the methods for discrete event-times are a straightforward choice in studies in which time is measured on a discrete scale, and are also useful for estimation of the continuous hazard. For a thorough treatment of the topics discussed in this paragraph, the readers are referred to: [8,10].

### 2.2 Simulation Plan

The whole simulation plan is described in the following paragraphs according to the ADEMP (aims/data-generating mechanisms/estimands/methods/performance measures) scheme [11].

**Aims.** The general aim is to evaluate to what extent the performances of discrete survival models with regression splines for flexible specification of the hazard function may "fail" in complex, realistic scenarios. We designed two distinct scenarios in which a biostatistician is asked to estimate the hazard function for a determined clinical event on data recorded either in a homogeneous population or a heterogeneous one. Namely, by homogeneous population we mean a population in which a single probability function of the time-to-event T is shared by all its subjects; for the latter, we mean a population in which the probability distribution varies according to one or more covariates. The biostatistician may safely use any type of smoothing method, but the strategy of analysis that he/she can adopt is not the same in both scenarios.

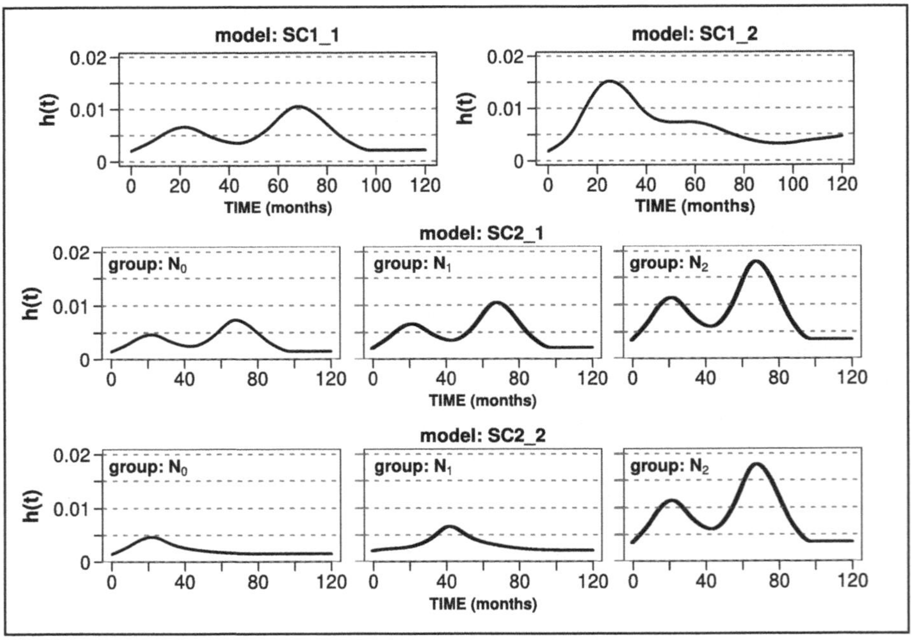


**Fig. 1. Target hazard functions adopted in simulations.** Upper panels: hazards for models $SC_{1.1}$ and $SC_{1.2}$ (homogeneous population); middle and lower panels: hazards for models $SC_{2.1}$ and $SC_{2.2}$ (heterogeneous population). PH = proportional hazards; NPH = non proportional hazards

To allow the scenarios to be representative of real data applications, we defined suitable expressions of the hazard function, able to reproduce the shape of estimated of hazards found in real studies. These functions are shown in Fig. 1. The first scenario ($SC_1$) includes two distinct hazard functions, shown in the upper layers of Fig. 1 (models: $SC_{1.1}$ and $SC_{1.2}$). In brief, these functions are characterized by a two-peaked shape, but for model $SC_{1.2}$, the second peak is less pronounced: note, in fact, the gentle slope of the function in the period between the 50-th and the 58-th month.

Under scenario $SC_2$ the time-to-event has a distinct distribution in three subgroups, identified by a categorical variable. For labeling the subgroups, reference was made to the variable lymph-nodal status, which is amongst the major prognostic variables in breast cancer studies. Thus, the adopted labels were: $N_0$ (no positive lymph nodes), $N_1$ (1 to 3 positive lymph nodes), and $N_2$ (4 or more). Note: the term "positive" means positive to the assessment of the presence of tumor cells within the lymph node.

According to clinical knowledge, the $N_0$ group is observed with higher frequency and is subjected to a lower risk of adverse events (such as death or tumor relapse), while $N_1$ and $N_2$ are relatively less frequent, and the risk in $N_2$ is substantially higher with respect to the other ones. Therefore, the frequencies of the three groups were set to 70%, 15%, and 15%, respectively, and the (conditional) distributions of the time-to-event were designed to have the worst (i.e., lower) survival probabilities in $N_2$. It is worth noting that under model $SC_{2.1}$ (Fig. 1, middle layer) the hazard functions for the $N_0$, $N_1$, and $N_2$ groups differ only by a multiplicative constant, so that the model satisfies the PH assumption. Under model $SC_{2.2}$ (bottom layer), the PH assumption is clearly violated, and the hazard function has a two-peaked shape only for the $N_2$ group. The readers interested in a formal discussion of the PH assumption and the related survival models are referred to [10].

**Data-Generating Mechanisms.** From the expressions of the hazard functions, the respective corresponding probability functions were obtained and used for data generation. Distinct simulations were performed, considering: four distributional models ($SC_{1.1}$, $SC_{1.2}$, $SC_{2.1}$, $SC_{2.2}$), two sample sizes ($N = 500$, 2000), and two regression splines (RCS, BS). In each simulation, 1000 samples were generated from the pertinent probability function and an independent censoring variable with uniform distribution. The proportion of censored times occurring within the follow-up period was fixed at 10%.

**Estimands.** The estimands consist in the sequence of 120 estimated values discrete hazard function over the partition of the follow-up period: $\hat{h}_j; j = 1, ..., 120$ months.

**Methods.** For the simulations performed under scenario $SC_1$, the methods are as follows. The follow-up period is partitioned into 120 one-month intervals. The generated dataset is reshaped in a format suitable for discrete-time survival models ("augmented" format). The new dataset includes $n_i$ records for each subject, $n_i$ being the number of months from the beginning of follow-up until the subject event-time or censoring time. Each record in the new dataset includes two variables: time (from 1 to 120 months) and subject status (1 if the target event has been observed within the corresponding month, 0 otherwise).

Several logistic regression models are fitted on the new dataset, with response variables corresponding to the above subject status, and with a logit link. For each model, time is included as an independent variable, using a regression spline for a flexible modelization. Each model is estimated by the generalized estimating equations (GEE) method to account for within-subject correlation [12]. Several models are estimated, with varying number of spline knots (up to a specific maximum value, see below), then the "best" model is selected according to the quasi-likelihood information criterion (QIC), which is well suited to GEE estimation [13]. Furthermore, to pursue additional protection against overfitting, the following rules are applied: 1) a sound maximum value is set for the number of knots: ten knots and eight internal knots, respectively, for RCS and BS; 2) every model with at least one estimated coefficient higher than 50 is discarded.

A similar estimation strategy is adopted under scenario $SC_2$, although with a necessary additional task. Within any simulation run, three general regression formulas are considered: 1) a "homogeneous population" model, in which only time is included as independent variable; 2) a model with additive effects of time and lymph-nodal status; and 3) a model with an interaction effect between time and lymph-nodal status. Several models are fitted for each formula, with varying number of spline knots (as in the previous case). The final model is chosen by pooling all the results and picking up the estimated model with the minimum QIC value. This implies that the QIC criterion is used for both tuning model complexity (by selecting the "optimal" number of knots) and variable selection.

**Performance Measures: Estimation Error.** MSE, bias, and SD were calculated for each estimand. Bias and SD were expressed both on an absolute scale and as a percentage of the respective "true" hazard value. To quantify the estimation error for the whole hazard function, the integrated square error (ISE) was adopted:

$$ISE = \int_0^{120} [h(s) - \hat{h}_j(s)]^2 \, ds \; ; \tag{5}$$

where $h(s)$ is the target hazard (expression (1)) and the $\hat{h}_j(s)$ are the estimands $\hat{h}_j$. The mean value of ISE over an entire set of 1,000 simulation runs gives an estimate of the mean integrated square error (MISE).

To better understand how much the estimates agree with the target hazard function, the $R^2$ coefficient was calculated from a regression of the discrete target hazard (expression (3)) on the estimated one. In each simulation run, the expression of the pertinent regression model was the following:

$$h_j = \beta_0 + \beta_1 \hat{h}_j + \epsilon \quad , \quad j = 1, \ldots, 120 \; ; \tag{6}$$

where $h_j$ are the values of the discrete hazard in the j-th month, $\hat{h}_j$ the corresponding estimates from the discrete-time model. and $\epsilon$ the residual.

**Performance Measures: Pattern Identification.** In each simulation run, the peaks in the series of estimated values: $\hat{h}_j; j = 1, \ldots, 120$ were identified by the following algorithm for detecting local maxima:

- for each $j = 3, \ldots, 118$, consider the sequence: $\hat{h}_{j-2}$, $\hat{h}_{j-1}$, $\hat{h}_j$, $\hat{h}_{j+1}$, $\hat{h}_{j+2}$, and calculate the finite differences: $\hat{\Delta}_k = \hat{h}_k$ - $\hat{h}_{k-1}$ for $k = j-1$, $j$, $j+1$ and $j+2$;
- a peak occurs within month $j$ if the first two differences are positive ($\hat{\Delta}_{j-1} > 0$, $\hat{\Delta}_j > 0$) and the remainder ones are negative ($\hat{\Delta}_{j+1} < 0$, $\hat{\Delta}_{j+2} < 0$).

To evaluate the departures of the timings of the estimated peaks from the target ones, the following criteria were adopted:

- total number of peaks within the whole follow-up period (to be compared between target and estimate);
- proximity in time between "estimated" and "true" peaks: this feature was evaluated by the frequency of simulation runs (over 1,000) in which, for each peak in the target hazard, there is at least one peak in the estimated hazard occurring within 15 months before or after it;
- closeness between peaks: evaluated by the frequency of simulation runs in which, for each peak in the target hazard, there is at least one peak in the estimated hazard in the proximity of the true one (previous criterion) and the difference between the values of the hazard at peak times (estimated versus target) is no larger than 15% of the value of the target;
- presence of spurious peaks, that is: of peaks in the estimated hazard within a time interval in which the target has a monotone behavior ("false positives");
- absence of peaks in the estimated hazard within an interval in which the target exhibits one peak ("false negatives").

To have a clean evaluation of the probability of occurrence of false positives and false negatives, the following rules were adopted: first, the simulations were performed only under the models $SC_{1.1}$ and $SC_{1.2}$. Second, false positives and false negatives were assessed within the same fixed time period, under two similar distributional models. The scheme for detection of false positives and false negatives is sketched in Fig. 2 for model $SC_{1.1}$. The results were reported in terms of frequency, over 1,000 simulation runs, of false positives and false negatives detection.

## 3 Results

The simulations were performed for two homogeneous population models and two heterogeneous population models (scenarios: $SC_1$ and $SC_2$, respectively), according to the type of flexible polynomial included in the regression model (RCS, BS) and the sample size (N = 500, 2000). Table 1 presents a synthetic report about the number of events in the generated datasets and the complexity of the fitted models – complexity assessed by the respective number of regression

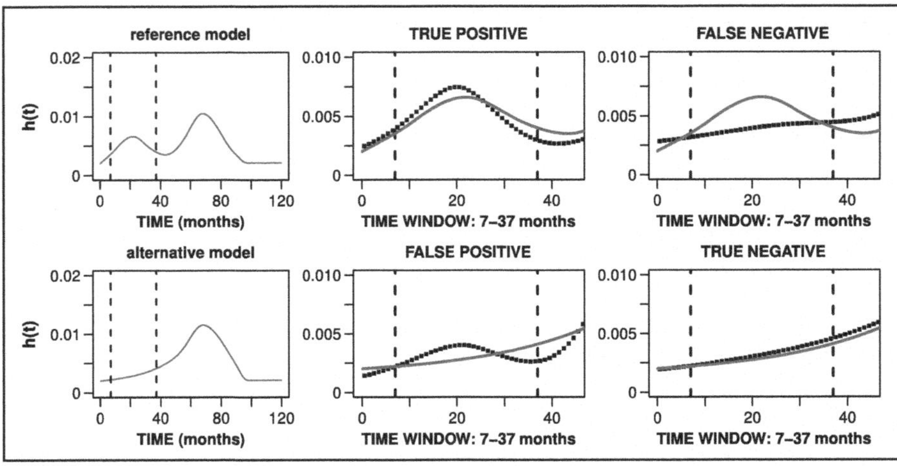


**Fig. 2. Detection of false negatives and false positives** 1) a time interval centered around a fixed peak time (shown by the vertical dashed lines) is considered; 2) for each simulation run under the reference distributional model, a false negative is identified by the estimated hazard showing zero peaks within the reference interval (upper right panel); 3) an alternative distributional model is designed, the hazard of which closely approximates the hazard function of model $SC_{1.1}$ outside the reference interval (lower left panel); 4) For each simulation run under the alternative model, a false positive is identified by the presence of one or more peaks of the estimated hazard within the reference interval.

coefficients (cfr: [9]). For what concerns the homogeneous population models, on average the number of generated events was between 200 and 300 for N = 500, and between 800 and 1100 for N = 2000. It may be seen that these values largely exceed the threshold of ten times the number of regression coefficients, which is generally adopted as a guideline for ensuring reliable estimation [14]. Due to this large discrepancy, one may safely expect that the performances of the estimates are not affected by issues typically associated with small sample sizes or a low number of events per variable (EPV) [15].

For the heterogeneous population models, the average number of generated events was between 100 and 200 for N = 500, and between 600 and 800 for N = 2000. Of note, for N = 500 and model $SC_{2.2}$, the average number of model coefficients was practically equal to the threshold above (approximatively 150 events versus 15 model coefficients).

### 3.1 Estimation Error and Agreement with the Target Hazard

A summary of the values of ISE and $R^2$ are shown in Table 2. The global estimation error (measured by the ISE) was comparable between smoothing strategies using either RCS or BS, for each scenario, model and sample size.

Concerning the agreement with the target hazard, the distribution of the $R^2$ coefficient (within each set of 1,000 simulation runs) was left-skewed, with a high

**Table 1. Generated events and model selection.** First row (Events): average number of generated events in each set of 1,000 simulation runs. Second row (Model width): average number of model coefficients.

| | SPLINE | N | MODEL | | | |
|---|---|---|---|---|---|---|
| | | | $SC_{1.1}$ | $SC_{1.2}$ | $SC_{2.1}$ | $SC_{2.2}$ |
| Events | RCS | 500 | 219 | 271 | 197 | 151 |
| | | 2000 | 874 | 1082 | 788 | 607 |
| | BS | 500 | 218 | 270 | 196 | 152 |
| | | 2000 | 879 | 1080 | 788 | 607 |
| Model width | RCS | 500 | 8 | 7 | 11 | 15 |
| | | 2000 | 9 | 9 | 12 | 23 |
| | BS | 500 | 8 | 6 | 10 | 15 |
| | | 2000 | 9 | 8 | 11 | 21 |

prevalence of values within the range 0.8–1.0. Reported in Table 2 there are the following indices: the minimum value, the 5-th centile: $q_{5\%}$; and the first and third quartiles: $Q_1$ and $Q_3$. For both the designed scenarios, the agreement was highly satisfactory when the sample size was largest (N = 2000). Considering the $q_{5\%}$ index, the worst case occurred for model $SC_{2.2}$ and smoothing using BS, where $q_{5\%}$ was 0.80. This means that the (estimated) probability of having $R^2 \geq 0.80$ in one random sample from the reference model was at least 95%.

For N=500 there were still non-trivial chances of obtaining an appreciable agreement in a random sample. For example, for model $SC_{1.1}$, $Q_1$ was equal to 0.82 and 0.84, when using, respectively, RCS and BS as smoothers. These values indicate that the probability of obtaining $R^2 \geq 0.80$ in a random sample was higher than 75.0% in both cases. Moreover, as for the ISE, no substantial differences emerged between RCS and BS.

For what concerns the pointwise estimation error, Table 4, reported in the Appendix, shows root MSE, bias and SD for the estimates of the hazard function at peak times. For what concerns the absolute error, the bias was, overall, negligible as compared to the root MSE, with few exceptions. For example, the bias obtained at the second peak under model $SC_{2.2}$ for the $N_2$ subgroup was about $-0.004$ ($-4.03 * 10^{-3}$) for N = 500 and $-0.001$ ($-1.03 * 10^{-3}$) for N = 2000, which are roughly one half and one third of the respective root MSE.

Further insights may be obtained by comparing bias and SD to the magnitude of the target hazard function (relative error). Under the homogeneous population scenario the relative bias was negligible: close to 1% of the respective "true" hazard for model $SC_{1.1}$ at both peaks times, and for the 1st peak time in model $SC_{1.2}$; below 5% for estimation of second peak time. For each of the above models, the relative SD was lower or equal than 20% for N = 500, and lower of equal than 10% for N = 2000 (BS performing slightly better than RCS). More in general, the amount of relative bias and SD for each month (120 time intervals) was comparable, overall, to the values above.

**Table 2.** Global estimation error and agreement with the target hazard.

| MODEL | SPLINE | N | root ISE mean, range | $R^2$ min, $q_{5\%}$, $Q_1$-$Q_3$ |
|---|---|---|---|---|
| $SC_{1.1}$ | RCS | 500 | 0.012, 0.005-0.029 | 0.56, 0.73, 0.82-0.91 |
| | | 2000 | 0.007, 0.004-0.012 | 0.84, 0.92, 0.94-0.97 |
| | BS | 500 | 0.011, 0.004-0.027 | 0.56, 0.74, 0.84-0.92 |
| | | 2000 | 0.006, 0.003-0.011 | 0.85, 0.93, 0.95-0.98 |
| $SC_{1.2}$ | RCS | 500 | 0.014, 0.007-0.026 | 0.61, 0.81, 0.88-0.94 |
| | | 2000 | 0.008, 0.003-0.015 | 0.89, 0.94, 0.96-0.98 |
| | BS | 500 | 0.013, 0.006-0.029 | 0.65, 0.82, 0.89-0.95 |
| | | 2000 | 0.008, 0.004-0.014 | 0.90, 0.94, 0.96-0.98 |
| $SC_{2.1}$ | RCS | 500 | 0.034, 0.013-0.112 | 0.26, 0.58, 0.81-0.91 |
| | | 2000 | 0.018, 0.008-0.054 | 0.70, 0.86, 0.94-0.97 |
| | BS | 500 | 0.032, 0.012-0.092 | 0.22, 0.64, 0.83-0.92 |
| | | 2000 | 0.016, 0.008-0.052 | 0.73, 0.92, 0.95-0.98 |
| $SC_{2.2}$ | RCS | 500 | 0.048, 0.021-0.110 | 0.23, 0.46, 0.59-0.79 |
| | | 2000 | 0.024, 0.014-0.047 | 0.73, 0.84, 0.89-0.94 |
| | BS | 500 | 0.048, 0.024-0.127 | 0.23, 0.43, 0.57-0.79 |
| | | 2000 | 0.023, 0.010-0.054 | 0.51, 0.80, 0.89-0.95 |

Concerning the heterogeneous population models: under model $SC_{2.1}$, the values of relative bias and SD were very similar to those previously described. This is not surprising, due to the similarity of the hazard shapes among the three models (Fig. 1). Moreover, it is worth noting that in this case the values of relative bias and SD did not show sensible differences among the sub-groups (that is: $N_0$, $N_1$ and $N_2$). For model $SC_{2.2}$, a higher extent of bias was shown for $N = 500$, especially for the first peak in the $N_0$ group and the second peak in $N_2$. The extent of relative SD was higher with respect to the homogeneous population models, especially for the smaller sample size.

### 3.2 Pattern Identification

Table 3 presents the evaluation of the following criteria (Data and Methods section): number of peaks in the target hazard, proximity in time between peaks, closeness between peaks, false negatives and false positives. For what concerns the homogeneous population models, the number of peaks was correctly detected in large percentages of simulation runs for model $SC_{1.1}$ (roughly 75% in the worst case) but not for model $SC_{1.2}$. For the former model, satisfactory results were obtained when considering both the correct number of peaks and the proximity in time (i.e. the presence of a peak of the estimated hazard within a range of 31 months centered at the "true" peak time), while the performances to detect both the correct number, the proximity in time and the distance between the peaks (closeness) were reduced, except for BS and $N = 2000$ (81.7%).

**Table 3.** Pattern identification.

| PATTERN | SPLINE | N | MODEL | | | | |
|---|---|---|---|---|---|---|---|
| | | | $SC_{1.1}$ | $SC_{1.2}$ | $SC_{2.1}$ | $SC_{2.2}$ | $SC_{2.2}$: group $N_2$ |
| number of peaks | RCS | 500 | 76.3% | 40.7% | 73.0% | 12.1% | 56.1% |
| | | 2000 | 88.7% | 60.4% | 81.7% | 21.5% | 80.9% |
| | BS | 500 | 91.3% | 27.1% | 87.9% | 17.4% | 66.7% |
| | | 2000 | 99.0% | 20.1% | 97.7% | 16.9% | 93.7% |
| number of peaks AND proximity | RCS | 500 | 76.3% | 35.0% | 72.2% | 10.2% | 54.9% |
| | | 2000 | 88.7% | 58.8% | 81.7% | 21.3% | 80.9% |
| | BS | 500 | 91.3% | 22.2% | 86.8% | 10.3% | 60.7% |
| | | 2000 | 99.0% | 19.0% | 97.6% | 16.1% | 93.6% |
| number of peaks AND closeness | RCS | 500 | 26.9% | 7.0% | 2.6% | 0.4% | 5.2% |
| | | 2000 | 69.0% | 41.9% | 30.4% | 3.5% | 28.7% |
| | BS | 500 | 38.6% | 9.4% | 6.5% | 0.0% | 7.0% |
| | | 2000 | 81.7% | 14.8% | 39.7% | 2.8% | 37.6% |
| FALSE NEGATIVES | RCS | 500 | 1.0% | 47.4% | - | - | - |
| | | 2000 | 0.0% | 23.6% | - | - | - |
| | BS | 500 | 0.4% | 72.4% | - | - | - |
| | | 2000 | 0.0% | 75.8% | - | - | - |
| FALSE POSITIVES | RCS | 500 | 15.8% | 10.3% | - | - | - |
| | | 2000 | 5.4% | 3.6% | - | - | - |
| | BS | 500 | 13.4% | 6.8% | - | - | - |
| | | 2000 | 4.0% | 0.7% | - | - | - |

For what concerns the detection of false negatives and false positives, appreciable performances were obtained for model $SC_{1.1}$. Specifically, the percentages of occurrence of a monotone behaviour of the estimate (i.e.: a false negative) in the nearby of the first peak time were all below 1.0%. The percentages of occurrence of one spurious peak (false positive) was assessed by simulating under an alternative distributional model (see Fig. 2). Here, the chances of occurrence of such pattern are roughly 15% and 5%, for N = 500 and 2,000 respectively, and BS showing slighlty better performances than RCS. For model $SC_{1.2}$, the reference interval for the evaluation of false negatives and false positives was taken in the nearby of the second peak of the target hazard (less pronounced: see the Data and Methods section). In this case, the estimates failed to detect the second peak with a substantially high frequency.

Of note is the relationship, for model $SC_{1.2}$, between the higher probability of detection of false negatives and the reduced ability to identify the correct number of peaks. In fact, the absence of peaks detected in the nearbly of the "true" peak reduces the ability to correctly identify the number of peaks in the target hazard, For illustratory purposes, the percentages of samples in which the numer of detected peaks was one (instead of two) were 41.4%, 22.0% for

RCS (N = 500, 2000, respectively) and 66.7% and 74.7% for BS (N = 500, 2000, respectively).

Concerning the heterogeneous population models: for model $SC_{2.1}$, BS and RCS showed an appreciable ability in recognizing the exact number of peaks and the peak times, with percentages of correct detection simular to those obtained for model $SC_{1.1}$. Again, this is likely due the fact that, under this model, the shape of the hazards for the groups $N_0$, $N_1$ and $N_2$ are similar to the shape of the hazard for model $SC_{1.1}$ (Fig. 2). Finally, for model $SC_{2.2}$ the percentages of pattern identification were, overall, below 25%. However, it may be noted from the last column of Table 3, that appreciable performances, similar to those previously illustrated, were obtained for the group $N_2$, especially for N=2,000 (where no issues are expected for model estimation, see the first sentences in the results paragraph). More in detail, even though the frequency of the group $N_2$ in the overall population is relatively low (15%), the model estimation strategy performed well in identifying the shape of the hazard. The "bad" performances obtained for the groups $N_0$ and $N_1$ could be justified by considering the shape of the respective target hazard functions, which are "flat" as compared to group $N_2$ (Fig. 1).

## 4 Conclusion

In clinical studies, investigating the hazard function is relevant for providing insights into the dynamic of events, for purposes such as monitoring the effectiveness of treatments, identifying patients at higher risk, and predicting the future course of a disease. To this end, accurate identification of the shape is a fundamental task. In this work, we assessed the performance of a commonly adopted strategy for estimating flexible regression models - namely: include regression splines with several knots and tune the complexity of the spline using AIC-like criteria - in the context of survival analysis, using realistic distributional models for data generation. It is worth noting that the investigation of the hazard function by the Cox proportional hazards model, for which joint estimation of the baseline hazard and the regression coefficients is not feasible, could provide an inaccurate evaluation of the relationship between hazard and covariates.

From the results reported here, it emerges that an accurate identification of the main patterns of the target function is not guaranteed, even in the presence of an optimal estimation error and a large sample size that may ensure an abundant ratio between observed events and model complexity. Specifically, the adopted regression splines (BS, RCS) gave non-optimal performances in estimating those hazards with a "flattened" shape, resulting in under-detection ("false negatives") or over-detection ("false positives") of the correct number of peaks. Of note, BS performed better than RCS in the pattern identification task, while estimation error and agreement were practically equivalent.

The evaluation of false positives and false negatives provided clear insights into the potential misspecification of the hazard function's behavior within a specific sub-period of the follow-up. The concepts of false positives and false negatives have been adopted from the theory concerning the accuracy of a binary diagnosis (healthy/ill) and are strictly linked to the problem of classification of a categorical outcome (see, e.g., [16]). This explains why we had to perform additional simulations and design additional models, to achieve a complete assessment of the possibility of (local) misspecification. Therefore, if we had assessed only the misspecification of the presence of one peak, our evaluation would have been incomplete. Furthermore, from the theories of diagnostic accuracy and binary classification, we know that false positives and false negatives are evaluated conditionally to the "true" status of the target. For this reason, these indices are not helpful for assessing the predictive performances of an estimator. This is in complete agreement with the fact that predictive performances are not among the aims of this work.

To the extent of our knowledge, the features addressed here, and the criteria used for evaluating the pattern identification ability have never been considered in the literature. However, we still need to fully characterize the shapes of the estimated hazard functions obtained throughout simulations. We doubt that the above criteria could achieve this goal, because they reflect single features of the target hazard. Alternative methods will be considered in future works. Furthermore, the simulations will be extended to further common estimation strategies, that is, generalized additive models (GAM) smoothing (for GAM methods in the discrete-time survival framework, see, e.g., [17]).

Finally, although realistic, the scenarios adopted for the simulation study have limitations regarding the degree of heterogeneity in the target population, which, in turn, depends on the number of covariates included in the data-generating model. In this respect, our conclusions are more tailored for clinical studies involving selected sub-groups of patients with similar clinical features, and do not extend easily to large studies on general populations. However, we are confident to state that, in the latter context, our conclusions would be overly optimistic. This assertion is based on technical considerations: in applications with a large number of potential "predictive" variables, model selection techniques must address the challenge of identifying both the correct set of covariates and the functional form of their effects, in addition to the functional form of the baseline hazard function. In the broader context of smoothed regression, AIC-like criteria were found insufficient for handling this issue, leading to the proposal of alternative techniques in the literature [18] For an evaluation of the performance of these alternative techniques in the framework of discrete-time survival models, see [19].

In conclusion, we emphasize that the authors' principal interest lies in evaluating the ability of discrete-time methods to identify the shape of target functions in survival analysis. Further research could be performed for the following aims: (1) extending the evaluations presented here to more complex frameworks, such as competing risks analysis, and (2) assessing the pattern recognition performances in higher-dimensional settings, eventually including machine learning techniques.

# A Appendix

**Table 4. Pointwise estimation error.** ABSOLUTE ERROR: root MSE, BIAS and SD multiplied by$10^3$. RELATIVE ERROR: root MSE, BIAS and SD expressed as percentage of the respective "true" value.

| MODEL | peak-time (months) | N=500 | | | | N=2000 | | | |
|---|---|---|---|---|---|---|---|---|---|
| | | ABSOLUTE ERROR root MSE, BIAS, SD | | RELATIVE ERROR % root MSE, BIAS, SD | | ABSOLUTE ERROR root MSE, BIAS, SD | | RELATIVE ERROR % root MSE, BIAS, SD | |
| | | RCS | BS | RCS | BS | RCS | BS | RCS | BS |
| SC1_1 | 22 | 1.22, 0.07, 1.22 | 1.25, -0.01, 1.25 | 18.6, 1.0, 18.6 | 19.1, -0.2, 19.1 | 0.58, 0.02, 0.58 | 1.01, 0.13, 1.00 | 8.8, 0.4, 8.8 | 9.1, 1.1, 9.0 |
| | 68 | 1.89, 0.12, 1.89 | 1.65, -0.12, 1.64 | 18.2, 1.2, 18.1 | 15.8, -1.2, 15.8 | 0,60, 0.07, 0.59 | 0.93, 0.11, 0.93 | 9.6, 1.2, 9.6 | 9.0, 1.1, 8.9 |
| SC1_2 | 22 | 2.18, 0.12, 2.17 | 1.92, -0.15, 1.91 | 14.3, 0.1, 14.3 | 13.0, 0.1, 13.0 | 1.12, 0.02, 1.12 | 0.77, 0.28, 0.72 | 8.2, 0.8, 8.1 | 7.8, 1.1, 7.7 |
| | 68 | 1.48, 0.02, 1.48 | 1.19, -0.33, 1.14 | 20.2, 0.3, 20.2 | 16.2, -4.5, 15.6 | 1.10, 0.11, 0.09 | 0.66, -0.24, 0.62 | 10.5, 3.8, 9.8 | 9.0, -3.2, 8.4 |
| SC2_1 | group: $N_0$ | | | | | | | | |
| | 22 | 0.94, 0.00, 0.94 | 0.90, -0.02, 0.90 | 20.2, 0.1, 20.2 | 19.4, -0.5, 19.4 | 0.48, 0.00, 0.48 | 0.50, 0.06, 0.49 | 10.3, 0.1, 10.3 | 10.7, 1.3, 10.7 |
| | 68 | 1.53, 0.18, 1.52 | 1.39, -0.10, 1.38 | 20.8, 2.5, 20.7 | 18.8, -1.3, 18.8 | 0.77, 0.03, 0.77 | 0.71, 0.06, 0.71 | 10.5, 0.5, 10.5 | 9.7, 0.9, 9.7 |
| | group: $N_1$ | | | | | | | | |
| | 22 | 1.88, 0.06, 1.88 | 1.84, 0.03, 1.84 | 28.6, 0.9, 28.6 | 28.0, 0.5, 28.0 | 0.91, 0.02, 0.91 | 0.87, 0.11, 0.86 | 13.8, 0.3, 13.8 | 13.2, 1.7, 13.1 |
| | 68 | 2.84, 0.28, 2.83 | 2.63, -0.02, 2.63 | 27.3, 2.7, 27.1 | 25.3, -0.2, 25.3 | 1.39, 0.05, 1.39 | 1.33, 0.12, 1.32 | 13.4, 0.4, 13.4 | 12.8, 1.2, 12.7 |
| | group: $N_2$ | | | | | | | | |
| | 22 | 2.82, 0.07, 2.81 | 2.64, -0.05, 2.64 | 25.0, 0.6, 25.0 | 23.5, -0.4, 23.5 | 1.40, 0.07, 1.40 | 1.40, 0.11, 1.40 | 12.5, 0.6, 12.5 | 12.5, 1.0, 12.4 |
| | 68 | 4.77, 0.47, 4.74 | 4.07, -0.27, 4.06 | 26.7, 2.6, 26.6 | 22.8, -1.5, 22.8 | 2.36, 0.15, 2.35 | 2.06, 0.07, 2.06 | 13.2, 0.8, 13.2 | 11.6, 0.4, 11.5 |
| SC2_2 | group: $N_0$ | | | | | | | | |
| | 22 | 1.23, -0.37, 1.17 | 1.12, -0.45, 1.03 | 26.5, -8.0, 25.2 | 24.2, -9.8, 22.2 | 0.62,-0.03, 0.62 | 0.56,-0.14, 0.54 | 13.3,-0.6, 13.3 | 12.1, -3.1, 11.7 |
| | group: $N_1$ | | | | | | | | |
| | 33 | 2.02, -0.02, 2.02 | 1.80, 0.21, 1.79 | 45.2, -0.5, 45.2 | 40.3, 4.7, 40.0 | 1.21, 0.10, 1.21 | 1.10, 0.15, 1.09 | 27.1, 2.2, 27.0 | 24.5, 3.4, 24.3 |
| | group: $N_2$ | | | | | | | | |
| | 22 | 4.10, 0.04, 4.10 | 3.77, 0.06, 3.77 | 36.5, 0.4, 36.5 | 33.5, 0.5, 33.5 | 2.08,-0.01, 2.08 | 2.05,-0.13, 2.05 | 18.5,-0.1, 18.5 | 18.3, -1.2, 18.2 |
| | 68 | 6.99, -2.96, 6.34 | 7.32, -4.03, 6.11 | 39.2,-16.6, 35.5 | 41.0,-22.6, 34.3 | 3.28, 0.34, 3.26 | 3.09,-1.03, 2.91 | 18.4, 1.9, 18.3 | 17.3, -5.8, 16.3 |

# References


1. Simes, R.J., Zelen, M.: Exploratory data analysis and the use of the hazard function for interpreting survival data: an investigator's primer. J. Clin. Oncol. **3**(10), 1418–1431 (1985)
2. Gray, R.J.: Flexible methods for analyzing survival data using splines, with applications to breast cancer prognosis. J. Am. Stat. Assoc. **87**(420), 942–951 (1992)
3. Murray, T.A., Hobbs, B.P., Sargent, D.J., Carlin, B.P.: Flexible Bayesian survival modeling with semiparametric time-dependent and shape-restricted covariate effects. Bayesian Anal. **11**(2), 381 (2015)
4. Demicheli, R., et al.: Recurrence and mortality according to estrogen receptor status for breast cancer patients undergoing conservative surgery. Ipsilateral breast tumour recurrence dynamics provides clues for tumour biology within the residual breast. BMC Cancer **10**(1), 1–9 (2010)
5. Yin, W., Di, G., Zhou, L., Lu, J., Liu, G., Wu, J., et al.: Time-varying pattern of recurrence risk for Chinese breast cancer patients. Breast Cancer Res. Treat. **114**, 527–535 (2009)
6. Feng, X.Y., Chen, Y.B., Wang, W., Guan, Y.X., Li, Y.F., Chen, S., et al.: Time-varying pattern of recurrence risk for gastric cancer patients. Med. Oncol. **30**, 1–5 (2013)

7. Demicheli, R., Abbattista, A., Miceli, R., Valagussa, P., Bonadonna, G.: Time distribution of the recurrence risk for breast cancer patients undergoing mastectomy: further support about the concept of tumor dormancy. Breast Cancer Res. Treat. **41**, 177–185 (1996)
8. Tutz, G., Schmid, M.: Modeling Discrete Time-to-Event Data. Springer (2016)
9. Harrell, F.E.: Regression modeling strategies: with applications to linear models, logistic regression, and survival analysis. Springer (2001)
10. Therneau, T.M., Grambsch, P.M.: Modeling Survival Data: Extending the Cox Model. Springer (2013)
11. Morris, T.P., White, I.R., Crowther, M.J.: Using simulation studies to evaluate statistical methods. Stat. Med. **38**(11), 2074–2102 (2019)
12. Zeger, S.L., Liang, K.Y., Albert, P.S.: Models for longitudinal data: a generalized estimating equation approach. Biometrics **44**(4), 1049–1060 (1988)
13. Pan, W.: Akaike's information criterion in generalized estimating equations. Biometrics **57**(1), 120–125 (2001)
14. Peduzzi, P., Concato, J., Feinstein, A.R., Holford, T.R.: Importance of events per independent variable in proportional hazards regression analysis II. Accuracy and precision of regression estimates. J. Clin. Epidemiol. **48**(12), 1503–1510 (1995)
15. Van Smeden, M., de Groot, J.A., Moons, K.G., Collins, G.S., Altman, D.G., Eijkemans, M.J., et al.: No rationale for 1 variable per 10 events criterion for binary logistic regression analysis. BMC Med. Res. Methodol. **16**, 1–12 (2016)
16. Pepe, M.S.: The statistical evaluation of medical tests for classification and prediction. Oxford University Press (2003)
17. Bender, A., Groll, A., Scheipl, F.: A generalized additive model approach to time-to-event analysis. Stat. Model. **18**(3–4), 299–321 (2018)
18. Marra, G., Wood, S.N.: Practical variable selection for generalized additive models. Comput. Stat. Data Anal. **55**(7), 2372–2387 (2011)
19. Rodríguez-Girondo, M., Kneib, T., Cadarso-Suárez, C., Abu-Assi, E.: Model building in nonproportional hazard regression. Stat. Med. **32**(30), 5301–5314 (2013)

# Nested Named Entity Recognition in Chinese Electronic Medical Records

Maolin Yang[1], Zeran Lu[2], Yucong Lin[1], Hong Song[3](✉), and Jian Yang[4]

[1] School of Medical Technology, Beijing Institute of Technology, Beijing 100081, China
3220212095@bit.edu.cn

[2] School of Statistics, Renmin University of China, Beijing 100872, China

[3] School of Computer Science and Technology, Beijing Institute of Technology, Beijing 100081, China
songhong@bit.edu.cn

[4] School of Optics and Electronics, Beijing Institute of Technology, Beijing 100081, China

**Abstract.** Nested named entity recognition (NER) is crucial in processing Chinese electronic medical records (EMRs). Recently, the BERT-based model using CNN and a multi-head Biaffine decoder has shown promising results in nested NER on news datasets. However, this model faces difficulties in dealing with the complex and unevenly distributed entities in Chinese EMRs, resulting in prediction errors. This paper proposes an MC-BERT-CGC model based on MC-BERT semantic features comprising Context-Gated Convolution and multi-head Biaffine decoder. Our model initially incorporates Chinese medical language knowledge by leveraging MC-BERT to represent medical descriptions as sentence vectors. We then use Context-Gated Convolution to accurately define the boundaries of nested entities by learning overlapping relationships between different entities. Finally, we use Focal Loss to classify difficult-to-distinguish entities. Experimental results tested on our Chinese EMRs and the CMeEE-V2 dataset show that our model performs better than existing baseline models in Chinese medical NER tasks. The impacts of this study on the life of patients are significant, as more accurate and detailed medical information can be extracted from EMRs, potentially leading to improved diagnoses, personalized treatment recommendations, and proactive identification of health risks. Our code is available at https://github.com/ymlmorning/MC-BERT-CGC.

**Keywords:** Chinese electronic medical records · Nested named entity recognition · MC-BERT · Context-Gated convolution · Focal Loss

## 1 Introduction

In recent years, the widespread use of Chinese electronic medical records (EMRs) has generated substantial textual data in hospitals. EMRs are comprehensive


M. Vettoretti et al. (Eds.): CIBB 2023, LNBI 14513, pp. 58–69, 2025.
https://doi.org/10.1007/978-3-031-90714-2_5

records containing valuable medical knowledge and patient health information. However, manual extraction of structured information from EMRs is time-consuming. Named entity recognition (NER) has emerged as a solution to automate the extraction of electronic medical record information.

A significant challenge in recognizing medical terms in EMRs is the presence of nested entities. The flat NER approach is limited in accurately identifying entities with such nested relationships. As a result, researchers have explored nested NER techniques. Early studies utilized traditional conditional random field (CRF)-based methods [1] to construct separate flat NER models for each hierarchical entity. However, CRF-based methods have the obvious disadvantage of over-reliance on rule-based post-processing, which limits the ability of the model to recognise complex structural entities. Later, region-based methods [2,3] attempted to recognise nested entity types using a single interpretable nested NER model by identifying all subsequences in the sentence. However, region-based methods rely on manually designed conversion actions and suffer from error propagation when conversion operations are performed word by word. Additionally, some scholars employed hypergraph-based methods [4,5] to model complex relationships between subsequences, storing nested sequence boundaries, types, and overlapping relationships in a hypergraph. While hypergraphs can represent all types of entity mentions, the edge and point representations of hypergraphs need to be carefully designed manually to avoid structural ambiguity problems. In addition, such models gradually generate hypergraphs based on words, which may lead to the problem of error propagation. Recent research has utilized seq2seq models [6,7], which encode input text into fixed-length word vectors and decode them into nested span sequences. Nevertheless, seq2seq approaches are confronted with the challenges of inefficient decoding and exposure bias. The latest trend involves span-based methods [8,9], which convert nested NER into a span classification task. Although span-based methods perform well in identifying nested entities, their enumeration process struggles to balance model performance and efficiency.

Nested NER exhibits significant promise in assisting patients. By leveraging nested NER to acquire detailed and well-structured information, we can subsequently develop a wider range of sophisticated medical intelligence models: Initially, we can develop phenotypic classification models to enhance the precision of patient diagnoses. Additionally, reinforcement learning models can provide personalized treatment recommendations tailored to each patient's specific needs. Furthermore, risk prediction models can proactively identify potential health risks in patients and assess them at an early stage.

Our principal contributions can be summarized as follows:

- Our method leverages MC-BERT to obtain embedded representations of EMRs, addressing the challenges posed by complex phrase combinations of Chinese medical terms.
- To recognise long and unevenly distributed nested entities more accurately, we employ Context-Gated Convolution that models the spatial relationship

between spans in the score matrix (combined with Focal Loss which adjusts the weight of difficult-to-classify samples).

- Our experimental results on two independent datasets (the Chinese EMRs and the CMeEE-V2 dataset) indicate that our model can accurately identify nested named entities, outperforming four existing models specifically on the task of nested NER.

## 2 Related Work

Previously, nested NER was tackled using various neural network techniques, including CRF, LSTM, and CNN. CRF-based approaches, such as the two-layer CRF model utilized by Alex et al. [1], identify internal entities with one CRF layer and subsequently recognise external entities based on the outcomes of the first layer using another CRF. However, this approach necessitates prior knowledge of the maximum depth of nested entities. Region-based approaches, such as the Transition approach used by Wang et al. [2], achieve sentence-to-forest mapping by obtaining an action sequence through a shift-reduce system. Although the tree construction process involves nested entity recognition, it relies on additional part-of-speech (POS) tagging results. Hypergraph-based approaches, like the LSTM-Hypergraph model introduced by Katiyar et al. [4], involve applying multiple-layer BIOES labeling to sentences. This approach combines identical characters tagged with different layers into a hypernode, with overlapping spans represented as hyperedges. Leveraging RNN for feature extraction enables the acquisition of nested entity hypergraph representations of sentences. Nonetheless, this methodology entails a complex inference and decoding process.

With the rise of pre-trained language models, BART and BERT have been increasingly adopted for nested NER. Seq2seq models, such as the BARTNER proposed by Yan et al. [6], involve constructing a seq2seq framework that utilizes the BART with a pointer mechanism to pinpoint entity positions in sentences. This model effectively integrates flat, nested, and discontinuous NER recognition tasks into a unified framework, albeit with the drawback of time-consuming decoding. Recently, Span-based models such as W2NER, proposed by Li et al. [8], have garnered significant attention. These models leverage BERT to generate embeddings for distance, words, and regions. They then employ multiple 2D dilated convolutions to capture the adjacent relationships between entity words. Finally, Biaffine and MLP are leveraged to decode and construct a unified framework for recognising the three tasks of NER. However, these methods have been predominantly developed and evaluated on general-domain datasets, with limited fine-tuning on Chinese EMRs. Consequently, their direct application to named entity recognition in Chinese EMRs may encounter challenges, potentially resulting in inaccuracies such as missed medical entities or imprecise entity boundary localization.

The investigation of NER is of significant importance in propelling research on Chinese EMRs. In previous work, Yin M et al. [10] devised the BiLSTM-CRF model, which incorporates radical features and a self-attention mechanism to capture long-term character dependencies. Qian Wan et al. [11] utilized fine-tuning techniques to enhance the performance of the ELMo model on a small medical corpus. They employed a Transformer encoder to mitigate the issue of context dependence and devised the ELMo-Transformer-CRF model. Jun Cao et al. [12] have converted the nested NER task into a question-answering task and have enhanced the performance of entity recognition in Chinese EMRs by employing an MRC neural network model with a Biaffine mechanism. These approaches outperform the traditional CRF-based flat NER model. However, current recognition of Chinese EMRs mainly relies on sequence labeling-based NER models, which do not consider nested entities, leading to incomplete recognition and missing information. Although some nested NER models exist for Chinese EMRs, they are typically compared to flat NER models, and the experimental results lack sufficient persuasiveness.

## 3 Data and Methods

### 3.1 Dataset

This study collected 200 Chinese EMRs from the Liver Disease Research Center of Beijing Friendship Hospital. These EMRs contain patient information like medical history, diagnosis, and test results. Five doctors then annotated the dataset by marking only the outermost entities with a triple form of start position, end position, and entity type. After that, the annotation results were transformed into a nested named entity dataset through data processing, as discussed in Sect. 3.2.

We segmented the dataset into three subsets, namely, the training, development, and test dataset, based on a document ratio of 6:2:2. The dataset comprises five categories of nested entities: position, disease, operation, place, and clinical manifestation. The total number of entities in the dataset is 39,048, of which

**Table 1.** The statistical description for five nested named entities in Chinese EMRs. Items: record and entity category quantity. Train, Dev and Test: training, development, and testing datasets. Nested: the Nested entities quantity in each category of entity.

| Items | Train | Dev | Test | Total | Nested |
|---|---|---|---|---|---|
| Documents | 120 | 40 | 40 | 200 | - |
| Sentences | 3,524 | 1,114 | 1,101 | 5,739 | - |
| Percentage | 60% | 20% | 20% | 100% | - |
| Position | 6,203 | 2,024 | 1,922 | 10,149 | 2,765 |
| Disease | 2,586 | 846 | 886 | 4,318 | 1,138 |
| Operation | 3,288 | 989 | 1,019 | 5,296 | 1,496 |
| Place | 642 | 234 | 208 | 1,084 | 392 |
| Clinical manifestation | 11,179 | 3,542 | 3,480 | 18,201 | 6,592 |
| Total | 23,898 | 7,635 | 7,515 | 39,048 | 9,894 |

9,894 entities exhibit nested relationships, accounting for roughly 25% of the overall dataset. Table 1 presents detailed statistics on five types of nested named entities found in Chinese EMRs.

We also conducted experiments on the CMeEE-V2 labelled dataset, which was divided into training, development and test datasets in a ratio of 6:2:2. As shown in Table 2, the dataset contains nine categories of entities: disease (Dis), clinical manifestation (Sym), medical procedure (Pro), medical equipment (Equ), drug (Dru), medical test item (Ite), body (Bod), department (Dep), and microbiology (Mic). In total, the dataset contains approximately 80,000 entities, of which over 15,000 are nested entities, accounting for approximately 18% of the total number of entities.

**Table 2.** The statistical description for nested named entities in CMeEE-V2 dataset. Items: entity category quantity. Train, Dev and Test: training, development, and testing datasets. Nested: the Nested entities quantity in each category of entity.

| Items | Train | Dev | Test | Total | Nested |
|---|---|---|---|---|---|
| Bod | 15,127 | 4,825 | 4,001 | 23,953 | 2,330 |
| Dis | 10,596 | 3,937 | 4,514 | 19,047 | 74 |
| Sym | 9,357 | 3,436 | 4,251 | 17,044 | 11,795 |
| Pro | 5,958 | 1,949 | 1,655 | 9,562 | 303 |
| Ite | 3,186 | 663 | 524 | 4,373 | 511 |
| Dru | 2,795 | 763 | 793 | 4,351 | 25 |
| Mic | 1,563 | 367 | 367 | 2,297 | 20 |
| Equ | 703 | 115 | 65 | 883 | 28 |
| Dep | 188 | 46 | 27 | 261 | 1 |
| Total | 49,473 | 16,101 | 16,197 | 81,771 | 15,087 |

### 3.2 Data Preprocessing

We follow the procedures outlined in Fig. 1 to construct the nested medical entities dataset. Initially, we utilized the *python-doc* python package to extract Chinese electronic medical record Word documents and transform them into plain text files. Next, medical practitioners employed the annotation system Doccano [13] to annotate the flat entities in the text files. The data processing method consists of three steps. Firstly, we establish a Chinese liver disease professional lexicon based on the annotations provided by doctors. Secondly, we utilize the NTLK toolkit [14] to process each sentence, including sentence segmentation, word segmentation, and POS tagging. Finally, we concatenate the segmented outcomes to obtain possible nested combinations that match the lexicon, thereby obtaining the nested vocabulary annotation outcomes.

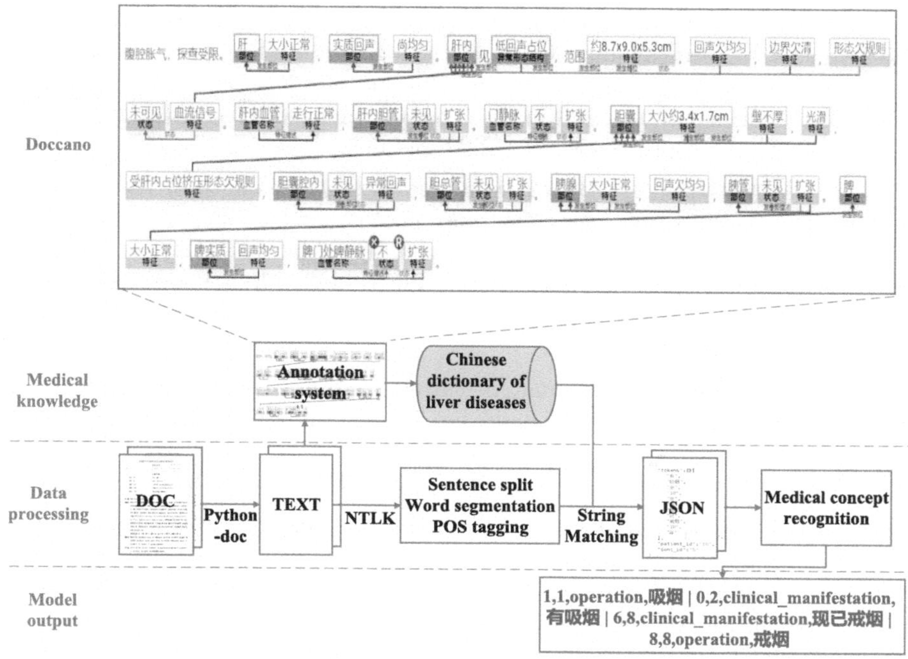


**Fig. 1.** Flowchart of the construction and recognition process for the medical nested named entity dataset. "6,8,clinical_manifestation,现已戒烟" denotes the string "现已戒烟" from the sixth character to the eighth (inclusive), represents a clinical manifestation term.

### 3.3 Nested Medical Concept Recognition

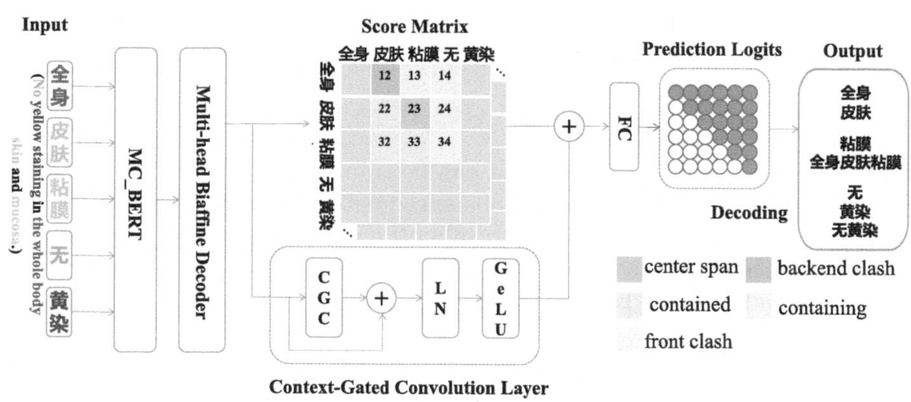


**Fig. 2.** An overview of the proposed MC-BERT-CGC nested NER model. In the model input, words with the same color in both Chinese and English are one-to-one corresponded. Additionally, we have included a bottom-right note to represent the spatial relationships between each span and the center span in the score matrix. Finally, we calculate probabilities only in the upper right triangle of the score matrix to improve efficiency.

Figure 2 illustrates the proposed model architecture. We choose MC-BERT [15], which injects biomedical knowledge into the representation model instead of BERT to embed input sentences. For an input sentence $\mathbf{x} = \{x_1, x_2, ..., x_n\}$ comprising $n$ tokens, we first convert each token $x_i$ into word pieces and then feed $\mathbf{x}^{'} = \{x_1, x_2, ..., x_{n^{'}}\}$ into the MC-BERT function $f_{mc}$, as follows: $\mathbf{H}' = f_{mc}(\mathbf{x}')$. Here, $\mathbf{H}'$ is a matrix in $\mathbb{R}^{n^{'} \times h}$, where $n^{'}$ represents the length of the sequence containing all the word pieces, and $h$ represents the hidden size.

After utilizing MC-BERT, we extend the dimensions of $\mathbf{H}^{'}$ to $n \times n^{'} \times h$, and subsequently employ the max-pooling function $f_{mp}$ to generate token-level representations. The process can be expressed as follows: $\mathbf{H}$=$f_{mp}(\mathbf{H}')$, where $\mathbf{H} \in \mathbb{R}^{n \times h}, \mathbf{H}' \in \mathbb{R}^{n \times n^{'} \times h}$.

Next, we obtain the score matrix through a multi-head Biaffine decoder $f_{mhb}$ [16], which incorporates the multi-head attention into the Biaffine decoder:

$$\mathbf{R} = f_{mhb}(f_{lr}(\mathbf{H}\mathbf{W}_s), f_{lr}(\mathbf{H}\mathbf{W}_e)) \tag{1}$$

where the hidden size is denoted by $d$, and $\mathbf{W}_s, \mathbf{W}_e \in \mathbb{R}^{h \times d}$. The output $\mathbf{R} \in \mathbb{R}^{n \times n \times r}$ and the feature size is represented by $r$. The function $f_{lr}$ is the LeakyReLU activation function.

To account for the spatial relations between various spans, we opt for Context-Gated Convolution $f_{cgc}$ [17] and perform layer normalization $f_{ln}$ in the feature dimension, as outlined below:

$$\mathbf{R}' = f_{cgc}(\mathbf{R}), \tag{2}$$

$$\mathbf{R}'' = f_{gl}(f_{ln}(\mathbf{R}' + \mathbf{R})), \tag{3}$$

where $f_{gl}$ represents the GeLU activation function.

Finally, we feed the sum of the initial score matrix and the processed one into a fully connected layer and apply the sigmoid function $f_s$ to generate the prediction logits, as follows:

$$P = f_s(\mathbf{W}_a(\mathbf{R} + \mathbf{R}'') + \mathbf{b}), \tag{4}$$

where $\mathbf{W}_a$ and $\mathbf{b}$ are parameters that can be trained.

In classification tasks involving imbalanced data, the model may exhibit bias towards the majority class, leading to suboptimal performance for the minority class. To overcome this challenge, we leverage Focal Loss $\mathcal{L}_{fl}$ [18] to compute the loss instead of Cross Entropy Loss:

$$\mathcal{L}_{fl} = -\frac{1}{n^2} \sum_{i=1}^{n} \sum_{j=1}^{n} \sum_{t=1}^{T} y_{ijt}(1 - p_{ijt})^{\gamma} \log(p_{ijt}), \tag{5}$$

where $T$ is the number of labels, and $p_{ijt}$ is the probability that the $i$-th row and $j$-th column belong to the $t$-th label.

The model training batch size is set to four, with a total of ten rounds of iterative training. The initial learning rate is set with the objective of optimising model training efficiency and performance. The number of attention heads of the Multi-head Biaffine decoder module is set to four, and the size of the hidden layer is set to 200, with the objective of enhancing the model's ability to capture and process complex features. Furthermore, the convolution kernel size is set to 3 and the channel dimension is set to 120, enabling more efficient capture of spatial features with local spans. Model parameters are updated using the Adam optimiser, with a warm-up factor of 0.1 to dynamically adjust the growth rate of the learning rate, thereby balancing the convergence speed and stability of the model.

## 4 Experimental Results and Analysis

The model utilizes Pytorch, fastNLP and transformers as three main Python packages. As shown in Table 3, we conducted repeated experiments on the test dataset. The neural transition-based model demonstrates the least effective recognition performance on Chinese EMRs, achieving an F1 score of 85.44% and an MCC score of 0.7449. These two scores are significantly lower than those of other models, indicating that training a nested NER model without utilizing a pre-training language model is challenging and may not yield significant results. The three models, BARTNER, W2NER, and BERT-CNN, exhibit remarkably similar F1 scores on the test dataset, hovering around 92%, and their MCC scores are above 0.85 but below 0.90. This observation underscores the impressive capacity of the pre-trained language model to represent text vectors with excellent efficacy, enabling it to deliver commendable performance even without fine-tuning. Boasting the highest F1 and MCC score, our model outperforms the other four nested NER models. Our model achieves an F1 score of 96.28% and an MCC score of 0.9149. Additionally, our model shows a significant improvement of 6% in the MCC score compared to the original BERT-CNN model. These comparisons demonstrate that the nested NER generalization ability is significantly enhanced after model adjustment.

Table 3 also displays the results of our model ablation experiment on Chinese EMRs. The F1-score and MCC columns demonstrate that improving each component has a beneficial effect on the model's predictive ability. MC-BERT significantly enhances the model performance, thereby highlighting the significance of integrating medical knowledge for medical entity recognition tasks. Although the CGC and Focal Loss components do not demonstrate a significant improvement in model performance, we believe they are pivotal for the Nested NER model of Chinese EMRs, particularly in the context of processing large-scale datasets in the future.

Table 4 reports experiments comparing our method with other nested named entity recognition models on the CMeEE-V2 dataset. The Transition model exhibits the lowest recognition performance, with the F1 score of less than 60% and the MCC metric of less than 0.5. The comparison of the Chinese EHR dataset results and the CMeEE-V2 dataset results indicates that the error propagation in tokenization has a significant impact on the model performance when

**Table 3.** Experimental results on Chinese EMRs. Evaluation Metrics (P: Precision, R: Recall, F1: F1-Score, SPE: Specificity, NPV: Negative Predictive Value, MCC: Matthews Correlation Coefficient)

| | Model | P | R | F1 | SPE | NPV | MCC |
|---|---|---|---|---|---|---|---|
| Baseline model | Transition [2] | 0.8569 | 0.8520 | 0.8544 | 0.9822 | 0.7022 | 0.7449 |
| | BARTNER [6] | 0.9172 | 0.9078 | 0.9133 | 0.9785 | 0.8569 | 0.8555 |
| | W2NER [8] | 0.8985 | 0.9345 | 0.9161 | 0.9916 | 0.8624 | 0.8879 |
| | BERT-CNN [16] | 0.9368 | 0.9195 | 0.9281 | 0.9935 | 0.8117 | 0.8534 |
| Ablation experiment | **MC-BERT-CGC** | **0.9704** | **0.9552** | **0.9628** | **0.9960** | **0.8871** | **0.9139** |
| | (w/o) MC-BERT | 0.9280 | 0.9131 | 0.9206 | 0.9951 | 0.7757 | 0.8283 |
| | (w/o) CGC | 0.9550 | 0.9547 | 0.9549 | 0.9956 | 0.8547 | 0.8896 |
| | (w/o) FocalLoss | 0.9659 | 0.9575 | 0.9617 | 0.9947 | 0.8878 | 0.9121 |

**Table 4.** Experimental results on CMeEE-V2 dataset. Evaluation Metrics (P: Precision, R: Recall, F1: F1-Score, SPE: Specificity, NPV: Negative Predictive Value, MCC: Matthews Correlation Coefficient)

| | Model | P | R | F1 | SPE | NPV | MCC |
|---|---|---|---|---|---|---|---|
| Baseline model | Transition [2] | 0.5817 | 0.5755 | 0.5786 | 0.8531 | 0.5456 | 0.4057 |
| | BARTNER [6] | 0.7311 | 0.7138 | 0.7223 | 0.9147 | 0.7191 | 0.6326 |
| | W2NER [8] | 0.7427 | 0.7744 | 0.7582 | 0.9372 | 0.7496 | 0.6948 |
| | BERT-CNN [16] | 0.7342 | 0.7130 | 0.7234 | 0.9249 | 0.7001 | 0.6292 |
| Ablation experiment | **MC-BERT-CGC** | **0.7680** | **0.7803** | **0.7741** | **0.9267** | **0.7519** | **0.6870** |
| | (w/o) MC-BERT | 0.7678 | 0.7660 | 0.7669 | 0.9299 | 0.7408 | 0.6785 |
| | (w/o) CGC | 0.7684 | 0.7711 | 0.7697 | 0.9294 | 0.7451 | 0.6825 |
| | (w/o) FocalLoss | 0.7650 | 0.7682 | 0.7666 | 0.9310 | 0.7431 | 0.6819 |

the entity distribution is more complex. Additionally, the BARTNER, W2NER and BERT-CNN models achieve an F1 score of 72%, which indicates that pre-trained language models can learn contextual information from samples in complex entity distribution situations. Our model achieves the highest F1 score of 77.41% and second highest MCC metric of 0.6870. The F1 score of our model is significantly improved by 5% compared to the original BERT-CNN baseline model, which demonstrates that the generalisation ability of the nested named entity recognition model is significantly enhanced after model optimization.

Table 4 also presents the results of the model ablation experiments on the CMeEE-V2 dataset. The five columns, apart from the specificity metric, show that the predictive power of the model increases to a certain degree with the improvement of each component. Incorporating medical knowledge through the use of MC-BERT improves the performance of the model, particularly in the MCC metrics, by approximately 1% compared to the ordinary BERT model. Additionally, the use of Context-Gated Convolution results in an improvement of approximately 0.4% in the MCC metrics compared to ordinary convolution, which indicates that adjusting the weights of the convolution kernel based on contextual information can effectively enhance the recognition accuracy of positive and negative sample spans. In addition, the use of FocalLoss improves the F1 score of the model by approximately 0.7% and the MCC metric by approximately 0.5% compared to the cross-entropy loss function. The CMeEE-V2 dataset contains a larger number of entity classes and a more uneven distribution of majority and minority classes compared to the Chinese medical EHR dataset, therefore the FocalLoss helps to address the issue of prediction results being biased towards the majority class.

**Table 5.** A case study of the predictions made by BERT-CNN and MC-BERT-CGC models on the same Mandarin Chinese sentence

| | Content |
|---|---|
| Mandarin Chinese | [ "个人", "史", ":", "出生", "并", "久", "居于", "本地", "否认", "疫水", "疫区", "接触", "史", "否认", "其他", "放射性", "物质", "及", "毒物", "接触", "史"] |
| English | Personal history : He was born and lived here for a long time. He denied the contact history of the epidemic area and other radioactive substances and poisons. |
| Accurate label | 并久居于本地 (and lived here for a long time): A,个人 (Personal): A,<br>接触(contact): C, 久居于本地 (lived here for a long time): A,<br>史 (history): A, 疫水 (epidemic water): B,<br>接触史 (the contact history): A,<br>疫区接触史 (the contact history of the epidemic area): A,<br>其他放射性物质 (other radioactive substances): A,其他 (other): A,<br>疫区 (the epidemic area): B,<br>出生并久居于本地 (born and lived here for a long time): A |
| BERT-CNN | Two fewer entities than the accurate label:<br>并久居于本地 (and lived here for a long time):A<br>and 久居于本地 (lived here for a long time):A |
| MC-BERT-CGC | One fewer entity than the accurate label:<br>出生并久居于本地(born and lived here for a long time): A |

The Chinese and English words with matching colors are one-to-one correspondences in the second and third lines. In the fourth line, the English text within parentheses corresponds to the Chinese entity preceding the parentheses one-to-one and label type:(A: 'clinical manifestation', B: 'operation', C: 'position').

Table 5 compares our proposed model with the BERT-CNN model through a case study. We evaluated the predictions of the two models on the same Chinese sentence. As demonstrated by the accurate labels, it is evident that most nested entities encompass various clinical manifestations, which serve as the fundamental components that constitute the structured electronic medical record.

Regarding the predictions made by the models presented in the last two rows of the table, our model predicts one fewer external entity than the accurate label. In contrast, BERT-CNN predicts two fewer internal entities. Although our model prediction omits an external entity, it is more suitable than BERT-CNN for populating the structured electronic medical record because it accurately predicts the internal entities representing the smallest information unit.

## 5 Conclusion

We propose the MC-BERT-CGC model to identify nested named entities in two independent Chinese datasets (the Chinese EMRs and the CMeEE-V2 dataset), and compare it with four other nested NER models: Neural Transition-based, BARTNER, W2NER, and BERT-CNN. Our evaluation metrics include entity-level Precision, Recall, F1, Specificity, NPV and MCC scores, and we trained the models for ten epochs. Our study shows that the MC-BERT-CGC model outperforms the others on the two independent datasets. We also conclude three main points of optimization: use MC-BERT component to emphasis the inclusion of knowledge of Chinese EHRs, utilize CGC component to consideration of the need to nest longer sentence information in the medical NER, and employ Focal Loss component to focus attention on indistinguishable entities. In the future, we plan to construct more extensive and higher-quality Chinese electronic medical record nested named entity datasets and enhance our model to simultaneously handle flat, nested, and non-continuous NER tasks in a single framework. Additionally, we will use this model to build artificial intelligence models to help the patients, including phenotypic classification for precise diagnoses, personalized treatment recommendations harnessing reinforcement learning, and early risk prediction models geared towards proactive health management.

**Acknowledgments.** This work was supported by National Key R&D Program of China (2021ZD0113201).

## References

1. Alex, B., Haddow, B., Grover, C.: Recognising nested named entities in biomedical text. In: Biological, Translational, and Clinical Language Processing, Prague, Czech Republic, pp. 65–72. Association for Computational Linguistics (2007). https://aclanthology.org/W07-1009
2. Wang, B., Lu, W., Wang, Y., Jin, H.: A neural transition-based model for nested mention recognition. In: Proceedings of the 2018 Conference on Empirical Methods in Natural Language Processing, Brussels, Belgium, pp. 1011–1017. Association for Computational Linguistics (2018). https://aclanthology.org/D18-1124
3. Ju, M., Miwa, M., Ananiadou, S.: A neural layered model for nested named entity recognition. In: Proceedings of the 2018 Conference of the North American Chapter of the Association for Computational Linguistics: Human Language Technologies, Volume 1 (Long Papers), pp. 1446–1459 (2018)

4. Katiyar, A., Cardie, C.: Nested named entity recognition revisited. In: Proceedings of the 2018 Conference of the North American Chapter of the Association for Computational Linguistics: Human Language Technologies, vol. 1 (2018)
5. Wang, B., Lu, W.: Neural segmental hypergraphs for overlapping mention recognition. In: Proceedings of the 2018 Conference on Empirical Methods in Natural Language Processing, Brussels, Belgium, pp. 204–214. Association for Computational Linguistics (2018). https://aclanthology.org/D18-1019
6. Yan, H., Gui, T., Dai, J., Guo, Q., Zhang, Z., Qiu, X.: A unified generative framework for various NER subtasks. In: Proceedings of the 59th Annual Meeting of the Association for Computational Linguistics and the 11th International Joint Conference on Natural Language Processing (Volume 1: Long Papers), pp. 5808–5822. Association for Computational Linguistics, Online (2021). https://aclanthology.org/2021.acl-long.451
7. Lewis, M., Liu, Y., et al.: BART: denoising sequence-to-sequence pre-training for natural language generation, translation, and comprehension. In: Proceedings of the 58th Annual Meeting of the Association for Computational Linguistics, pp. 7871–7880. Association for Computational Linguistics, Online (2020). https://aclanthology.org/2020.acl-main.703
8. Li, J., et al.: Unified named entity recognition as word-word relation classification. In: Proceedings of the AAAI Conference on Artificial Intelligence, vol. 36, pp. 10965–10973 (2022)
9. Eberts, M., Ulges, A.: Span-based joint entity and relation extraction with transformer pre-training. arXiv preprint arXiv:1909.07755 (2019)
10. Yin, M., Mou, C., Xiong, K., Ren, J.: Chinese clinical named entity recognition with radical-level feature and self-attention mechanism. J. Biomed. Inform. **98**, 103289 (2019)
11. Wan, Q., Liu, J., Wei, L., Ji, B.: A self-attention based neural architecture for Chinese medical named entity recognition. Math. Biosci. Eng. **17**(4), 3498–3511 (2020)
12. Cao, J., et al.: Electronic medical record entity recognition via machine reading comprehension and biaffine. Discrete Dyn. Nat. Soc. **2021** (2021)
13. https://doccano.github.io/doccano/
14. https://www.nltk.org/index.html
15. Zhang, N., Jia, Q., Yin, K., Dong, L., Gao, F., Hua, N.: Conceptualized representation learning for Chinese biomedical text mining. arXiv preprint arXiv:2008.10813 (2020)
16. Yan, H., Sun, Y., Li, X., Qiu, X.: An embarrassingly easy but strong baseline for nested named entity recognition. In: Proceedings of the 61st Annual Meeting of the Association for Computational Linguistics (Volume 2: Short Papers), Toronto, Canada, pp. 1442–1452. Association for Computational Linguistics (2023). https://aclanthology.org/2023.acl-short.123
17. Lin, X., Ma, L., Liu, W., Chang, S.-F.: Context-gated convolution. In: Vedaldi, A., Bischof, H., Brox, T., Frahm, J.-M. (eds.) ECCV 2020. LNCS, vol. 12363, pp. 701–718. Springer, Cham (2020). https://doi.org/10.1007/978-3-030-58523-5_41
18. Lin, T.Y., Goyal, P., Girshick, R., He, K., Dollár, P.: Focal loss for dense object detection. In: Proceedings of the IEEE International Conference on Computer Vision, pp. 2980–2988 (2017)

# Transformers for Interpretable Classification of Histopathological Images

Domenico Amato[1], Salvatore Calderaro[1(✉)], Giosué Lo Bosco[1], Giuseppe Marino[1], Filippo Vella[2], and Riccardo Rizzo[2]

[1] Department of Mathematics and Computer Science, University of Palermo, 90123 Palermo, Italy
{domenico.amato01,salvatore.calderaro01,giosue.lobosco}@unipa.it, giuseppe.marino24@community.unipa.it

[2] Institute for High-Performance Computing and Networking, National Research Council of Italy, 90146 Palermo, Italy
{filippo.vella,riccardo.rizzo}@icar.cnr.it

**Abstract.** Recent studies have shown how machine learning techniques can help and support the analysis of expert physicians' digitalized biopsy images. In particular, these techniques can improve the process related to some repetitive tasks, provide information, and speed up physician decision-making. Classifiers based on neural networks, particularly those that use Vision Transformers, have proven to be state-of-the-art in several classification tasks. In this paper, we apply vision transformers to classify histological images provided in the BreakHis and GasHisSDB datasets, which collect images from breast and gastric biopsies, respectively. The transformer architecture allows classification performances close to the state-of-the-art, also providing an explainable mechanism based on attention maps that localize and highlight the image regions relevant to the classifier's decision.

**Keywords:** Vision Transformer · Classification · Interpretability · Breast Cancer · Gastric Cancer

## 1 Introduction

According to the World Health Organization, cancer is the leading cause of death worldwide and counted 10 million deaths in 2020, or nearly one in six deaths[1]. The clinical pathology of cancer involves examining of high-resolution microscopic pictures of biopsy material, which enable the observation of specific areas of tissues. However, the invasive nature of tissue removal in pathology can lead to significant clinical effects. The presence of a wide range of data, such as images and their corresponding molecular features, has allowed machine learning scientists to create advanced computational methods that aid in automatic diagnostic assistance [29]. Deep Neural Network techniques can effectively process

[1] https://www.who.int/news-room/fact-sheets/detail/cancer.


M. Vettoretti et al. (Eds.): CIBB 2023, LNBI 14513, pp. 70–88, 2025.
https://doi.org/10.1007/978-3-031-90714-2_6

visual content and support physicians in repetitive image analysis tasks. One such application is the analysis and classification of histopathological images. A widely used dataset in the contest of histological image classification is BreakHis (Breast Cancer Histopathological Database) [36], a benchmark dataset containing histological images of breast cancer acquired at different magnification factors and divided into two main classes: benign and malignant, each of them divided into four subclasses. A more detailed description of the dataset is provided in Sect. 3. These images are available at four different magnification levels, so their analysis can be carried out in a magnification-specific (MS) or magnification-independent (MI) fashion. The authors of the BreakHis dataset made available a five-fold train-test split to compare the performance of different classification systems. These folds are patient-based, so images of the same patient are never in both the training and the test set. In this paper, we are only interested in the benign versus malignant dichotomy and have considered the MS setup. We also used the same split (as defined by the authors in [36]) for an adequate comparison with other state-of-the-art works [36]. The other dataset we used in our experimental activity was GasHisSDB (Gastric Histopathology Sub-size Image Database) [20]. It contains square patches of gastric cancer pathology divided into abnormal and normal classes and organized into three sub-datasets according to their dimensions. For the experimental part, we considered the patches with a $160 \times 160$ pixels resolution. A more detailed description of the dataset is in Sect. 3. This paper proposes a Vision Transformer network (ViT) for histopathology image classification that highlights the most relevant image regions the neural network uses in the labelling process, providing explainability to the model. This is a significant advancement in histopathology as it offers practical implications that physicians can directly apply and evaluate in their work. In particular, ViT is distinguished by a key component, the *attention* mechanism that enables the selection of the most relevant information in the input sequence and facilitates the analysis of the mutual relevance among the different portions of the image. Artificial Intelligence and Deep Learning tools used in real-world critical applications should always offer interpretable insights; in the medical domain, users should understand the mechanism behind the decision-making process. Some tools based on Convolutional Neural Networks(CNN) provide a mask superimposed to the input image that indicates important details involved in the classification decision, some of these visualizations are connected with attention mechanisms in neural networks, for example, the ones described in [30,32], other on some saliency mechanism typical of the image processing field. The impact of artificial intelligence tools on clinical decision-making is an active field of study. For example, Kim et al. [22] analyze and classify Clinical Decision Support Systems based on explainable AI tools, while Evans et al. [17] examine the impact of AI errors on clinical decisions. Usually, AI-based methods are tested as proof-of-concept and even for papers that deal with clinical trial. This is the case of the paper by Juan Ramon et al. [21] where the evaluation is performed as a Retrospective Validation to assess the algorithm performance in the case it would be implemented in the clinical workflow. To the best of our knowledge, there is no analysis of the use of histopathological image analysis in the field of AI-based systems.

Our proposal leverages ViT's attention mechanism to create intuitive visual explanations of model predictions, addressing a critical requirement for interpretability in AI applications within clinical settings. Unlike CNN-based attention models, our ViT-based approach has the potential to offer finer granularity in highlighting relevant image regions, potentially enhancing the trust of medical personnel in AI tools. The paper is organized as follows. Section 2 describes an overview of the State of the Art in the classification of the used datasets, the data used for the classification, the ViT model and the explainability mechanism are described in Sect. 3. The results of the experiments are shown in Sect. 4.

## 2 Related Works

Many works use the BreakHis dataset as a benchmark; for example, multiple approaches have been suggested for Magnification Specific Binary Classification (MSB). To ensure a consistent comparison, we have focused on methods that use the split training-test introduced in [36]. This ensures that all accuracy values are based on the same data split. In addition to the data division provided for training and testing, Spanhol et al. explored two methods for extracting smaller image sections called "patches" [37]. Another contribution by Spanhol et al. was a new approach [35], called DeCAF, that extracts features by taking the outputs of the top 3 layers from a pre-trained network similar to AlexNet. Then, these features are used as input for a separate CNN classifier. The concept of DeCAF features was further explored by Benhammou et al. [5] using a pre-trained Inception v3 network [42]. In their approach, only the fully connected layer before the softmax layer is adjusted during training, while the rest of the network remains unchanged. Two studies by Song et al. utilized pre-defined features for image classification. Both approaches use Fisher Vector encoding [28] to represent the data. In the first paper [34], authors use a VGG-D network, pre-trained on the ImageNet dataset, to extract features and represent them with Fisher Vector encoding. Then, an adaption layer, composed of two locally connected layers, and a classification layer extract descriptors used to train a support vector machine with a linear kernel. The second paper [33] proposed a supervised method using a neural network to improve the discriminator power of Fisher Vectors. Feature vectors of size 512 were extracted with the last layer of a pre-trained VGG-VD ConvNet, using as input images re-scaled to multiple sizes. These vectors are combined to form the Fisher vector based on ConvNet. Also, a support vector machine with a linear kernel was used as a classifier in this case. Sudharshan et al. [39] explored another machine learning approach called Multiple Instance Learning (MIL) for analyzing medical images with CNNs. During training, they randomly extracted 1000 small image patches ($64 \times 64$ pixels) from each patient's image. Each patch was then converted into a specific type of feature vector (162 elements long) based on Parametric-Free Threshold Adjacency Statistics (PFTAS). Sun et al. [40] proposed a new approach based on the computation of similarity embeddings that seems to be a more effective form of representation for histopathological images. Calderaro et al. [8] and Amato et al. [1] adopted the same approach based on metric learning and triplet loss. In the first paper, the authors proposed a method based on a fuzzy clustering algorithm, i.e. C-Means, to classify the embedding trained using a ResNet CNN and Metric Learning with Triplet Loss. The same architecture was used in the second paper, but the classification was performed using a k-NN classifier following an approach similar to those reported in [10,11]. A complete survey about the application of metric learning in medical imaging is in [3]. It is essential to point out that despite a decreasing PLA (Patient Level Accuracy) performance compared to those presented so far, those methods enable the model to provide information supporting the decision-making of the model itself. This information makes them more reliable as decision support systems in a critical field such as medical science. As mentioned, all the papers discussed above use the same 5-fold split to evaluate classifier performance. However, in the literature, other splits, not actually considered as benchmarks, can be found, e.g., the one proposed by Kumar and Rao [25]. An interested researcher could adopt it in their benchmark studies. It is also essential to emphasize that the images of the same patient must all be either in the training or the test set. Otherwise, the accuracy results are higher and do not simulate a real setting. Examples of works of CNN architecture that maintain separate training and test sets that can have images of the same patient in the two sets are [45] and [4]. In addition, Graph Neural Networks [12], GoogleNet [14], VGG networks [13], and Restricted Boltzmann Machine [26] are used in other papers. The GasHisSDB dataset, which is less common than BreakHis, is focused on gastric cancer and binary classification. The dataset was released by Hu et al. [20] for the development of systems aimed at diagnosing gastric cancer and is divided into three sub-datasets. Additionally, the study tests the effectiveness of a variety of classifiers. Yong et al. propose using transfer learning to classify histopathological images using the GasHisSDB dataset in several works. In their first paper [50], they compared different convolutional networks pre-trained on ImageNet and then fine-tuned them on the reference dataset. Notably, the best results were achieved by the DenseNet121 network by eliminating "Empty" patches with a high concentration of white pixels. Then, this work is extended in other two papers [51,52] that focus on using ensemble learning to improve transfer learning results. In particular, they chose the best three or five pre-trained neural networks as base models and combined them with three ensemble strategies. In [19], the dataset image is represented by extracting seven features and then classifying the histopathological images. Commonly used hand-prepared features are extracted. The GasHisSDB was also used in [46] as a reference for the identification of RoIs in pathological images and in the paper [18] as train/test dataset for the development of a system composed by a neural network for representation, an Extreme Learning Machine for classification and a meta-heuristic algorithm for optimization of parameters. The results of this system seem interesting, but the proposed methodology has been introduced previously. However, in digital pathology, deep learning tools that offer interpretable insights are essential. These tools help pathologists understand the rationale behind the model's decision-making process, which is critical for integrating AI systems into clinical workflows and enhancing trust in AI-generated outcomes. Recently, CNN with attention mechanisms have been developed to analyze the predictions, providing region-level supervision to identify errors and uncertainties in model decisions [30,32]. In [49], Yang et al. introduced the Guided Soft Attention Network (GuSA) for breast cancer classification, designed to focus on diagnostically significant regions of an image. The architecture consists of a Region of Interest (RoI) prediction branch for localizing key features and a classification branch for categorizing the images. A CNN backbone handles feature extraction, while GuSA produces activation maps that serve as visual explanations of the network's focus areas. In [6], BenTaieb et al. proposed a recurrent visual attention model for cancer prediction from histopathological images. This model incorporates Recurrent Neural Networks (RNN) paired with a visual attention mechanism that iteratively analyzes image regions. The architecture features convolutional layers for feature extraction and recurrent layers for learning spatial context. The attention module sequentially focuses on specific image regions, generating attention maps that highlight the most relevant areas for diagnosis. In [24], Krishna et al. introduced a novel interpretable decision-support model leveraging a CNN with a trainable attention mechanism tested on breast cancer histopathological images. The architecture fuses a DarkNet19 [31] variant as a feature extractor with an attention branch, creating the Branch Attention Network (ABN). This model provides diagnostic interpretability by generating attention maps that highlight RoIs, utilizing the convolutional layers of DarkNet19 and Global Average Pooling (GAP) to model the context of the visual features. Recently, new techniques have been developed that do not rely on attention mechanisms. For instance, in [2], Amato et al. proposed a granular computing methodology using deep metric learning for histopathological image classification. This approach enhances classification confidence by identifying and highlighting critical image fragments, which contributes to a significant reduction in false negatives.

## 3 Material and Methods

In this section, we first briefly describe the datasets involved in our experimental activity. Then, we describe the overall mechanism and the single components of the ViT. In the last part, we discuss an approach to identify the image regions relevant to the labelling process, starting from the transformer's attention. Using this approach, we create a heatmap that we superimpose onto the original image, thus providing an explainability mechanism that can be useful to medical personnel.

### 3.1 The BreakHis Dataset

The BreakHis dataset was proposed in [36], consisting of 7909 images with $700 \times 460$ pixels obtained from the Whole Slide Image of 82 breast cancer patients.

The images belong to two main classes, Malignant (M) and Benign (B), each divided into four subclasses of tumours and acquired at four magnification factors (40×, 100×, 200×, and 400×), but in this paper we only consider the primary classes. In Table 1, we report the sample distribution for each magnification factor. Figure 1 reports some samples of the dataset. The dataset is imbalanced because the malignant images are 5429 and the benign 2480; this imbalance is solved using data augmentation with rotation, flipping, and mirroring; it also used mild deformations and colour alterations (see Fig. 2 for an example). We did not apply pre-processing to the images, such as stain-normalization or whitening, since this processing does not ensure a sensible improvement in the classification performance. As mentioned above, the authors of [36] also propose a 70% – 30% split between train and test, with a five-fold structure.

**Table 1.** BreakHis dataset, samples distribution for each magnification factor.

| Magnification Factor | Benign | Malignant | Total |
|---|---|---|---|
| 40× | 625 | 1370 | 1995 |
| 100× | 644 | 1437 | 2081 |
| 200× | 623 | 1390 | 2013 |
| 400× | 588 | 1232 | 1820 |
| Total | 2480 | 5429 | 7909 |

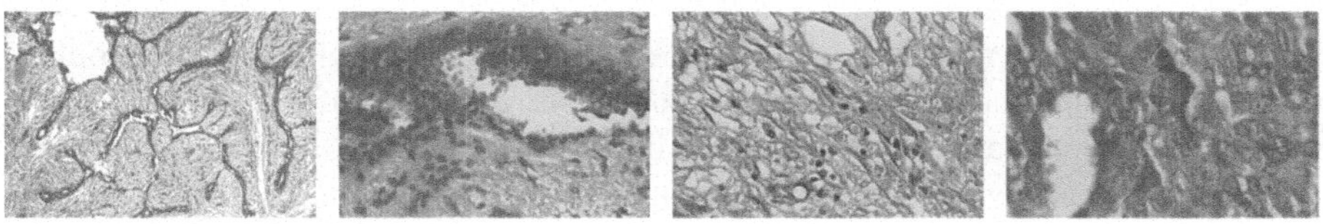

**Fig. 1.** Samples of BreakHis Images: the first two images are from the benign class, and the last two are from the malignant class.

### 3.2 The GasHisSDB Dataset

Another dataset we used to test the effectiveness of our proposed approach is the so-called GasHisSDB [20]. The dataset comprises 245.196 images (97.076 for the "abnormal" class and 148.120 for the "normal" class) obtained from 600 images of gastric cancer pathology with resolution 2048 × 2048. Starting from these images, the authors create three sub-datasets, A, B, and C, with square patches using three cropping sizes: $160 \times 160$, $120 \times 120$, and $80 \times 80$. An overview of the sample distribution for each sub-dataset is in Table 2. We considered the dataset

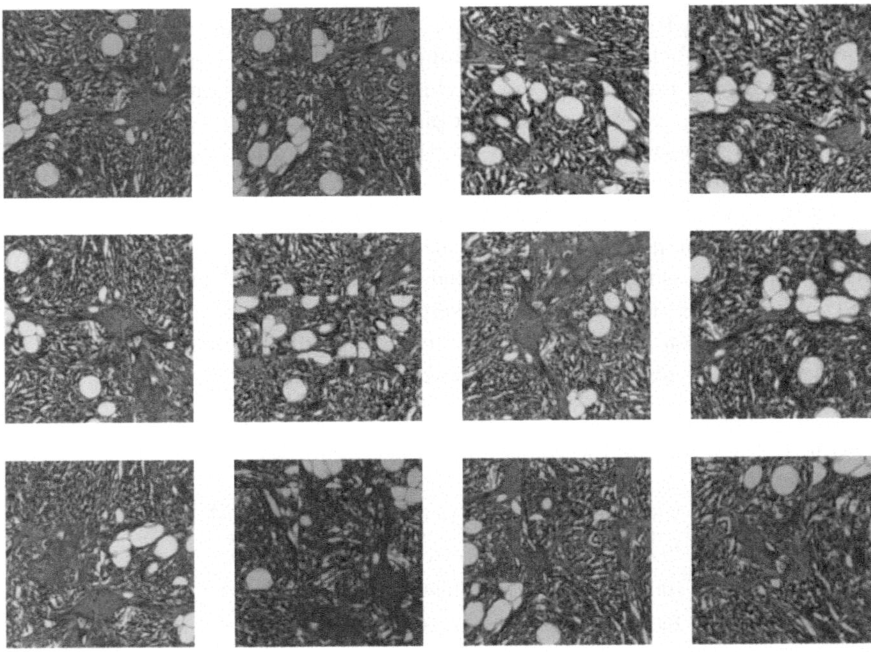

**Fig. 2.** Example of augmentation applied to the original (upper left) image (dye, saturation, contrast, rotation, scaling and cropping transformations).

with the patches of dimension 160 × 160 for our experiments, including 13.124 abnormal images and 20.160 normal ones. The images of the "abnormal" class contain gastric cancer, while the images of the "normal" class do not contain cancerous regions; the dataset does not contain any patient information. In Fig. 3, we report some dataset samples. Also, in this case, we did not apply any preprocessing technique, such as stain normalization or whitening, to the dataset images. Our preliminary experiments indicated that such techniques did not yield improvements in the classification results. Unlike the BreakHis dataset experiments, we did not need to apply any data augmentation procedure due to the absence of significant class imbalances.

**Table 2.** GasHisSDB dataset, samples distribution for each sub-dataset. The sub-dataset used for our experiments is A.

| Sub-Dataset | Abnormal | Normal | Total |
|---|---|---|---|
| A (160 × 160) | 13124 | 20160 | 33284 |
| B (120 × 120) | 24801 | 40460 | 65261 |
| C (80 × 80) | 59151 | 87500 | 146651 |
| **Total** | 97076 | 148120 | 245196 |

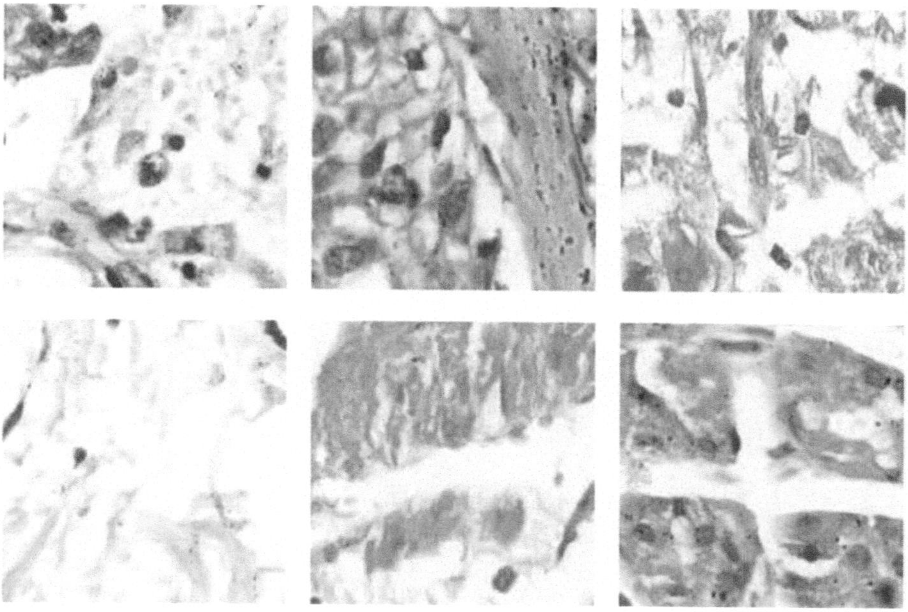

**Fig. 3.** Samples of GasHisSDB dataset: in the first row we have abnormal images while in the second normal ones.

### 3.3 Vision Transformer

ViT [16] are customized versions of the original Transformer [44]. The transformer architecture was proposed for word sequence processing and is an encoder-decoder architecture characterized by the so-called *self-attention layer*. It can provide higher accuracy than classical recurrent networks when replicated and opportunely stacked with a MultiLayer Perceptron (MLP). The ViT customization (see Fig. 4) inherits the specific transformer property of being able to model long-term relationships in the input data, which is an important feature for the case of histopathological image classification.

The ViT does not include a decoder part; it is based only on an encoder incorporating *multi head* self-attention (MHA). A single self-attention scheme starts from a tokenized input sequence $Z = [\boldsymbol{z}_1, \ldots, \boldsymbol{z}_t]$, where $z_i \in \mathbb{R}^m$, is used to compute a hidden representation $[\boldsymbol{h}_1, \ldots, \boldsymbol{h}_t]$ by the following equation:

$$\boldsymbol{h}_i = [\boldsymbol{W}_v \boldsymbol{z}_1, \ldots, \boldsymbol{W}_v \boldsymbol{z}_t]\, \mathrm{softmax}\left[\frac{(\boldsymbol{W}_q \boldsymbol{z}_i \cdot \boldsymbol{W}_k \boldsymbol{z}_1)}{\sqrt{d}}, \ldots, \frac{(\boldsymbol{W}_q \boldsymbol{z}_i \cdot \boldsymbol{W}_k \boldsymbol{z}_t)}{\sqrt{d}}\right] \tag{1}$$

where $\boldsymbol{W}_q, \boldsymbol{W}_k, \boldsymbol{W}_v$ are the *queries, key and values* matrices, respectively, and represent the learning parameters of the self-attention layer. The rationale behind this computation is that the hidden component $h_i$ represents the relevance of the token $z_i$ to generate the corresponding target. The MHA is (1)

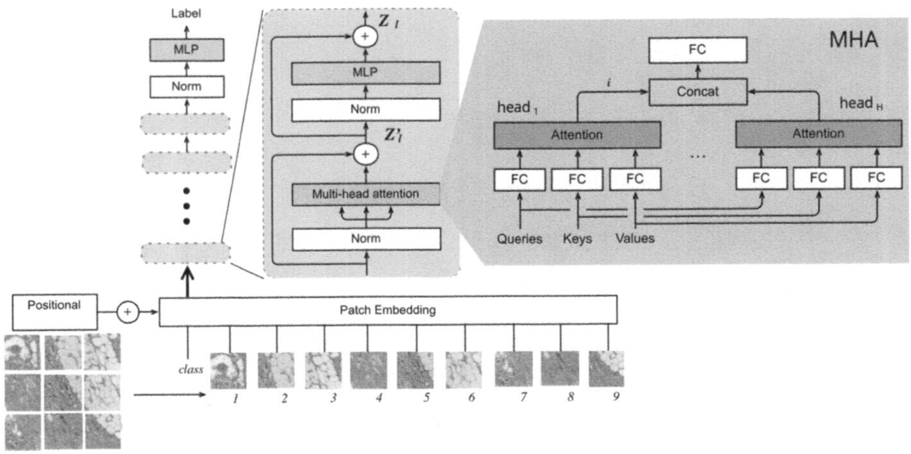


**Fig. 4.** A sketch of the transformer structure.

the parallel computation of several single self-attentions, (2) the concatenation of the corresponding hidden representations, and (3) their computation by a linear Fully Connected layer (FC). Applying the MHA enables us to derive different kinds of relationships between tokens. The permutation-invariant nature of self-attention is a drawback in sequence processing. To spatially contextualize a token in a sequence, with a relative position and an absolute position, a *positional encoding* scheme is used. To each component of the tokenized input sequence $[\boldsymbol{x}_1, \ldots, \boldsymbol{x}_t]$, a positional encoding vector is added. In the context of ViT, an image $\mathbf{X} \in \mathbb{R}^{r \times c \times p}$ of height $r$, width $c$ and $p$ channels is tiled in square patches with sides equal to $s$, each representing a token of the transformer input sequence. Each token $\boldsymbol{x}_i$ is the linearization of a $p$ channels $s \times s$ patch, and the length of the sequence is $t = \frac{r \times c}{s^2}$. Each layer of the Transformer receives and returns vectors of the same dimension $d$. For this reason, the embedding of the patches must project a linearized patch of length into a vector of dimension $d$. For the case of ViT, a linear learned embedding of the input sequence is used together with the positional encoding information. The transformer input sequence can be finally represented as $\boldsymbol{Z} = [\boldsymbol{x}_{class}, \boldsymbol{E}\boldsymbol{x}_1, \boldsymbol{E}\boldsymbol{x}_2, \ldots, \boldsymbol{E}\boldsymbol{x}_t] + \boldsymbol{E}_{pos}$, where $\boldsymbol{E} \in \mathbb{R}^{(s^2 \times p) \times d}$ is the learned embedding matrix, and $\boldsymbol{E_{pos}} \in \mathbb{R}^{(t+1) \times d}$ is the positional encoding matrix. Note that the input sequence starts with an additional embedding $\boldsymbol{x}_{class}$ that captures the representation of the whole sequence, such as a weighted average of the tokens in the sequence. ViT is characterized only by an encoder part composed of the concatenation of $n$ encoding blocks, each consisting of MHA followed by an MLP with one hidden layer. Layer normalization (LN) is applied to each of these, and a residual connection is added at the output. The Gaussian Error Linear Unit (GELU) activation function is used within the MLP. In formulas:

$$\begin{aligned} \mathbf{Z}'_\ell &= \mathrm{MHA}(\mathrm{LN}(\mathbf{Z}_{\ell-1})) + \mathbf{Z}_{\ell-1} \\ \mathbf{Z}_\ell &= \mathrm{MLP}_{GELU}(\mathrm{LN}(\mathbf{Z}'_\ell)) + \mathbf{Z}'_\ell \end{aligned} \qquad \ell = 1, \ldots, n. \tag{2}$$

In the general case of image classification by deep models, transfer learning is always profitable and even mandatory for datasets with few images. For the case of ViT, a pre-trained model on a large dataset is considered, followed by fine-tuning on a specific task (using a dataset with fewer examples). In both the pre-training and fine-tuning phases, MLP is used. Both processings take as input $z_n^0$, which is the vector corresponding to $x_{class}$ at the end of the $n$ encoder blocks. In the pre-training step, an MLP is trained on a huge dataset, such as ImageNet. In the fine-tuning phase, another MLP returns a vector of the same size as the number of classes.

### 3.4 Explainability

The proposed explainability mechanism is based on the visualization of the attention, which is not trivial due to the complex structure of ViTs, consisting of various concatenated blocks. The last block is certainly the one with the highest level of abstraction and is, therefore, considered in the present visualization. Although the analysis is limited to the last block's attention, it is impossible to directly evaluate a single attention distribution. Each block has $h$ different attention heads, each evaluating $t+1$ different attentions, one for each patch plus

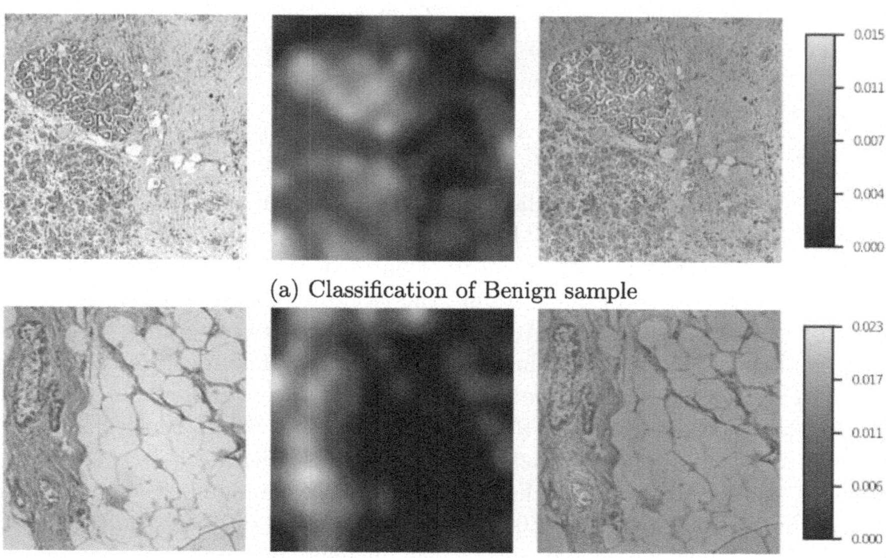


(a) Classification of Benign sample

(b) Classification of Malignant sample

**Fig. 5.** Visualization of Vision Transformer's attention for the classification of images from the BreakHis dataset.

one for the token $\boldsymbol{x}_{class}$. We decided to consider softmax attention $s_h$ relative to $\boldsymbol{x}_{class}$ as the query and the embeddings of the patches as the keys, i.e.:

$$s_h = \text{softmax}\left[\frac{(\boldsymbol{W}_q\boldsymbol{x}_{class} \cdot \boldsymbol{W}_k\boldsymbol{z}_1)}{\sqrt{d}}, \ldots, \frac{(\boldsymbol{W}_q\boldsymbol{x}_{class} \cdot \boldsymbol{W}_k\boldsymbol{z}_t)}{\sqrt{d}}\right]. \quad (3)$$

Consequently, one softmax attention vector was obtained for each head attention, then aggregated by mean, obtaining a vector $S$ of $t$ components, one related to each token and, consequently, to each patch. Rearranging these $t$ values in a grid of $H/P$ and $W/P$, then scaling it to a resolution of $H \times W$, resulted in an attention map that could be overlaid on the input image of the ViT (see Fig. 5 for an example). In the resizing process, bilinear interpolation was used, followed by a median filter with a structural element of radius $P/2$ and a Gaussian blur with $\sigma = P/4$. We discuss this approach in more detail in [9].

## 4 Experiments

The experimental activities involved different kinds of transformers, the "Base" model (ViT-B) introduced by [16], and two additional configurations, named the "Tiny" (ViT-T) and "Small" (ViT-S) models, proposed in [43]. Details about the models are reported in Table 3. The images in both datasets involved in our experimental activity have been rescaled to $224 \times 224$ using the nearest neighbours interpolation technique to fit the ViT input dimension. The number of considered patches is $t = 196$, so the size of the single patch is $16 \times 16$. For the BreakHis dataset, the performances of the considered ViTs are measured by using the PLA [36] defined as:

$$PLA = \frac{\sum_{p=1}^{N_{pat}} ACC_p}{N_{pat}} \quad (4)$$

where $ACC_p$ is the fraction of the correctly classified image of a patient $p$ and $N_{pat}$ is the number of patients. For the GasHisSDB, the PLA metric cannot be computed because the patient information is unavailable. For this reason, to assess the performance obtained, we use the Accuracy metric, a commonly used measure in supervised learning. Accuracy is the ratio between correctly classified images and the total number of images in the test set. The Inception ResNetV2 defined in [41] was selected as the baseline for comparison, as made in [48]. Note that the ResNetV2 has about $56 \times 10^6$ parameters. The vision transformers were trained using the Adam optimization algorithm [23] with a batch size of 32 (larger or smaller sizes do not affect performances); the number of epochs was 25, and the learning rate was 0.002; a greater number of epochs and different learning rate values were used without an increase on the performances. All the code was written in Python using the PyTorch library [27]; the ViTs were implemented using the Timm library [47] starting from default parameters. The models are pre-trained using the ImageNet-21k dataset [15] and fine-tuned with two classes, Benign and Malignant for the Breakhis dataset and Normal and

Abnormal for the GasHisSDB one. In the present work, any reference to ViT training is precisely intended as the execution of such fine-tuning. According to the authors of Timm, they referred to Dosovitskiy et al. [16], Steiner et al. [38], and Beyer et al. [7] for the implementation details of the architectures and the configuration of the pre-training phase. For the BreakHis dataset, we performed the experiments using a workstation equipped with an Intel Core i7-8700 3.2 GHz CPU and an Nvidia Titan V GPU. The total system memory is 32 GB of DDR4, and the GPU supplies 12 GB of DDR5 memory. The operating system is Ubuntu LTS 20.04. For the GasHisSDB dataset, we use a different workstation equipped with a 12th Gen Intel CoreTM i9-12900KS and an Nvidia 3090 Ti GPU. The total system memory is 32 GB of DDR5, and the GPU supplies 24 GB of DDR6 memory. The operating system is Ubuntu 23.10. Both workstations adopt a CUDA parallel computing platform.

### 4.1 Results with the BreakHis Dataset

For the BreakHis dataset, the PLA values of the binary classification Benign vs Malignant, for all the considered models and all the magnification values, are reported in Table 4. Among all ViTs considered, ViT-S obtained the best results across multiple magnification values. In comparison to the baseline model, Inception ResNet v2 (denoted as IRNV2 in the table), ViT-S achieves comparable results at 40× magnification, superior results at 200× and 400×, but performs slightly worse at 100×. Figure 5 reports examples of images and attention maps. The leftmost portion of the image shows the sample to be classified; the central part is the activation map; the rightmost image is the superposition of the attention map to the sample. The legend added to each image shows the pseudo-colour representation of the activation level. The examples in Fig. 5a and Fig. 5b are correctly classified. An analysis of the pattern in the regions highlighted by the method shows a specific pattern that appears significant for discriminating the label. Conversely, the first three samples identify areas in the input images with a pattern that can be associated with the pathology label. However, the authors underline that the analysis is purely on the informatics stance. A detailed and complete review of these results requires the suggestion and evaluation of medical personnel that can provide insight into the relevance of the retrieved patterns.

**Table 3.** ViT configurations. $n$ is the number of model layers, $d$ is the dimension of the input token embeddings, *MLP* is the number of features used in the MLP part, head is the number of attention heads and *parameters* is the total number of trainable parameters of the model.

| Model | n | d | MLP | heads | parameters |
|---|---|---|---|---|---|
| ViT-T | 12 | 92 | 768 | 3 | $5 \times 10^6$ |
| ViT-S | 12 | 384 | 1536 | 6 | $22 \times 10^6$ |
| ViT-B | 12 | 768 | 3072 | 12 | $86 \times 10^6$ |

**Table 4.** Patient Level Accuracy (PLA) and Training Time (TT) in seconds for each ViT configuration for the BreakHis dataset.

| Model | 40× | | 100× | |
|---|---|---|---|---|
| | PLA | TT | PLA | TT |
| ViT-T | 0.8769 ± 0.03 | 1178 ± 65 | 0.8559 ± 0.03 | 1301 ± 78 |
| ViT-S | 0.8898 ± 0.04 | 941 ± 370 | 0.8552 ± 0.02 | 1320 ± 78 |
| ViT-B | 0.8599 ± 0.05 | 1319 ± 68 | 0.8321 ± 0.01 | 1440 ± 82 |
| IRNV2 | 0.892 ± 0.02 | 1267 ± 72 | 0.8756 ± 0.02 | 1418 ± 76 |

(a) Values of PLA and TT for images with magnification factor 40× and 100×

| Model | 200× | | 400× | |
|---|---|---|---|---|
| | PLA | TT | PLA | TT |
| ViT-T | 0.8589 ± 0.04 | 1262 ± 57 | 0.8015 ± 0.04 | 1139 ± 46 |
| ViT-S | 0.8781 ± 0.03 | 1286 ± 57 | 0.8249 ± 0.03 | 1163 ± 46 |
| ViT-B | 0.85415 ± 0.04 | 1437 ± 85 | 0.7893 ± 0.06 | 1281 ± 48 |
| IRNV2 | 0.8715 ± 0.03 | 1379 ± 51 | 0.8005 ± 0.04 | 1248 ± 45 |

(b) Values of PLA and TT for images with magnification factor 200× and 400×

**Table 5.** BreakHis dataset: comparison of PLA values between our approach with state-of-the-art methods.

| Method | 40× | 100× | 200× | 400× |
|---|---|---|---|---|
| Sun et al. [40] | 87.51 ± 4.07 | 89.12 ± 2.86 | 90.82 ± 3.31 | 87.10 ± 3.80 |
| Spanhol et al. [37] | 90.0 ± 6.7 | 88.4 ± 4.8 | 84.6 ± 4.2 | 86.1 ± 6.2 |
| Spanhol et al. [35] | 84.0 ± 6.9 | 83.9 ± 5.9 | 86.3 ± 3.5 | 82.1 ± 2.4 |
| Benhammou et al. [5] | 87.6 ± 3.9 | 82.4 ± 2.7 | 86.1 ± 0.7 | 79.7 ± 3.2 |
| Song et al. [34] | 90.0 ± 3.2 | 88.9 ± 5.0 | 86.9 ± 5.2 | 86.3 ± 7.0 |
| Song et al. [33] | 90.2 ± 3.2 | 91.2 ± 4.4 | 87.8 ± 5.3 | 87.4 ± 7.2 |
| Sudharshan et al. [39] | 92.1 ± 5.9 | 89.1 ± 5.2 | 87.2 ± 4.3 | 87.8 ± 5.6 |
| Amato et al. [1] | 87.60 ± 3.92 | 88.17 ± 3.86 | 89.37 ± 3.26 | 85.98 ± 2.78 |
| Proposed method | 88.98 ± 4.00 | 87.56 ± 2.00 | 87.81 ± 3.00 | 82.49 ± 3.00 |

In Table 5, we reported the PLA values of our approach and the state-of-the-art methodologies using the same patient-based split. It is essential to point out that the values reported for our approach in the table refer to the ViT that obtains the best results regarding the PLA values. In this case, we received the best results for all the magnification factors using the ViT-S architecture.

### 4.2 Results with the GasHisSDB Dataset

For the GasHisSDB dataset, a predefined split into training and test sets is unavailable. Furthermore, patient information is not provided, so a patient-based split is not possible. For this reason, we decided to use a classical five-fold cross-validation protocol to obtain reliable results. In Table 6, we report the mean and standard deviation of the accuracy values and training times for all models considered. We obtained the best results with the ViT-T architecture regarding the transformer architectures. At the same time, we have a performance drop of about eight percentage points using the ViT-B, the most complex transformer model we used. Among all the considered models, IRNV2 achieved the best results (about 96%) but with the limitation that it is not directly interpretable. During our experimental activity for the GasHisSDB dataset, we performed a further experiment that removed the so-called "empty patches" from the considered datasets. A similar approach has been used in other works that used the same dataset: [46,50–52]. The results of this further experiment are in the Table 7. Analyzing the results reported in Tables 6 and 7, it is evident that the removal of the empty patches does not allow us to obtain better results using Vision Transformers as a classifier. The comparison - for both experiments - of the results between our approach and state-of-the-art methods are in Table 8. Figure 6 shows some examples of images and related attention maps for normal and abnormal cases. In particular, the original image is shown on the left, while the same image with the superimposed attention map is shown on the right. Both samples are correctly classified, and the visual analysis of the image pattern in the regions highlighted by the method shows a specificity that appears important to discriminate between the classes..

**Table 6.** Accuracy (Acc.) and Training Time (TT) in seconds for each ViT configuration for the GasHisSDB dataset.

| Model | Acc. | TT |
|---|---|---|
| ViT-T | $0.895 \pm 0.010$ | $663 \pm 4$ |
| ViT-S | $0.884 \pm 0.006$ | $1401 \pm 4$ |
| ViT-B | $0.815 \pm 0.019$ | $4050 \pm 97$ |
| IRNV2 | $0.962 \pm 0.014$ | $2257 \pm 102$ |

**Table 7.** Accuracy (Acc.) and Training Time (TT) in seconds for each ViT configuration for the GasHisSDB dataset with the removal of empty patches.

| Model | Acc. | TT |
|---|---|---|
| ViT-T | $0.896 \pm 0.007$ | $576 \pm 0.89$ |
| ViT-S | $0.878 \pm 0.014$ | $1304 \pm 1$ |
| ViT-B | $0.804 \pm 0.025$ | $3729 \pm 3$ |

**Table 8.** GasHisSDB dataset: comparison between our approach and state-of-the-art methodologies.

| Method | Accuracy |
|---|---|
| (a) Experiment with all the patches | |
| Hu et al. [19,20] | 96.09% |
| Proposed approach | 89.5% |
| (b) Experiment without the empty patches | |
| Yong et al. [50] | 98.68% |
| Yong et al. [51,52] | 99.10% |
| Wei et al. [46] | 96.29% |
| Proposed approach | 89.6% |

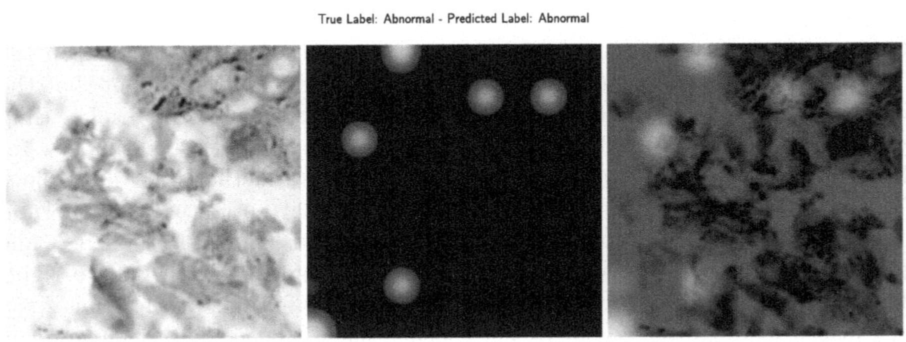


(a) Example of classification of an image from the *Abnormal* class of the GasHisSDB dataset

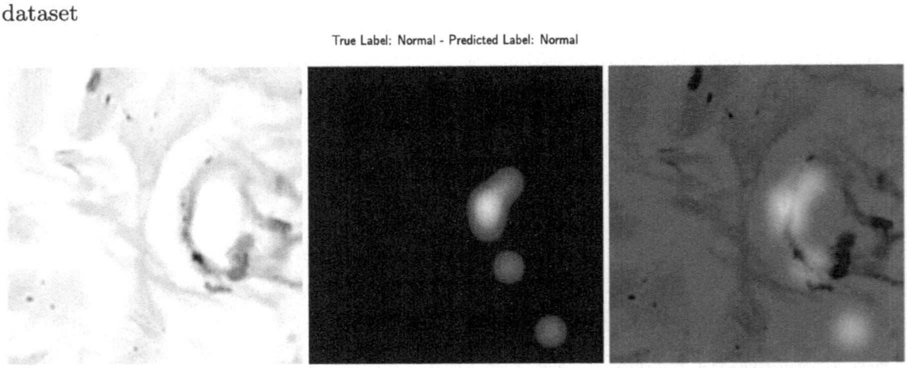


(b) Example of classification of an image from the *Normal* class of the GasHisSDB dataset

**Fig. 6.** Visualization of Vision Tranformer's attention for the classification of images from the GasHisSDB dataset.

## 5 Conclusions

This paper shows that applying ViT in histopathology image classification is both effective and provides an explainable mechanism for physicians. The explainability of the model is based on the aggregated information of the attention level, particularly focusing on attention maps of the first token in the image path sequences. Obviously, only a pathologist, with specialized expertise, can evaluate the significance of high attention values in the image regions; a positive result would constitute a strong argument in favour of the model, indicating a promising direction for future research.

**Data Availibility Statement.** The BreakHis dataset is publicly available at the following Link. The GasHisSDB is publicly available at the following Link. The source code can be found in the following GitHub repository.

## References

1. Amato, D., Calderaro, S., Lo Bosco, G., Rizzo, R., Vella, F.: Metric learning in histopathological image classification: opening the black box. Sensors **23**(13), 6003 (2023)
2. Amato, D., Calderaro, S., Lo Bosco, G., Rizzo, R., Vella, F.: Explainable histopathology image classification with self-organizing maps: a granular computing perspective. Cogn. Comput. 1–21 (2024)
3. Amato, D., Calderaro, S., Lo Bosco, G., Rizzo, R., Vella, F.: Leveraging deep embeddings for explainable medical image analysis. In: Machine Learning and Granular Computing: A Synergistic Design Environment, pp. 225–261. Springer (2024)
4. Bardou, D., Zhang, K., Ahmad, S.M.: Classification of breast cancer based on histology images using convolutional neural networks. IEEE Access **6**, 24680–24693 (2018)
5. Benhammou, Y., Tabik, S., Achchab, B., Herrera, F.: A first study exploring the performance of the state-of-the art CNN model in the problem of breast cancer. In: proceedings of the International Conference on Learning and Optimization Algorithms: Theory and Applications, pp. 1–6 (2018)
6. BenTaieb, A., Hamarneh, G.: Predicting cancer with a recurrent visual attention model for histopathology images. In: Frangi, A.F., Schnabel, J.A., Davatzikos, C., Alberola-López, C., Fichtinger, G. (eds.) Medical Image Computing and Computer Assisted Intervention - MICCAI 2018, pp. 129–137. Springer, Cham (2018)
7. Beyer, L., et al.: FlexiViT: one model for all patch sizes. In: Proceedings of the IEEE/CVF Conference on Computer Vision and Pattern Recognition, pp. 14496–14506 (2023)
8. Calderaro, S., Lo Bosco, G., Rizzo, R., Vella, F., et al.: Fuzzy clustering of histopathological images using deep learning embeddings. In: CEUR Workshop Proceedings, vol. 3074, pp. 1–9 (2022). http://ceur-ws.org/Vol-3074/
9. Calderaro, S., Lo Bosco, G., Rizzo, R., Vella, F., et al.: Visualization and analysis of transformer attention. In: CEUR Workshop Proceedings, vol. 3563 (2023). https://ceur-ws.org/Vol-3563/paper_11.pdf

10. Calderaro, S., Lo Bosco, G., Rizzo, R., Vella, F.: Deep metric learning for histopathological image classification. In: 2022 IEEE Eighth International Conference on Multimedia Big Data (BigMM), pp. 57–64. IEEE (2022)
11. Calderaro, S., Lo Bosco, G., Rizzo, R., Vella, F.: Deep metric learning for transparent classification of COVID-19 X-ray images. In: 2022 16th International Conference on Signal-Image Technology & Internet-Based Systems (SITIS), pp. 300–307. IEEE (2022)
12. Calderaro, S., Lo Bosco, G., Vella, F., Rizzo, R.: Breast cancer histologic grade identification by graph neural network embeddings. In: International Work-Conference on Bioinformatics and Biomedical Engineering, pp. 283–296. Springer (2023)
13. Cascianelli, S., et al.: Dimensionality reduction strategies for CNN-based classification of histopathological images. In: Intelligent Interactive Multimedia Systems and Services 2017, pp. 21–30. Springer (2018)
14. Das, K., Karri, S.P.K., Roy, A.G., Chatterjee, J., Sheet, D.: Classifying histopathology whole-slides using fusion of decisions from deep convolutional network on a collection of random multi-views at multi-magnification. In: 2017 IEEE 14th International Symposium on Biomedical Imaging (ISBI 2017), pp. 1024–1027. IEEE (2017)
15. Deng, J., Dong, W., Socher, R., Li, L.J., Li, K., Fei-Fei, L.: ImageNet: a large-scale hierarchical image database. In: 2009 IEEE Conference on Computer Vision and Pattern Recognition, pp. 248–255. Ieee (2009)
16. Dosovitskiy, A., et al.: An image is worth 16×16 words: transformers for image recognition at scale. In: 9th International Conference on Learning Representations, ICLR 2021, Virtual Event, Austria, 3–7 May 2021 (2021)
17. Evans, H., Snead, D.: Understanding the errors made by artificial intelligence algorithms in histopathology in terms of patient impact. NPJ Digit. Med. **7**(1), 89 (2024)
18. Fan, D., Liang, H., Qu, C., Ma, J., Hasani, R.: A novel approach for early gastric cancer detection using a hybrid of AlexNet, extreme learning machine, and adjusted gorilla troops optimization. Biomed. Signal Process. Control **93**, 106126 (2024)
19. Hu, W., et al.: A comparative study of gastric histopathology sub-size image classification: from linear regression to visual transformer. Front. Med. **9**, 1072109 (2022)
20. Hu, W., et al.: GashisSDB: a new gastric histopathology image dataset for computer aided diagnosis of gastric cancer. Comput. Biol. Med. **142**, 105207 (2022)
21. Juan Ramon, A., et al.: Development and deployment of a histopathology-based deep learning algorithm for patient prescreening in a clinical trial. Nat. Commun. **15**(1), 4690 (2024)
22. Kim, S.Y., Kim, D.H., Kim, M.J., Ko, H.J., Jeong, O.R.: XAI-based clinical decision support systems: a systematic review. Appl. Sci. **14**(15), 6638 (2024)
23. Kingma, D., Ba, J.: Adam: a method for stochastic optimization. In: International Conference on Learning Representations (ICLR), San Diego, CA, USA (2015)
24. Krishna, S., Suganthi, S., Bhavsar, A., Yesodharan, J., Krishnamoorthy, S.: An interpretable decision-support model for breast cancer diagnosis using histopathology images. J. Pathol. Inform. **14**, 100319 (2023)
25. Kumar, K., Rao, A.C.S.: Breast cancer classification of image using convolutional neural network. In: 2018 4th International Conference on Recent Advances in Information Technology (RAIT), pp. 1–6. IEEE (2018)

26. Nahid, A.A., Mikaelian, A., Kong, Y.: Histopathological breast-image classification with restricted Boltzmann machine along with backpropagation. Biomed. Res. (India) **29**(10), 2068–2077 (2018)
27. Paszke, A., et al.: PyTorch: an imperative style, high-performance deep learning library. In: Advances in Neural Information Processing Systems, vol. 32 (2019)
28. Perronnin, F., Dance, C.: Fisher kernels on visual vocabularies for image categorization. In: 2007 IEEE Conference on Computer Vision and Pattern Recognition, pp. 1–8. IEEE (2007)
29. Pinello, L., Lo Bosco, G., Hanlon, B., Yuan, G.C.: A motif-independent metric for DNA sequence specificity. BMC Bioinform. **12**, 408 (2011)
30. Pocevičiūtė, M., Eilertsen, G., Lundström, C.: Survey of XAI in digital pathology. In: Artificial Intelligence and Machine Learning for Digital Pathology: State-of-the-Art and Future Challenges, pp. 56–88 (2020)
31. Redmon, J., Farhadi, A.: YOLO9000: better, faster, stronger. In: 2017 IEEE Conference on Computer Vision and Pattern Recognition (CVPR), pp. 6517–6525 (2016)
32. Ribeiro, M.T., Singh, S., Guestrin, C.:"Why should i trust you?": explaining the predictions of any classifier. In: Proceedings of the 22nd ACM SIGKDD International Conference on Knowledge Discovery and Data Mining, KDD 2016, pp. 1135–1144. Association for Computing Machinery, New York (2016)
33. Song, Y., Chang, H., Huang, H., Cai, W.: Supervised intra-embedding of fisher vectors for histopathology image classification. In: Descoteaux, M., Maier-Hein, L., Franz, A., Jannin, P., Collins, D.L., Duchesne, S. (eds.) MICCAI 2017. LNCS, vol. 10435, pp. 99–106. Springer, Cham (2017). https://doi.org/10.1007/978-3-319-66179-7_12
34. Song, Y., Zou, J.J., Chang, H., Cai, W.: Adapting fisher vectors for histopathology image classification. In: 2017 IEEE 14th International Symposium on Biomedical Imaging (ISBI 2017), pp. 600–603 (2017). https://doi.org/10.1109/ISBI.2017.7950592. iSSN 1945-8452
35. Spanhol, F.A., Oliveira, L.S., Cavalin, P.R., Petitjean, C., Heutte, L.: Deep features for breast cancer histopathological image classification. In: 2017 IEEE International Conference on Systems, Man, and Cybernetics (SMC), pp. 1868–1873. IEEE (2017)
36. Spanhol, F.A., Oliveira, L.S., Petitjean, C., Heutte, L.: A dataset for breast cancer histopathological image classification. IEEE Trans. Biomed. Eng. **63**(7), 1455–1462 (2015)
37. Spanhol, F.A., Oliveira, L.S., Petitjean, C., Heutte, L.: Breast cancer histopathological image classification using convolutional neural networks. In: 2016 International Joint Conference on Neural Networks (IJCNN), pp. 2560–2567. IEEE (2016)
38. Steiner, A.P., Kolesnikov, A., Zhai, X., Wightman, R., Uszkoreit, J., Beyer, L.: How to train your ViT? Data, augmentation, and regularization in vision transformers. Trans. Mach. Learn. Res. (2022). https://openreview.net/forum?id=4nPswr1KcP
39. Sudharshan, P., Petitjean, C., Spanhol, F., Oliveira, L.E., Heutte, L., Honeine, P.: Multiple instance learning for histopathological breast cancer image classification. Expert Syst. Appl. **117**, 103–111 (2019)
40. Sun, Y., Huang, X., Wang, Y., Zhou, H., Zhang, Q.: Magnification-independent histopathological image classification with similarity-based multi-scale embeddings (2022)
41. Szegedy, C., Ioffe, S., Vanhoucke, V., Alemi, A.: Inception-v4, inception-ResNet and the impact of residual connections on learning. In: Proceedings of the AAAI Conference on Artificial Intelligence, vol. 31 (2017)

42. Szegedy, C., Vanhoucke, V., Ioffe, S., Shlens, J., Wojna, Z.: Rethinking the inception architecture for computer vision. In: Proceedings of the IEEE Conference on Computer Vision and Pattern Recognition, pp. 2818–2826 (2016)
43. Touvron, H., Cord, M., Douze, M., Massa, F., Sablayrolles, A., Jégou, H.: Training data-efficient image transformers & distillation through attention. In: International Conference on Machine Learning, pp. 10347–10357. PMLR (2021)
44. Vaswani, A., et al.: Attention is all you need. In: Advances in Neural Information Processing Systems, vol. 30 (2017)
45. Wei, B., Han, Z., He, X., Yin, Y.: Deep learning model based breast cancer histopathological image classification. In: 2017 IEEE 2nd International Conference on Cloud Computing and Big Data Analysis (ICCCBDA), pp. 348–353. IEEE (2017)
46. Wei, Z., et al.: Deep learning-based stratification of gastric cancer patients from hematoxylin and eosin-stained whole slide images by predicting molecular features for immunotherapy response. Am. J. Pathol. **193**(10), 1517–1527 (2023)
47. Wightman, R.: PyTorch image models (2019). https://github.com/rwightman/pytorch-image-models, https://doi.org/10.5281/zenodo.4414861
48. Xie, J., Liu, R., Luttrell, J., IV., Zhang, C.: Deep learning based analysis of histopathological images of breast cancer. Front. Genet. **10**, 80 (2019)
49. Yang, H., Kim, J.Y., Kim, H., Adhikari, S.P.: Guided soft attention network for classification of breast cancer histopathology images. IEEE Trans. Med. Imaging **39**(5), 1306–1315 (2019)
50. Yong, M.P., Hum, Y.C., Lai, K.W., Goh, C.H., Yap, W.S., Tee, Y.K.: Histopathological gastric cancer detection using transfer learning. In: 2023 11th International Conference on Bioinformatics and Computational Biology (ICBCB), pp. 123–129 (2023)
51. Yong, M.P., et al.: Histopathological gastric cancer detection on GasHisSDB dataset using deep ensemble learning. Diagnostics **13**(10) (2023)
52. Yong, M.P., et al.: Histopathological cancer detection using intra-domain transfer learning and ensemble learning. IEEE Access **12**, 1434–1457 (2024)

# Breast Cancer Malignancy Prediction through Explainable Models based on a Multimodal Signature of Features

Francesco Prinzi[1], Carmelo Militello[2](✉), Tommaso Vincenzo Bartolotta[1], and Salvatore Vitabile[1]

[1] Department of Biomedicine, Neuroscience and Advanced Diagnostics (BiND), University of Palermo, 90127 Palermo, Italy

[2] Italian National Research Council, Institute for High-Performance Computing and Networking (ICAR-CNR), 90146 Palermo, Italy

carmelo.militello@cnr.it

**Abstract.** Breast cancer classification through ultrasound imaging poses a significant challenge due to the inherent noise present in ultrasound images. The radiologist's reporting process aims to assess the lesions within the images following the Breast Imaging-Reporting and Data System (BI-RADS). This work investigates whether the medical knowledge, represented by the BI-RADS information, augmented by pixel-based quantitative features, can improve breast cancer classification. Machine learning classifiers, including XGBoost, Random Forest, and Support Vector Machine, were trained with an intelligible multimodal signature composed of the BI-RADS and radiomic features. Exploiting the intrinsic interpretability of our model input, the work aims to obtain an explainable predictive model using *post-hoc* explanation methods. A proprietary dataset composed of 237 B-mode ultrasound scans was acquired, and a total of 103 radiomics features were extracted. Before the training of classifiers, a pipeline for selecting an informative and non-redundant signature was implemented. A 10-fold Cross-Validation repeated 20 times was considered for the training in 80% of the dataset, and the best model in terms of accuracy was selected for testing on the remaining 20%. Our results prove how the medical knowledge, represented by the BI-RADS information, is enhanced with the use of radiomic features. XGBoost was the best model, showing an AUROC of $0.977 \pm 0.029$ and 0.956 in the training and test phases, respectively. In addition, the implemented global explanation using the SHAP method and exploiting the intelligibility of radiomic features, allowed us to confirm some important model findings.

**Keywords:** Explainable model · Interpretable features · Breast cancer · Multimodal signature · Clinical features · Radiomic features


M. Vettoretti et al. (Eds.): CIBB 2023, LNBI 14513, pp. 89–100, 2025.
https://doi.org/10.1007/978-3-031-90714-2_7

## 1 Introduction

Breast cancer remains one of the most prevalent cancers worldwide, with a substantial impact on public health and clinical outcomes [18]. Accurate diagnosis and precise characterization of breast cancer are crucial for effective treatment and improved patient prognosis. The diagnosis of invasive breast cancer typically involves a comprehensive approach that combines clinical evaluation with advanced imaging techniques and biopsy, ensuring that treatment strategies are tailored to the individual needs of patients. Diagnosis is typically made through screening examinations like mammography [13], often supplemented by ultrasound. Once a diagnosis is established, more advanced imaging techniques, such as MRI, provide improved characterization of the lesion and aid in planning treatment interventions. Given the high cost of advanced exams like MRI [7], the ability to predict malignancy using simpler, minimally invasive methods, such as ultrasound, could significantly impact the timeliness and efficiency of subsequent treatment interventions

In the breast cancer diagnostic process, the Breast Imaging-Reporting and Data System (BI-RADS) lexicon plays a pivotal role in standardizing the radiological assessment. The BI-RADS is a score (which can take values from 0 to 6) used to classify a breast lesion/mass, allowing the results of a diagnostic examination (*e.g.*, mammography, ultrasound) to be presented in a standardized and understandable way, helps to reduce dependence on the operator at the reporting stage. The final BI-RADS value takes into account the following features: *i)* breast composition, *ii)* shape margin orientation, *iii)* echo pattern, *iv)* posterior features, *v)* calcifications, *vi)* architectural distortion, *vii)* duct changes, *viii)* skin thickening, *ix)* skin retraction, and *x)* edema. At the end of the evaluation of all these features, the radiologist assigns the BI-RADS score and decides on future interventions. For this reason, the BI-RADS score represents an initial assessment by the physician, which then needs to be possibly deepened through advanced examinations.

In recent years, radiomics has demonstrated the ability to provide valuable quantitative information that complements the qualitative assessments made by physicians. Through radiomic feature extraction, it is possible to quantify the relationships between gray levels and lesion texture, providing pixel-by-pixel details that are impossible for the human eye to see [1,3]. This approach facilitates the correlation of informative radiomic signatures, comprising sets of features, with clinical outcomes [8].

Although radiomic-based models have proliferated and demonstrated remarkable performance, the development of explainable methodologies remains relatively limited [5,11]. In clinical settings, the need for model explainability has become increasingly pronounced. Indeed, explainable models afford clinicians the ability to validate or refute model predictions, fostering trust and facilitating their integration into clinical workflows.

Within this context, deep learning techniques often lack transparency. While saliency map computation attempts to explain the features extracted by deep architectures, their outcomes have been inconsistent and, in some cases, unsatisfactory [15]. Leveraging intelligible radiomic features alongside shallow learning models aligns with the requisites of *i)* explainable modeling, and *ii)* training in small dataset scenario [13].

This retrospective study proposes an interpretable multimodal signature for breast cancer classification. We aim to analyze whether physician assessments, represented by the BI-RADS score, combined with radiomic features can enhance the performance of predictive machine learning models. In addition, through explainable AI methods, the trained model will be analyzed to evaluate the impact of both BI-RADS score and radiomic signature.

## 2 Material and Methods

### 2.1 Dataset Description

A cohort comprising 237 patients diagnosed with breast cancer was collected between September 2021 and December 2021 at the Breast Unit of the Policlinico Universitario "P. Giaccone" in Palermo, Italy. Images concerning Focal Breast Lesions (FBLs) were acquired utilizing the B-mode ultrasound (US) modality. Subsequently, the acquired dataset was divided into two subsets, considering 80% for training and 20% for testing. An expert breast radiologist (over 30 years of expertise in breast imaging) conducted the high-resolution breast US acquisitions. The RS85 ultrasound unit, manufactured by Samsung-Medison, Co. Ltd., equipped with a 3–12 MHz linear transducer, facilitated the acquisition process. Each FBL image was meticulously classified in accordance with BI-RADS criteria [14]. Table 1 presents a summary of the main dataset characteristics, along with those about the training and testing subsets. Figure 1 shows some examples of malignant (first two images and left) and benign (last two images on the right) breast lesions belonging to the dataset used.

**Table 1.** Overall characteristics of the FBLs dataset, and details concerning its subdivision into the training and test subsets.

| Characteristic | Whole Dataset | Training | Test |
|---|---|---|---|
| number of lesions | 237 | 189 | 48 |
| malignant | 100 | 77 | 23 |
| benign | 137 | 112 | 25 |
| range size [$mm$] | 4–90 | 4–90 | 5–32 |
| mean size [$mm$] | 15.12 ± 9.44 | 15.19 ± 9.99 | 14.85 ± 6.90 |
| age range [$years$] | 17–88 | 19–88 | 17–85 |
| mean age [$years$] | 53.15 ± 15.00 | 51.95 ± 14.63 | 57.95 ± 15.68 |

### 2.2 Analysis Workflow

The procedures for defining the signature and training the predictive model are covered in the following subsection.

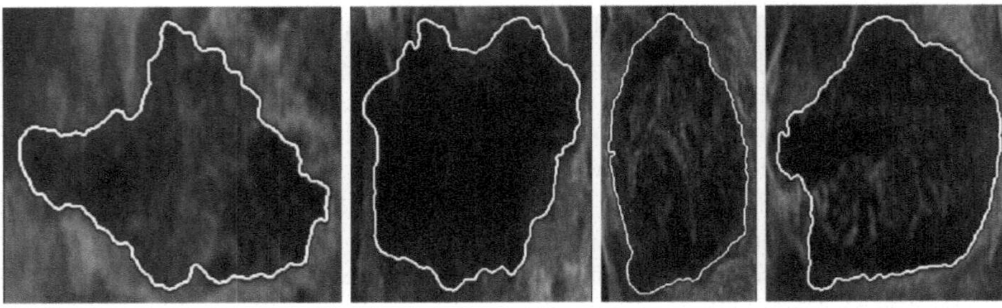

**Fig. 1.** Examples of malignant (first two images on the left) and benign (last two images on the right) breast lesions belonging to the dataset used.

*ROI Segmentation:* ROIs were segmented and validated by an expert radiologist using the S-DetectTM software, commercially available and integrated for clinical use. The physician uses the software on the RS85 ultrasound machine, to extract the ROI from each B-mode image of the breast lesion. If needed, the segmentations obtained by S-DetectTM were manually verified and any artifacts were removed.

*Feature Extraction:* Pyradiomics [16] was used for feature extraction, following the Image Biomarkers Standardization Initiative (IBSI) [17,19]. In particular, 103 radiomic features were extracted, including first order histogram statistics and texture features computed from the following matrices: Gray Level Run Length Matrix (GLRLM); Gray Level Co-occurrence Matrix features (GLCM); Neighboring Gray Tone Difference Matrix (NGTDM) categories; Gray Level Size Zone Matrix (GLSZM); Gray Level Dependence Matrix (GLDM). Also, 2D shape features were considered.

*Features Preprocessing and Selection:* The dataset was divided into the training and test sets, using an 80-20 ratio, respectively. Only the training set was used for feature selection and hyperparameter tuning. Regarding feature selection, three model-independent methods were employed to achieve a balanced ratio between samples and model inputs [8]. In particular:

- *near-zero variance*: features with low information content were discarded considering a variance of 0.005;
- *correlation analysis*; correlated features were removed using the Spearman correlation coefficient (considering 0.9 as a threshold);
- *statistical test*: the Mann-Whitney U-test was used to compare the 'Benign' and 'Malignant' distributions. This is the only supervised step, and no test subset was involved.

At the end of the feature selection process, the obtained radiomic signature was concatenated with the BI-RADS information, representing the knowledge-based feature of the physician fed into the models. For this reason, we call our signature multimodal considering the quantitative radiomic information, and the physician-based feature.

*Predictive Model Setup:* Considering their ability to manage small datasets [9], XGBoost (XGB), Random Forest (RF), and Support Vector Machine (SVM) were implemented as classifiers. Models were After selecting features, the training subset was used for tuning hyperparameters. A grid search and cross-validation were employed for each of the three ML methods. A search for the decision tree-based models was performed to optimize the number of estimators and the importance function, as well as the learning rate for XGB and the kernel for SVM. The dataset was divided into the training and test sets, using an 80-20 ratio, respectively. A stratified 10-fold cross-validation (repeated 20 times to reduce selection bias) was implemented in the training set, and the best model in terms of accuracy was selected for the test. For this reason, training performance is presented considering the mean and standard deviation computed across each fold.

*Model Explanation:* In the context of machine learning models, interpretability is crucial for understanding which features drive model predictions. It was shown that a model explanation should support several stakeholders involved in the clinical process, including engineers, physicians, and patients [11]. To provide explainability, we computed SHAP (SHapley Additive exPlanations) values, which provide a unified measure of feature importance based on cooperative game theory, evaluating the contribution of each feature for model predictions [4]. In our study, the SHAP method was particularly valuable for interpreting the contributions of the different components of the multimodal signature. Values are computed on the test set considering the best-trained model.

## 3 Experimental Results

RF and XGB were trained using 100 estimators. In addition, a learning rate of 0.3 and gain as the importance type were employed for XGB training, and Gini as the criterion for RF. The RBF was used as a kernel for SVM training, and feature standardization was performed before training. The metrics employed for evaluation included Balanced Accuracy, Accuracy, Specificity, Sensitivity, Positive Predictive Value (PPV), Negative Predictive Value (NPV), and Area Under the Receiver Operating Characteristic curve (AUROC).

### 3.1 Features Preprocessing

Table 2 shows the number of features selected after each preprocessing/selection step. The multimodal signature - comprising radiological and clinical data - was composed of the features represented in Table 3 and by the *BI-RADS* information.

**Table 2.** Results of feature preprocessing and selection steps. (*) The initial feature set consists of radiomic features (103) and clinical features (BI-RADS).

| Preprocessing Step | Remaining Features | Discarded Features |
|---|---|---|
| initial features | 104 (*) | N.A. |
| near-zero variance ($\sigma^2 \geq 0.005$) | 80 | 24 |
| correlation analysis (correlation $\geq 0.90$) | 20 | 60 |
| statistical analysis (p-value $\leq 0.05$) | 9 | 11 |

**Table 3.** Features belong to the radiomic signature and corresponding categories

| Feature Name | Feature Category |
|---|---|
| *Elongation* | Shape 2D |
| *Entropy* | First Order |
| *InterquartileRange* | First Order |
| *Kurtosis* | First Order |
| *Skewness* | First Order |
| *MaximumProbability* | GLCM |
| *Busyness* | NGTDM |
| *DependenceEntropy* | GLDM |

### 3.2 Predictive Model Performance

The work focuses on assessing the difference in model performance when the BI-RADS score, which represents the knowledge-based feature of the physician in the model, is enhanced by radiomic features. To this aim, RF, XGB, and SVM classifiers were trained *i)* using only the BI-RADS features, and *ii)* considering the addition of the radiomic features. The former configuration can be approximated to the radiologist's performance evaluation in classification. The latter configuration, on the other hand, assesses the radiologist's performance enhanced with the quantitative radiomic information. Table 4 shows the performances calculated within the cross-validation procedure using only BI-RADS and the multimodal signature, respectively, considering the three implemented classifiers. The test metrics were shown in Table 5 and the related ROC curve and confusion matrix in Figs. 2 and 3 for the unimodal and multimodal signature, respectively.

Comparing the results between the two signatures in Table 5, which represent predictive models using different sets of features (BI-RADS alone *versus* BI-RADS combined with radiomic features), several interesting observations can be made. The models trained with BI-RADS alone achieved balanced accuracies of 0.893 across all algorithms, whereas when radiomic features were included, the balanced accuracy improved significantly, reaching 0.916 for XGB and 0.896 for RF. This indicates that incorporating radiomic features enhances the model's ability to correctly classify both tumor classes, leading to better overall performance. The sensitivity improved notably when radiomic features were added, especially for the XGB and RF models, where it increased from 0.826 to 0.913. This signifies an enhanced ability to correctly identify true positives, which is crucial in medical diagnosis tasks. In addition, radiomic significantly enhances the ratio of specificity to sensitivity, and hence, PPV to NPV.

**Table 4.** Training results computed using the cross-validation procedure. B-Acc: balanced accuracy; Acc: accuracy; Spec: specificity; Sens: sensitivity;

| Evaluation Metric | BI-RADS | | | BI-RADS + radiomic features | | |
|---|---|---|---|---|---|---|
| | XGB | RF | SVM | XGB | RF | SVM |
| B-Acc | 0.855 ± 0.066 | 0.857 ± 0.067 | 0.855 ± 0.067 | 0.896 ± 0.073 | 0.893 ± 0.078 | 0.865 ± 0.080 |
| Acc | 0.862 ± 0.062 | 0.866 ± 0.064 | 0.863 ± 0.062 | 0.901 ± 0.069 | 0.901 ± 0.071 | 0.881 ± 0.071 |
| Spec | 0.894 ± 0.142 | 0.911 ± 0.081 | 0.899 ± 0.141 | 0.924 ± 0.080 | 0.935 ± 0.081 | 0.956 ± 0.067 |
| Sens | 0.817 ± 0.189 | 0.802 ± 0.184 | 0.812 ± 0.189 | 0.867 ± 0.128 | 0.851 ± 0.144 | 0.774 ± 0.157 |
| PPV | 0.890 ± 0.141 | 0.908 ± 0.137 | 0.895 ± 0.140 | 0.897 ± 0.104 | 0.912 ± 0.103 | 0.935 ± 0.096 |
| NPV | 0.898 ± 0.099 | 0.890 ± 0.096 | 0.896 ± 0.099 | 0.916 ± 0.075 | 0.909 ± 0.081 | 0.867 ± 0.081 |
| AUROC | 0.978 ± 0.021 | 0.977 ± 0.021 | 0.896 ± 0.140 | 0.977 ± 0.029 | 0.975 ± 0.031 | 0.963 ± 0.036 |

**Table 5.** Test results. B-Acc: balanced accuracy; Acc: accuracy; Spec: specificity; Sens: sensitivity;

| Evaluation Metric | BI-RADS | | | BI-RADS + radiomic features | | |
|---|---|---|---|---|---|---|
| | XGB | RF | SVM | XGB | RF | SVM |
| B-Acc | 0.893 | 0.893 | 0.893 | 0.916 | 0.896 | 0.853 |
| Acc | 0.896 | 0.896 | 0.896 | 0.917 | 0.895 | 0.854 |
| Spec | 0.960 | 0.960 | 0.960 | 0.920 | 0.880 | 0.880 |
| Sens | 0.826 | 0.826 | 0.826 | 0.913 | 0.913 | 0.826 |
| PPV | 0.950 | 0.950 | 0.950 | 0.913 | 0.875 | 0.863 |
| NPV | 0.857 | 0.857 | 0.857 | 0.920 | 0.917 | 0.846 |
| AUROC | 0.966 | 0.966 | 0.966 | 0.956 | 0.956 | 0.948 |

To statistically prove the improvement obtained employing radiomic features, McNemar's test was applied between the two feature configurations (BI-RADS *vs.* BI-RADS + radiomics) for each of the 3 models (*i.e.* RF, SVM, XGB) and no statistically significant differences ($p < 0.05$) were shown. This result is in accordance with the distribution of Shapley values shown in Fig. 4, suggesting that the BI-RADS score is predominant for prediction. However, the results

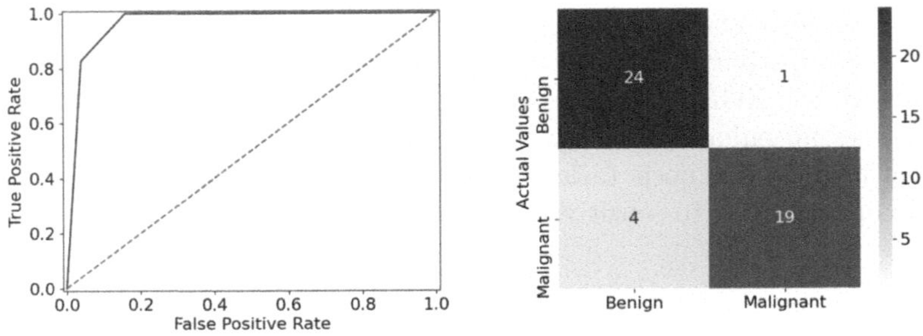


**Fig. 2.** (on the left) ROC curve obtained in the testing phase by the model based on XGB considering only the BI-RADS (0.966); (on the right) the corresponding confusion matrix.

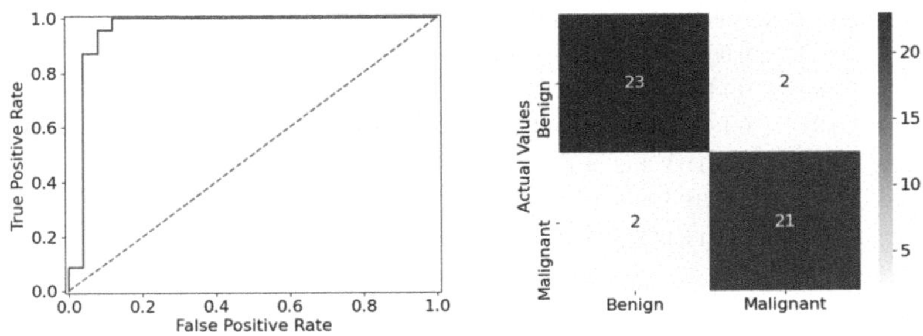


**Fig. 3.** (on the left) ROC curve obtained in the training phase by the model based on XGB considering the radiomic features and the BI-RADS ($0.977 \pm 0.029$); (on the right) the corresponding confusion matrix.

calculated in cross-validation (Table 4) highlight improved accuracy when the BI-RADS score is augmented with radiomic features. In fact, applying, the Wilcoxon test to compare the distribution of accuracy values calculated in cross-validation, a statistical difference was observed between the two feature configurations (BI-RADS *vs.* BI-RADS + radiomics) considering RF, SVM, and XGB classifiers.

### 3.3 Features Interpretability and Failure Analysis

The implemented radiomic analysis enabled the identification of an interpretable signature [13] that combined with shallow learning algorithms increased the model's explainability [12]. In particular, a global explanation allows us to compare the model findings with the clinical literature. The SHAP analysis [4] was used to interpret our XGB model, considered the most accurate compared to the other classifier. The impact of each feature within the multimodal signature to train the XGB model is shown in Fig. 4. Features are sorted in order of importance from top to bottom. The obtained plot shows that several crucial aspects comply with some radiologic aspects used to discriminate between malignant and benign FBLs.

- *busyness*: represents the difference between a pixel and its neighbor. Malignant FBLs are linked to lower values than benign ones, according to an analysis of the class distributions. This is related to the fact that malignant lesions typically show a heterogeneous pattern, meaning that the lesion has areas with varying densities (such as habitats, physiologically distinct regions within and composing the tumor [2]);
- *elongation*: shows the connection between the tumor's two biggest main axes. It is feasible to observe that, despite the lack of a distinct dichotomy, high values are primarily linked to malignant FBLs, while low values are indicative of benign ones. This is consistent with clinical observations that benign tumors typically have a less elongated and more oval shape;
- *maximum probability*: determines the frequency of the most common pairs of adjacent intensity values, which reveals a consistent pattern inside the lesion. Higher levels are primarily associated with FLBs that are malignant;
- *skewness*: evaluates the asymmetry of the pixel value distribution *versus* the image's mean value; skewness can be either positive or negative, depending on the location of the distribution's mass concentration and prolonged tail. Elevated values correspond to hyperintense FBLs, which are more compact; this is a condition commonly observed in malignant tumors;
- *entropy*: describes the pixel values' uncertainty or randomness. It is a metric for the heterogeneity of tumors. Greater heterogeneity compared to benign FBLs is linked to higher values in malignant FBLs;

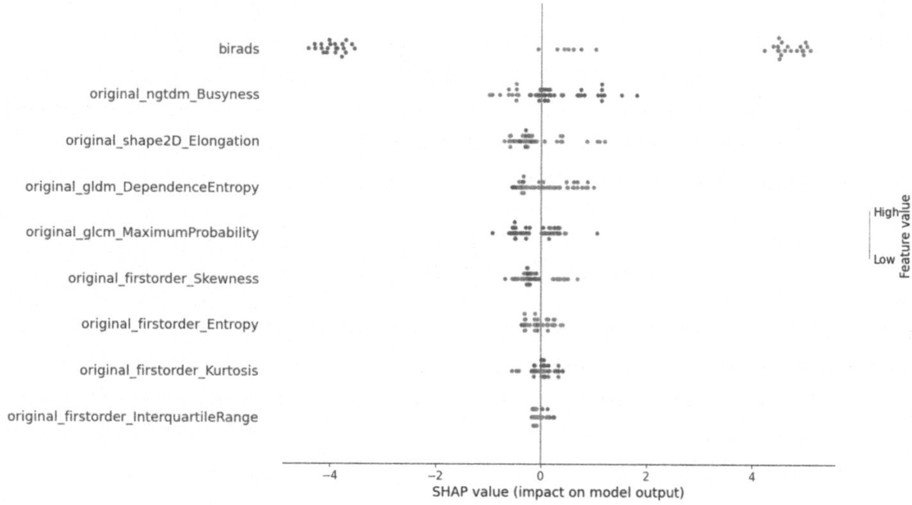


**Fig. 4.** SHapley analysis providing the global explanation of the predictive model exploiting the XGB classifier and trained with the multimodal signature.

To assess false positives (FP) and false negatives (FN), we performed a failure analysis considering the best-trained model (XGB). It was shown that the errors of the combined model (exploiting BI-RADS + radiomics) are 4 (FP = 2; FN = 2, three with BI-RADS = 4 and one with BI-RADS = 5), while those of the model based only on BI-RADS score are 5 (FP = 1; FN = 4, four with BI-RADS = 4 and one with BI-RADS = 5). From a clinical perspective, it is preferable to employ a conservative system that minimizes FNs (*i.e.*, patients with malignant lesions incorrectly classified as benign). Therefore, the enhanced model incorporating radiomic features aligns more closely with clinical practice. Furthermore, consistent with the meaning of the BI-RADS lexicon, model errors are concentrated on lesions identified by the clinician as BI-RADS 4, which by definition represent 'suspicious' cases ($2\% < likelihood\,of\,cancer < 95\%$). From this perspective, the contribution of radiomic features in correcting the prediction is evident.

## 4 Conclusion

This retrospective investigation aimed to define a predictive model utilizing a multimodal signature - comprising radiological and clinical data - to discern the malignancy or benignancy of FBLs, solid tumors focused in one area of the breast. By incorporating both BI-RADS and radiomic features, we enhanced predictive accuracy compared to models relying solely on BI-RADS information. A comparative analysis shown in Tables 4 and 5 for the training and testing phases respectively, reveals a more pronounced enhancement in predictive performance for models utilizing XGB and RF, with a less pronounced improvement observed for the SVM-based model.

Following literature guidelines [8], we implemented a meticulously structured processing pipeline to extract and select resilient and interpretable signatures for the training of predictive models. Leveraging shallow learning methodologies and interpretable radiomic features, a superior model introspection was established, thereby enhancing the explainability of the predictive model, a crucial prerequisite for its integration into clinical practice.

As prospective avenues for further exploration within this research domain, it would be pertinent *i)* to analyze the stability and robustness of features through the assessment of various segmentation methods [6] and pre-filtering strategies [10] applied to ultrasound images; *ii)* to explore different feature selection techniques and machine learning classifiers to enhance model efficacy. Furthermore, to evaluate the model's generalization capabilities, it is imperative to validate the trained model on external datasets, thereby assessing its performance across different data distributions.

**Funding Information.** This research has been supported by Piano Nazionale per gli investimenti Complementari al PNRR, project INNOVA "Italian network of excellence for advanced diagnosis", CUP B73C22001770006; and by European Union - Next Generation EU - MUR D.M. funds 737/2021, research project"A Trustworthy Clinical Decision Support System for non-communicable diseases detection and prediction".

## References

1. Gillies, R.J., Kinahan, P.E., Hricak, H.: Radiomics: images are more than pictures, they are data. Radiology **278**(2), 563–577 (2016). https://doi.org/10.1148/radiol.2015151169
2. Jardim-Perassi, B.V., et al.: Multiparametric MRI and coregistered histology identify tumor habitats in breast cancer mouse models. Can. Res. **79**(15), 3952–3964 (2019). https://doi.org/10.1158/0008-5472.CAN-19-0213
3. Lambin, P., et al.: Radiomics: the bridge between medical imaging and personalized medicine. Nat. Rev. Clin. Oncol. **14**(12), 749–762 (2017). https://doi.org/10.1038/nrclinonc.2017.141
4. Lundberg, S.M., Lee, S.I.: A unified approach to interpreting model predictions. In: Guyon, I., et al. (eds.) Advances in Neural Information Processing Systems, vol. 30, pp. 4765–4774. Curran Associates, Inc. (2017)
5. Militello, C., Prinzi, F., Sollami, G., Rundo, L., La Grutta, L., Vitabile, S.: CT radiomic features and clinical biomarkers for predicting coronary artery disease. Cogn. Comput. 1–16 (2023). https://doi.org/10.1007/s12559-023-10118-7
6. Militello, C., et al.: Robustness analysis of DCE-MRI-derived radiomic features in breast masses: assessing quantization levels and segmentation agreement. Appl. Sci. **12**(11), 5512 (2022). https://doi.org/10.3390/app12115512
7. Militello, C., et al.: 3D DCE-MRI radiomic analysis for malignant lesion prediction in breast cancer patients. Acad. Radiol. **29**(6), 830–840 (2022). https://doi.org/10.1016/j.acra.2021.08.024
8. Papanikolaou, N., Matos, C., Koh, D.M.: How to develop a meaningful radiomic signature for clinical use in oncologic patients. Cancer Imaging **20**, 1–10 (2020). https://doi.org/10.1186/s40644-020-00311-4
9. Prinzi, F., Currieri, T., Gaglio, S., Vitabile, S.: Shallow and deep learning classifiers in medical image analysis. Eur. Radiol. Exp. **8**(1), 26 (2024). https://doi.org/10.1186/s41747-024-00428-2
10. Prinzi, F., Militello, C., Conti, V., Vitabile, S.: Impact of wavelet kernels on predictive capability of radiomic features: a case study on COVID-19 chest X-ray images. J. Imaging **9**(2), 32 (2023). https://doi.org/10.3390/jimaging9020032
11. Prinzi, F., Militello, C., Scichilone, N., Gaglio, S., Vitabile, S.: Explainable machine-learning models for COVID-19 prognosis prediction using clinical, laboratory and radiomic features. IEEE Access **11**, 121492–121510 (2023). https://doi.org/10.1109/ACCESS.2023.3327808
12. Prinzi, F., Orlando, A., Gaglio, S., Vitabile, S.: Breast cancer classification through multivariate radiomic time series analysis in DCE-MRI sequences. Expert Syst. Appl. **249**, 123557 (2024). https://doi.org/10.1016/j.eswa.2024.123557
13. Prinzi, F., Orlando, A., Gaglio, S., Vitabile, S.: Interpretable radiomic signature for breast microcalcification detection and classification. J Imaging Inf. Med. (2024). https://doi.org/10.1007/s10278-024-01012-1
14. Sickles, E.A., D'Orsi, C.J.: How should screening breast US be audited? The BI-RADS perspective. Radiology **272**(2), 316–320 (2014). https://doi.org/10.1148/radiol.14140634
15. Signoroni, A., et al.: BS-Net: learning COVID-19 pneumonia severity on a large chest X-ray dataset. Med. Image Anal. **71**, 102046 (2021). https://doi.org/10.1016/j.media.2021.102046
16. Van Griethuysen, J.J., et al.: Computational radiomics system to decode the radiographic phenotype. Can. Res. **77**(21), e104–e107 (2017). https://doi.org/10.1158/0008-5472.CAN-17-0339

17. Whybra, P., et al.: The image biomarker standardization initiative: standardized convolutional filters for reproducible radiomics and enhanced clinical insights. Radiology **310**(2), e231319 (2024). https://doi.org/10.1148/radiol.231319
18. World Health Organization: International Agency for Research on Cancer. The Global Cancer Observatory (2020). https://gco.iarc.fr/. Accessed 15 Mar 2024
19. Zwanenburg, A., et al.: The image biomarker standardization initiative: standardized quantitative radiomics for high-throughput image-based phenotyping. Radiology **295**(2), 328–338 (2020). https://doi.org/10.1148/radiol.2020191145

# Exploring the Conformational Odorant Space in the Olfactory Receptor Binding Region

Anuththara Lokubandara[1,2], Vasanth Mandla[1,3], and Chiquito Crasto[1](✉)

[1] Center for Biotechnology and Genomics, Texas Tech University, #116, Experimental Sciences Building I, Lubbock, TX 79409, USA
chiquito.crasto@ttu.edu

[2] Department of Psychology, Texas Tech Health Science Center, Lubbock, TX, USA

[3] Center for DNA Fingerprinting and Diagnostics (CDFD), Hyderabad, Telangana, India

**Abstract.** We have ideated a novel mechanism of how odorants with differing and electronic-structural features can excite the same olfactory receptor. We explored the conformational space of an odorant and identified its likely "activating" or "inhibiting" conformers. The eventual challenge is to provide a molecular glimpse into how a few hundred olfactory receptors can discern several thousand odorants and odors—of various structures and functional groups. For a functionally characterized mouse olfactory receptor, mOR-EG, we tested odorant molecules with experimentally determined activating strengths: strong and weak. Vanillin—an odorant that strongly activates mOR-EG was used as a reference. The conformational space of each test odorant was explored and superimposed over the structure of the reference odorant. By narrowing our studies to mOR-EG's strong activator, eugenol and weak activator, allyl benzene, we identified specific molecular features among conformations that were reproduced even for disparate-looking odorants. It is likely then that a specific conformation catalyzes OR activation leading to the perception of an odor. We explore activating conformers as opposed to an odorant's gross chemical features such as aromatic rings, aliphatic chains, degree of bond-saturation, etc. Our studies can be extended to drug-design, primarily because olfactory receptors have a putative structure that is typical of type-A G- Protein Coupled Receptor. To accomplish the tasks represented in this work the BioSolveIT suite of software was used.

**Keywords:** Olfactory Receptors · Odorants · Deorphanization · Conformational analysis · odorant docking · preferred binding

## 1 Introduction and Background

Olfactory receptors (OR) have come into prominence relatively recently. Their discovery was announced in 1991 [1] and the discoverers were awarded the Nobel Prize in Physiology and Medicine in 2004 [2]. In mammals, ORs constitute superfamilies of genes numbering a few hundred. In humans, approximately 400 functional genes have been

A. Lokubandara and V. Mandla—Both authors contributed equally to this paper.


M. Vettoretti et al. (Eds.): CIBB 2023, LNBI 14513, pp. 101–116, 2025.
https://doi.org/10.1007/978-3-031-90714-2_8

identified out of approximately 1000 [3–6]–the remaining being pseudogenic. This large number of pseudogenes could be because the olfactory senses diminished as other senses evolved [7]. Alternatively, one might also speculate that olfactory repertoires and olfactory perceptions vary across different populations due to evolutionary and epigenetic factors, and the current perspectives drawn from the human genome do not sufficiently and comprehensively capture the olfactory variability in in human. Projects such as the 1000 Genome Project [8] have endeavored to address this.

The variability within the olfactory receptor gene family is necessitated because a few hundred receptors must bind (and catalyze olfactory perception) several thousand odorants or combinations of odorants that result in specific odors. Elucidating the mechanism of this variability at a molecular level is fraught with challenges. Olfactory receptors are membrane bound proteins [9]. They are poorly behaved in terms of ease of expression and the levels of the purity required to attempt to crystallize these proteins for structure determination [10, 11]. Also, each OR might be activated (or inhibited) by several different odorants bearing disparate electronic and structural features. Conversely, an odorant will bind different ORs with varying levels of activation strengths—promiscuous interactive behavior [12, 13]. A mechanism for these interactions is not understood; thus, identifying a rational panel of odorants that are likely to bind and activate an OR is cost prohibitive. This hinders functional characterization (or deorphanization) of ORs [14, 15].

A 2023 publication has published the results of structural studies of a human olfactory receptor, OR51E2 and its binding interactions with propionate, using cryo-electron micrography, a methodology that is being increasingly used for complex or poorly behaved proteins (from a structural studies standpoint) [16]. The olfactory receptor protein was fused with miniGs399, a protein that has been used often in reducing solubility issues associated with the protein being studied [17].

In seeking to understand the mechanism of olfactory receptor-odorant receptors at the molecular level, several computationally driven studies have been published [18–21]. Each of these studies began by developing a computational model of the OR protein. ORs are Class A GPCRs (G[TP-binding] Protein Coupled Receptors), whose structures are presumptively composed of seven helical regions (some anomalies exist) that traduce the plasma membrane, an extracellular N-terminus, a cytoplasmic C-terminus, and three extracellular and intracellular loops each.

The crystal structure of rhodopsin [22] and β-adrenergic receptor [23] have often been used as templates during the model building process. While the model building protocols in references 18–21 vary, the foundational idea remains the same: build a model of the OR protein and computationally dock an odorant into one of two binding pockets of the olfactory receptor.

Some of the challenges in developing a computational structure of the OR protein involved poor sequence homologies with the template proteins. We and others used a threading protocol that relied only on the template protein superstructure to construct the OR scaffold, while fine-tuning each of the helical regions [18, 19].

Computational OR-odorant interaction studies originally drew inferences from static docking of odorants in a pre-determined binding region to identify receptor amino acids that are responsible for odorant binding. These studies evolved to include all-atom, unrestricted molecular dynamics simulations of these interactions over nano-second timescales [19, 24]. The aim was to provide a molecular basis for olfactory-odorant interactions—a possible structural and functional excitation of the single olfactory sensory neuron that expressed that olfactory receptor. This would catalyze a process that would result in an olfactory response. The aim was to identify a rationale by which odorants that were likely to bind an olfactory receptor could be identified.

Our previous computational studies stemmed from selecting an olfactory receptor that was experimentally deorphanized, where the activating, non-activating, and inhibiting odorants were identified. The earliest such results were provided by Malnic et al., [26] who studied mouse olfactory receptor neuronal excitations with odorants at various concentrations. This paper reasoned that the promiscuous interactions between one olfactory receptor and several odorants were because olfactory receptors had binding-region amino-acid profiles that were complementary to specific structural and electronic features of the odorant. This notion was formalized as the odotope theory [27, 28]. A more comprehensive combinatorial study was performed by Saito et al. [14] who studied the excitations of 63 mouse and human ORs with 62 odorants at concentrations ranging from 100 nM to 1 mM.

In this study, we approach a molecule-level assessment of olfactory receptor-odorant interactions from the perspective of the odorant, irrespective of the olfactory receptor with which it interacts. To accomplish this, we leveraged the notions involved in pharmacophore discovery in the drug-design process. We used the BioSolveIT software [29] made available through success at an international research competition in which the authors competed.

The primary question that we sought to answer was, why do seemingly structurally (and electronically) disparate odorants bind and putatively activate the same olfactory receptor. Our hypothesis here is that while dynamically interacting within the binding region of an olfactory receptor, an odorant adopts an "activating" conformation that is likely different from its native stable configuration.

Our novel methodology used the FlexS software of the BioSolveIT suite to perform a comprehensive exploration of the conformational space of a panel of odorants that are known to activate or inhibit mOR-EG. We then performed a superimposition of all odorant conformers to identify likely activating conformers.

How does one address the seemingly promiscuous interactions between ORs and odorants? A structural change occurs in ORs following binding–a prelude to activation and odorant behavior in the binding region [30]. A computational model of the OR protein is first developed. An odorant ligand is docked in one or more putative binding regions of the OR. Simulation studies illustrate the dynamic behavior of the OR-odorant complex during these interactions [18, 19, 24]. We approach the OR-odorant complex formation starting from the perspective of the odorant shown to experimentally activate the OR. If our computational results can provide a mechanistic glimpse into experimental results, they can serve as models to hypothesize the nature of interactions for other OR proteins.

Our work here attempts to address earlier mentioned challenges by adopting a novel, computationally driven perspective. By assessing odorant conformations, we seek to identify specific structures adopted by an odorant that facilitates OR activation, especially within an OR binding pocket. We present an analysis and results of mOR-EG (ORL828 [ORDB] (https://ordb.biotech.ttu.edu/ORDB/Data/2775) [15]. This is a well characterized OR whose activation has been experimentally studied with the odorants, six of which are listed in Table 1.

We performed an extensive conformational scan (FlexS) for the odorants listed in Table 1. These conformations were then superimposed against vanillin (in red in Table 1) to assess the exact conformational similarities that resulted in OR activation. To explore the conformational space for odorants within the putative OR binding pockets, we also built a computational model for mOR-EG using an established protocol [19, 24]. The LeadIT software was then used to rapidly dock the odorant molecules into the putative binding pockets of the receptor protein models using two strategies: unrestricted docking, and targeted docking into previously known binding regions. The energy of binding and conformational freedom within the binding regions was used as the determinant for binding strengths.

In a typical computational pharmacology experiment, the conformers of putative drugs are tested by superimposing on the structure a central drug molecule. This "central" molecule is selected because of proven efficacy (having a positive effect on a condition or disease). Here, we tested six molecules with activations levels identified as strong and weak. Within each group, there is no evidence of relative strength and relative weakness. To select a central molecule, we selected a molecule that had the most variability in terms of structure and number of side chains. This would ensure that the maximum possible conformational superimpositions would be possible.

**Table 1.** The Table shows the odorants used in this analysis. Vanillin (red) is used as the reference on which the other odorant conformation structures are superimposed.

| **mOR-EG activating odorants** | |
|---|---|
| Strong | Weak |
| Vanillin<br>Eugenol | 2-methoxy-4-ethyl phenol,<br>2-methoxy-4-methyl phenol<br>Allyl benzene<br>Eugenol ethyl ether |

## 2 Methodology

The protocol used in this work is illustrated in Fig. 1.

## 2.1 Determining the Reference Odorant Ligand

Two criteria for determining the reference odorant for mOR-EG were used. 1) A strong activator; and 2) the largest number of functional groups and variations in structural features. As mentioned in the previous section, in drug design experiments, reference ligand is one that has been known to be efficacious as a pharmaceutical product. For mOR-EG, vanillin was used as a reference (highlighted red in Table 1). It was experimentally shown to be a strong activator while possessing in addition to a benzene ring, methoxy, hydroxy and aldehyde substituents.

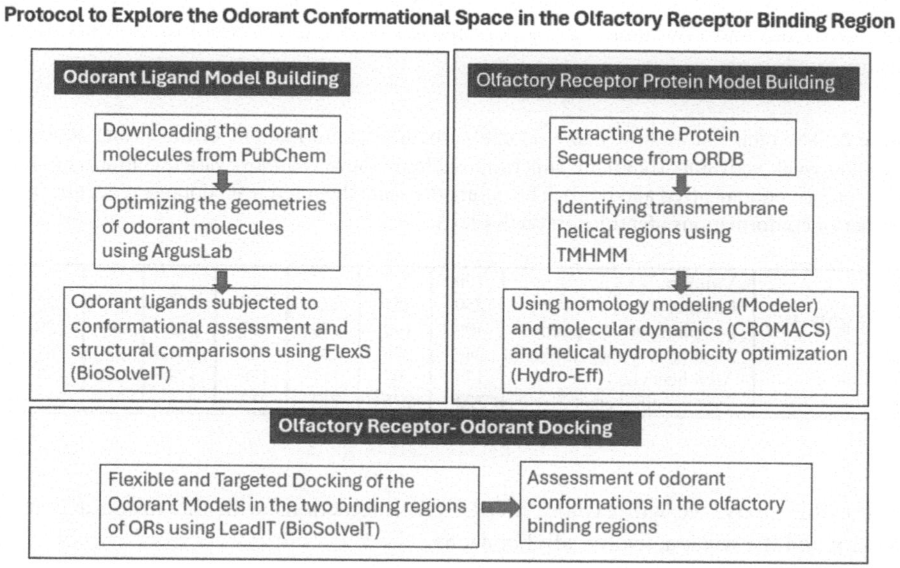


**Fig. 1.** The figure represents the protocol used that culminates in the assessment of conformational freedom of an odorant in the olfactory receptor binding pocket. The peach-colored box represents the procedure to explore the conformational space of the odorant (FlexS). The light green box represents the creation of the olfactory receptor protein model. The light blue box represents odorant-ligand docking in the binding regions of the olfactory receptor (LeadIT).

## 2.2 Determining the Reference Odorant Ligand

Two criteria for determining the reference odorant for mOR-EG were used. 1) A strong activator; and 2) the largest number of functional groups and structural features. As mentioned in the previous section, in drug design experiments, reference ligand is one that has been known to be efficacious as a pharmaceutical product. For mOR-EG, vanillin was used as a reference (highlighted red in Table 1). It was experimentally shown to be a strong activator while possessing in addition to a benzene ring, methoxy, hydroxy and aldehyde substituents.

### 2.3 Odorant Conformations Scan and Molecular Superposition with Reference Odorant

The FlexS software of BioSolveIT [30] (https://www.biosolveit.de/) was used to comprehensively explore the conformational space for each molecule in Table 1. 1000 conformations were identified for vanillin, which following manual inspection were divided into seven groups. A representative conformation was identified for each group. The rationale for this was to ensure that superimpositions of the conformations of other molecules would comprehensively explore the conformational space of the other odorant molecules. FlexS was also used to superimpose all the other odorants on the seven conformations of the references. Table 2 shows the total number of conformations for each molecule and how many are superimposed with the seven conformations of the reference odorants.

**Table 2.** The table shows how many odorant-structural conformations of the strong activators (red) and weak activators (blue) are superimposable on the seven most varied conformations of the reference odorants (red and bold). The numbers beside the reference odorants reflect the total number of conformations identified through FlexS.

| | | | | | | | | |
|---|---|---|---|---|---|---|---|---|
| mOR-EG | **Vanillin** | **1000** | | | | | | |
| | Eugenol | 250 | 252 | 226 | 237 | 254 | 225 | 232 |
| | 2-methoxy-4-ethylphenol | 199 | 190 | 190 | 191 | 213 | 209 | 229 |
| | 2-methoxy-4-methyl phenol | 246 | 276 | 238 | 238 | 261 | 277 | 267 |
| | Allyl benzene | 132 | 202 | 117 | 178 | 173 | 202 | 227 |
| | Eugenol ethyl ether | 222 | 225 | 213 | 213 | 215 | 215 | 224 |

For this study, we will explore mOR-EG's interactions with the strong activator, eugenol and the weak activator, allyl benzene.

### 2.4 Protein Model Building

Computational model building uses a template of the high-resolution, x-ray crystal structure of bovine rhodopsin (PDB: 1U19) [22] and molecular dynamics to remove rhodopsin-specific structural artifacts that are not likely to be found in olfactory receptors. We used it as a template to create the computational model of mOR-EG as a prelude to our docking studies. The template was created by employing a previously published rigorous protocol.

- We used a Hidden Markov Model transmembrane topology web-based software TMHMM [31] (highly rated in independent tests of transmembrane predicting tools [32]) to identify the helical regions, the N- and C-termini, the extracellular loop regions, and the cytoplasmic loop regions of mOR-EG. The results of TMHMM were validated through another membrane topology web-based software HMMTOP [33].
- We used the high-resolution structure of rhodopsin (PDB ID = 1U19) [22] to align the helical regions of OR2W1 sequentially and structurally with those from the crystal structure of rhodopsin.

- We performed homology modeling using the software Modeler [34] to create a preliminary model, primarily to correctly position the helical regions of the ORs.
- We abstracted out the loop regions from this preliminary model. Rhodopsin loops tend to be shorter than OR loops (based on the topology prediction when compared to rhodopsin's structure), while OR helices are shorter than rhodopsin helices.
- We computationally refined each helical region independently to ensure that rhodopsin- specific structural features do not influence the TM regions of the OR.
- We manually rotated each helix such that the effective hydrophilic side of the helix is pointed in binding region of the protein, while the effective hydrophobic side is pointed into the plasma membrane's lipid bilayer [35].
- We reintroduced the loops back into protein model.
- We optimized the geometry of the entire model.

## 2.5 Ligand Model Building

- Each molecule that would serve as the docking ligands (Table 1) were downloaded from the PubChem resource (https://pubchem.ncbi.nlm.nih.gov/) in the SDF (Structure Data Format).
- The SDF file was converted to the MOL format using the OpenBabel resource [36].
- The geometries of the ligand molecules were optimized using the Universal Force Field methodology using the ArgusLab software [37].
- These ligands' structure files were then used for further docking analysis with the computational protein model of mOR-EG using LeadIT.

## 2.6 Protein-Ligand Docking

To determine the conformational space within a binding region, two docking strategies were adopted. The LeadIT software in BioSolveIT provides rapid docking results. Another advantage of using LeadIT is that it allows unrestricted docking, where the software tracks the accessible regions of the protein and docks the ligand in that region. LeadIT also allows the users to specify putative binding regions.

In unrestricted docking, the results of the docking are filtered to remove odorants docked in the regions that would be biochemically non-feasible. These include the sides of the protein where it would be embedded in the lipid bilayer of the plasma membrane, or the cytoplasmic side of the protein to which the odorant ligand would have no access. Unrestricted docking has the advantage of exploring regions previously not considered in other computational docking experiments. These regions could be at the boundaries of the helical TM domains and the extracellular inter-helical loops.

Through targeted docking, depending on the nature of binding region in terms of its amino acid profiles as well as conformational freedom, a preferential binding site can be identified. Previous studies have shown that odorant binding in olfactory receptors typically occurs in two specific regions: between transmembrane domains 3, 4, 5, and 6, and between domains 1, 2, 3, and 7, with the first being preferred binding pocket [19] Once docking was complete, the conformational space and the energetics for an odorant in the docked position were explored within the binding region.

## 3 Results

A manual inspection of the superimpositions (Table 2) was undertaken to identify the best overlapping (most extensive in terms of the molecules superstructure and side chains) structures. The top overlapping molecules were identified for further assessment.

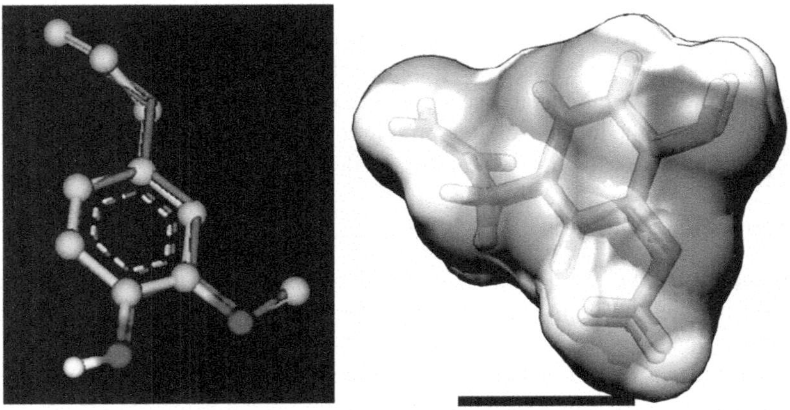

*a* *b*

**Fig. 2.** Shows the manually identified best superimposition between eugenol (beige) and vanillin. a) the superimposition in the stick figures shows overlap between the aromatic ring, the hydroxyl and the ether group. There is conformational-structural (but not functional group) overlap between the propenyl group of eugenol and aldehyde group of vanillin. b) The surface representation of the superimposition of vanillin and eugenol shows a clear bipolarity. The polar side is represented by the oxygen of the overlapping hydroxyl and ether oxygens

### 3.1 Eugenol Superimposed with Vanillin

Both odorants elicit a strong excitatory response with mOR- EG. Figure 2a shows a superimposition of both molecules. The structures are similar other than an isopropenyl group in eugenol (beige) and an aldehyde group in vanillin (blue). The bulk of the molecule and the positions of the side chains on the benzene ring in addition to the charged nature of these side chains are similar. The surface structure of the superimposed molecular shows distinct polarity, which is a better indicator of the similarities between the interactabilities of eugenol and vanillin (Fig. 2b).

Typically, two criteria drive odorant binding resulting in OR activation: that the molecule must easily fit into the OR's binding region—increased conformational freedom; and that the interactions between odorant and amino acids of the OR binding region have to be specific and sustained. For most odorants that have a charged functional group, e.g., vanillin and eugenol, the charged portion of the odorant interacts with a polar amino acid residue forming a Schiff's base tether. Additionally, the non-polar part of the odorant is stabilized by Van der Waals' interactions with non-polar amino acid residues.

### 3.2 Ally Benzene Superimposed with Vanillin

Allylbenzene is a weak activator of mOR-EG. Figure 3a shows that there is a strong coincidence in the benzene ring for both molecules. However, vanillin (blue) has two side chains that are absent in allyl benzene (beige) which only has a benzene ring and a propenyl group. The surface view of this superimposition also shows two charged regions, opposite each other. Following the argument made in the previous paragraph, the ligand stability cannot be sustained by Van der Waal's forces because of the lack of a distinct bipolar nature of the molecule. Table 1 also shows that the number of possible superimposable conformations for allyl benzene are far fewer than that of eugenol and is only comparable in one of the conformations of vanillin.

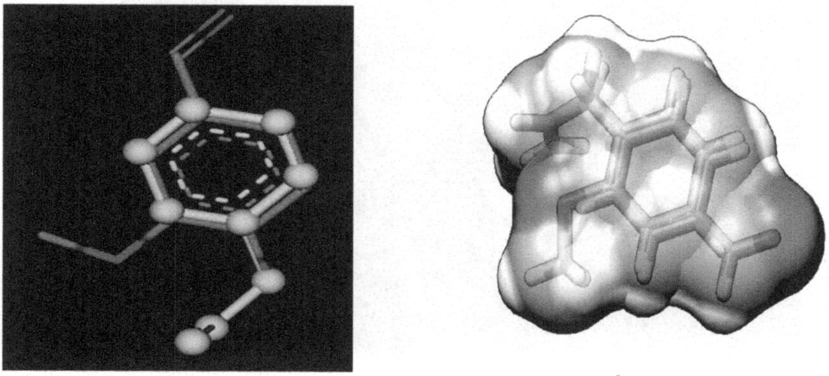

*a* *b*

**Fig. 3.** Shows the manually identified best superimposition between allyl benzene (beige) and vanillin. a) the superimposition in the stick figures shows overlap between the aromatic ring, and a conformational overlap between the allyl group and the hydroxyl group. The strongly electro negative aldehyde group and the ether group of vanillin find no overlap with allyl benzene b) The surface representation of the superimposition of vanillin and allyl benzene two distinct polar regions opposite each other, that arise from the charged functional groups of vanillin.

### 3.3 Energy, Conformations, and Preferred Binding Pocket

As mentioned in the previous section, targeted and flexible docking was carried out on the ligands with mOR-EG. Figure 4 shows a model of the OR created using our previously established protocols. The transmembrane helical regions are labeled, as are the putative binding pockets 1 and 2, henceforth abbreviated P1 and P2 in the graphs (Figs. 5a - 7b).

Figures 5a and 5b show the result of flexible binding of the strong activating odorant eugenol in terms of the RMSDs of each conformation and the accompanying energy of binding for pockets 1 and 2. The results show that number of viable conformations is higher in pocket 1 with lower overall energy. The opposite is true for pocket 2. While the same pattern is observed for the weakly binding allylbenzene, when compared with

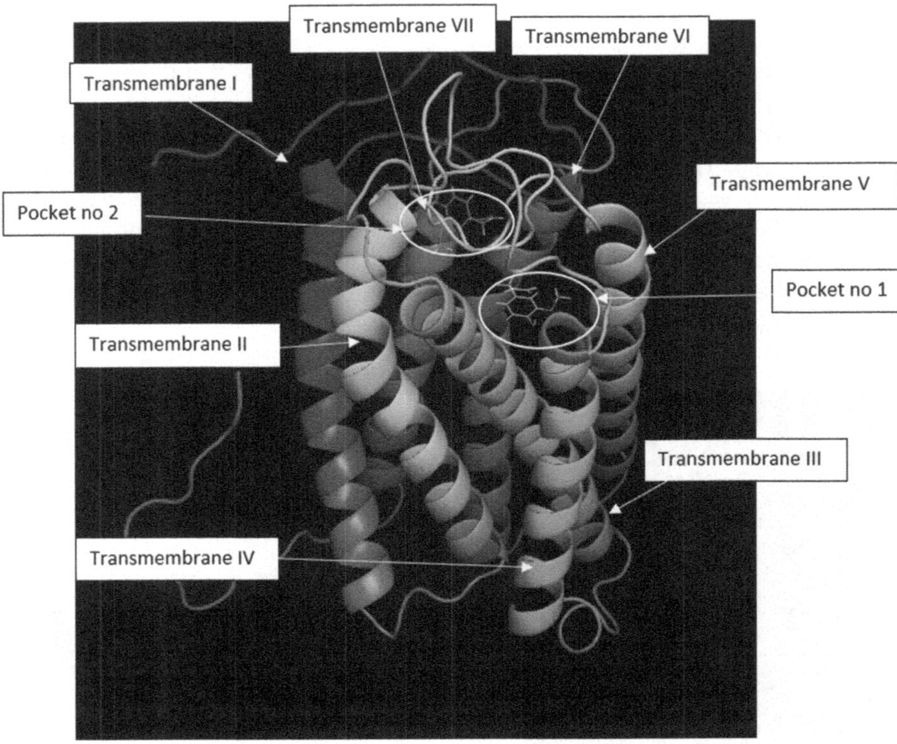


**Fig. 4.** A computational model of mOR-EG determined by combining homology modeling and molecular dynamics. The transmembrane (TM) helices are labeled and variably colored. The two putative binding pockets: P1 (preferable) and P2 are identified with a single conformation of an odorant in each binding pocket.

eugenol, the RMSDs (Root Mean Square Deviation) for conformations for the weakly binding allylbenzene are lower with corresponding higher energies. This means that the size of the pocket and the electronic character of the pocket is not conducive for binding for allylbenzene–which has no charged functional groups (Fig. 6a and 6b).

### 3.4 Targeted Docking Versus Flexible Docking

When the relative energies for targeted docking versus flexible docking for both pockets are compared, the flexibly docked ligands bind within pockets 1 and pockets 2, but with lower energy. When comparing the same metrics for a strong activator versus a weak activator, the energy patterns are the same–flexible docking occurs at a lower energy versus targeted docking. Figure 6a and 6b show that the overall energies of docking for a strong activator are far lower than those for docking with a weak activator.

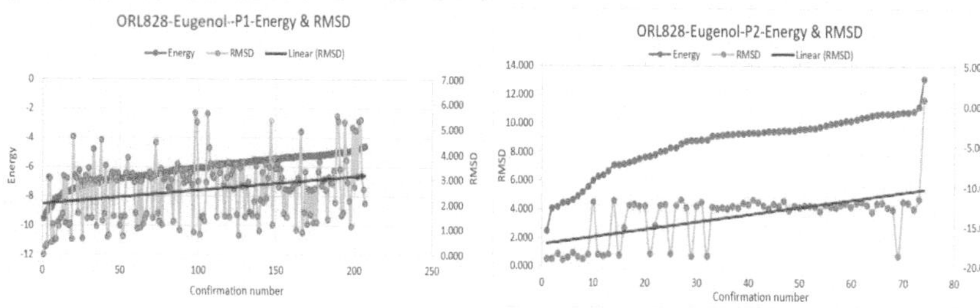


*a* *b*

**Fig. 5.** The conformational space for eugenol in pockets P1 (a) and in pocket (P2) (b). In P1 (a), the small variation in odorant RMSD shows that the ligand is stable within the binding region and the entire conformational space is centered around fewer structural variations than in P2. The overall energy is also lower in P1 and does not vary over the conformational space explored. For eugenol's behavior in P2 (b), the RMSD ranges over 2 Å, there are fewer conformations. It appears that the conformations vary in two distinct phases, though there are far fewer conformations. Conformations 33 to 68 appear as if the eugenol molecule is stabilized in a metastable configuration, which is of significantly high energy. Conformation 76 is a high energy conformation that depicts a configuration that is destabilized.

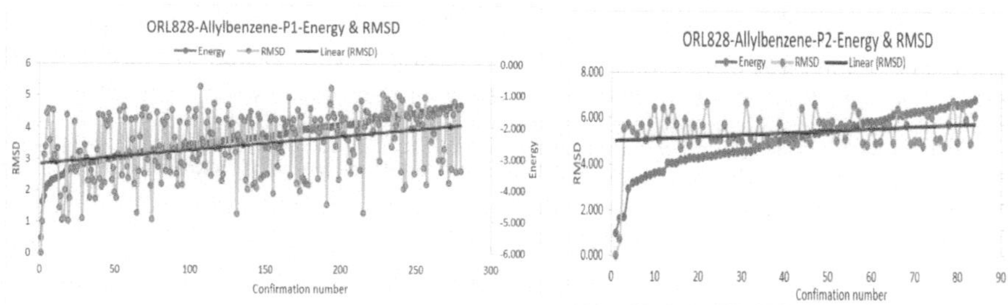


*a* *b*

**Fig. 6.** The conformational space for allyl benzene in pocket number 1 (P1) (a) and in pocket number 2 (P2). In P1 (a), the small variation in odorant RMSD shows that the ligand is stable within the binding region and the entire conformational space is centered around fewer structural variations than in P2. The overall energy is also lower in P1 and does not vary significantly over the conformational space explored. Allyl benzene's initial conformation is at a lower energy, but like eugenol (4b) the energy rises and there are smaller conformational variations in this unstable high energy state. Allyl benzene is stabilized in P1 with a larger number of low energy conformations and is less so in P2.

## 4 Discussion and Conclusions

Two novel notions are advanced in this paper. One, that we are considering olfactory receptor-odorant interactions from the perspective of the odorant. Two, we have used the metrics of energy and conformational space to determine the strength of interaction, and possible OR activation. The aim of this theoretical-computational study has been

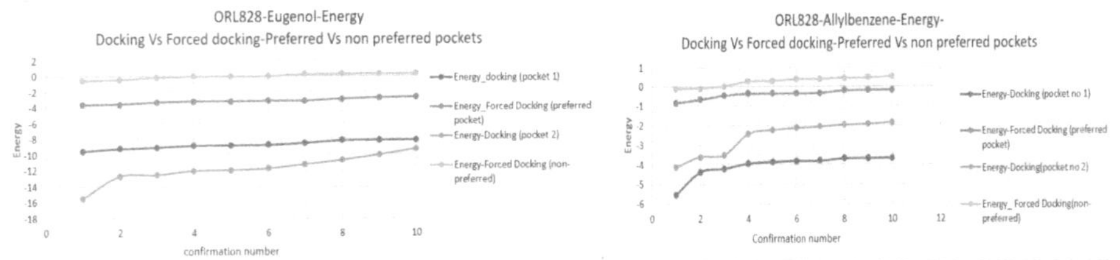


*a* *b*

**Fig.7.** Comparative energies of docking over 10 conformations show that flexible docking in pocket 1 has a lower energy than in pocket 2 for eugenol. Targeted docking energies are higher. Overall docking energies are lower for eugenol than for allyl benzene

to lay the mechanistic foundations of OR-odorant interactions. The eventual aim is to create a paradigm that allows researchers to identify a rational panel of odorants that are likely to bind specific olfactory receptors, thus mitigating some of the issues associated with deorphanizing an OR, as articulated in the Introduction and Background section.

We have borrowed notions associated with novel drug design, leveraging the software accessed through success at an international competition organized by the creators of BioSolveIT suite of software. The two programs that were specifically used were FlexS and LeadIT. FlexS explores the conformational space of a molecule and allows the superimposition of two molecules. LeadIT rapidly docks ligands in the receptor binding region. Here, a user can define putative binding regions (which we have termed targeted docking) or allow the software to identify ligand-accessible regions without user-intervention (flexible docking).

Novel drug design relies on structure-function similarities with known efficacious compounds. Our methodology leverages these notions by selecting a molecule to serve as the central compound—vanillin. This is because this compound is a known strong activator of mOR-EG while possessing the most variability in structure and number of functional groups. For this limited cohort of tested molecules, vanillin offers the most structural permutations—the conformational space.

Each of these conformations were manually and computationally inspected to filter and identify the conformations that had the largest superimposed overlap between eugenol and vanillin, and allyl benzene and vanillin. These results from docking experiments for these conformations are illustrated in Figs. 5a-7b. Superimposition of molecules indicates that only molecules with similar structures and electronic character will elicit similar activation responses. But scanning the conformational space also shows how some conformations temporarily mimic the structural characteristics of a strong activator, even though the native configuration of the ligand molecule looks different from that of a strong activator. The results of the superimposition of vanillin and eugenol versus that of vanillin and allyl benzene show that it is not just the size of odorant but also its electronic character that is critical to binding. Also important is the complementarity between the electronic character of the odorant ligand and that of the OR binding region.

LeadIT allows the rapid and customized docking of ligands in the binding region. Odorant binding in pockets 1 and 2, even when unrestricted, confirms that indeed two binding pockets exist. Flexible binding is preferred to forced binding in predetermined regions. The binding energies and the conformation freedom (from the number of conformations and the RMSDs) indicate that pocket 1 (TMs 3, 4, 5, 6) is preferable to pocket 2 (TMs 1, 2, 6, and 7). This was explored in our previous computational work [19]. The potential for activation strength can also be seen by determining the conformational space and energetics of the ligand in the binding pocket. Ligands that are weak activators are conformationally restricted within the binding region when compared to the RMSDs of conformations for strong activators.

One can put the results into the context of the odotope theory [28]. The odotope—the odorant has electronic structural features that makes it conducive to binding because of the complementary part of the OR's binding region. In addition to the functional groups that allow an electrostatic interaction with a charged amino acid or Van der Waals interactions with the non-polar parts of the odorant, one can also surmise that the odotope is a specific conformation that is conducive to binding, which is why seemingly structurally disparate odorants are known activators of the same ORs. This is a novel notion, which can be considered as an extension of the odotope theory.

This work was carried out using a known strong activator of mOR-EG olfactory receptor and four non-activators. The odors for each are as published in The Good Scents Company resource are in parentheses: Vanillin (vanilla, sweet, balsamic, spicy odors), eugenol (clove and spicy odor), and 2-methoxy-4-ethyl phenol (sweet, spicy, medicinal odor), 2-methoxy-4-methyl phenol (smoky, sweet, burnt, vanilla-like odor), allyl benzene (aromatic odor), and eugenol ethyl ether (floral, spicy, mild, vanilla-like) [38]. Of these, eugenol and allyl benzene were selected for further exploration, and the results are reflected in Figs. 2a-7b. The primary reason for seeking out experimentally well characterized ORs and their activities for such computational work is because our theoretical work is invariably founded on experiments. The computational results can provide a mechanistic basis for these interactions extended to OR activity that cannot be experimentally characterized.

One of the challenges of carrying out exploration of the olfactory and odorant space is that the olfactory response is driven by olfactory perception. Perception is not uniform or consistent between individuals, and the factors that drive perception have been explored, without conclusive results [39–41]. This is borne out by the odors for each of the molecules explored in this work. Allyl benzene is the only odorant that has a different smell from the others tested. Even the experimentally determined weak activators have the same vanilla-like, spicy odor of vanillin. This potentially means that for the vanilla-like perception more than one OR is involved. Though vanillin and eugenol are strong activators with overlapping smells, the vanilla-like smell is not found in eugenol. This indicates that the process of perception of specific odorants and odors is more complicated.

We have recently explored a different approach to elucidating OR-odorant interactions, also from the perspective of the odorant. We grouped molecules based on smell and taste perceptions [42]. We explored molecular fingerprints in these molecules—electronic-structural similarities in each group of molecules. These fingerprints involve atom-pairs (bonded or non-bonded) that have the same (or very close) electronic environments (determined by nuclear magnetic resonance chemical shifts) and interatomic distances [43].

This work shows that it is likely that conformational freedom without large structural variations within the binding pocket of an OR facilitates binding and activation. Figures 3a-7b illustrate this for a known strong activator and a weak activator. This, however, is based on an exploration of only two molecules. Clearly more work needs to be done to ascertain this, starting with the remaining non-activating molecules and extending to other well-characterized olfactory receptors.

The implications for this work have consequences not only in enhancing our understanding of the first step in the process that leads to olfactory perception, but they also have implications in the discovery of novel pharmaceutical products. Indeed, most protein receptors that respond to ligands that have pharmacophores are GPCRs and have similar putative structures to those of olfactory receptors.

## References

1. Buck, L., Axel, R.: A novel multigene family may encode odorant receptors: a molecular basis for odor recognition. Cell **65**(1), 175–187 (1991)
2. Buck, L., Axel, R.: Nobel prize in physiology or medicine, 2004. Indian J. Physiol. Pharmacol. **49**(1), 122 (2005)
3. Trimmer, C., et al.: Genetic variation across the human olfactory receptor repertoire alters odor perception. Proc. Natl. Acad. Sci. U. S. A. **116**(19), 9475–9480 (2019)
4. Zozulya, S., Echeverri, F., Nguyen, T.: The human olfactory receptor repertoire. Genome Biol. **2**(6), research0018.1-0018.12 (2001)
5. Glusman, G., Yanai, I., Rubin, I., Lancet, D.: The complete human olfactory subgenome. Genome Res. **11**(5), 685–702 (2001)
6. Niimura, Y., Nei, M.: Evolution of olfactory receptor genes in the human genome. Proc. Natl. Acad. Sci. U. S. A. **100**(21), 12235–12240 (2003)
7. Prieto-Godino, L.L., et al.: Olfactory receptor pseudo-pseudogenes. Nature **539**(7627), 93–97 (2016)
8. Auton, A., et al.: 1000 genomes project consortium–a global reference for human genetic variation. Nature **526**(7571), 68–74 (2015)
9. Gaillard, I., Rouquier, S., Giorgi, D.: Olfactory receptors. Cell. Mol. Life Sci. **61**(4), 456–469 (2004)
10. Topiol, S.: X-ray structural information of GPCRs in drug design: what are the limitations and where do we go? Expert Opin. Drug Discov. **8**(6), 607–620 (2013)
11. McCusker, E.C., Bane, S.E., O'Malley, M.A., Robinson, A.S.: Heterologous GPCR expression: a bottleneck to obtaining crystal structures. Biotechnol. Prog. **23**(3), 540–547 (2007)
12. Mombaerts, P.: Genes and ligands for odorant, vomeronasal and taste receptors. Nat. Rev. Neurosci. **5**(4), 263–278 (2004)
13. Oka, Y., Omura, M., Kataoka, H., Touhara, K.: Olfactory receptor antagonism between odorants. EMBO J. **23**(1), 120–126 (2004)
14. Saito, H., Chi, Q., Zhuang, H., Matsunami, H., Mainland, J.D.: Odor coding by a Mammalian receptor repertoire. Sci Signal. **2**(60), ra9 (2009)
15. Oka, Y., Nakamura, A., Watanabe, H., Touhara, K.: An odorant derivative as an antagonist for an olfactory receptor. Chem. Senses **29**(9), 815–822 (2004)

16. Billesbølle, C.B., et al.: Structural basis of odorant recognition by a human odorant receptor. Nature **615**(7953), 742–749 (2023)
17. Harris, J.A., et al.: Selective G protein signaling driven by substance P-neurokinin receptor dynamics. Nat. Chem. Biol. **18**(1), 109–115 (2022)
18. Crasto, C.J.: Computational biology of olfactory receptors. Curr. Bioinform. **4**(1), 8–15 (2009)
19. Lai, P.C., Guida, B., Shi, J., Crasto, C.J.: Preferential binding of an odor within olfactory receptors: a precursor to receptor activation. Chem. Senses **39**(2), 107–123 (2014)
20. Vaidehi, N., et al: Prediction of structure and function of G protein-coupled receptors. Proc. Natl. Acad. Sci. U. S. A. **99**(20), 12622–12627 (2002)
21. Floriano, W.B., Vaidehi, N., Goddard, W.A., 3rd., Singer, M.S., Shepherd, G.M.: Molecular mechanisms underlying differential odor responses of a mouse olfactory receptor. Proc. Natl. Acad. Sci. U. S. A. **97**(20), 10712–10716 (2000)
22. Okada, T., Sugihara, M., Bondar, A.N., Elstner, M., Entel, P., Buss, V.: The retinal conformation and its environment in rhodopsin in light of a new 2.2 Å crystal structure. J. Mol. Biol. **342**(2), 571–583 (2004)
23. Rasmussen, S.G., et al.: Crystal structure of the human beta2 adrenergic G-protein-coupled receptor. Nature **450**(7168), 383–387 (2007)
24. Lai, P.C., Crasto, C.J.: Beyond modeling: all-atom olfactory receptor model simulations. Front. Genet. **3**, 61 (2012)
25. Katada, S., Touhara, K.: A molecular basis for odorant recognition: olfactory receptor pharmacology. Nihon Yakurigaku Zasshi **124**(4), 201–209 (2004)
26. Malnic, B., Hirono, J., Sato, T., Buck, L.B.: Combinatorial receptor codes for odors. Cell **96**(5), 713–23 (1999)
27. Mori, K., Shepherd, G.M.: Emerging principles of molecular signal processing by mitral/tufted cells in the olfactory bulb. Semin. Cell Biol. **5**(1), 65–74 (1994)
28. Turin, L.: A method for the calculation of odor character from molecular structure. J. Theor. Biol. **216**(3), 367–385 (2002)
29. BiosolveIT. https://www.biosolveit.de/. Accessed 2017–2018
30. Moini, J., LoGalbo, A., Ahangari, R.: The senses of smell and taste. In: Moini, J., LoGalbo, A., Ahangari, R. (eds.) Foundations of the Mind, Brain, and Behavioral Relationships, pp. 161–174. Academic Press (2024)
31. Krogh, A., Larsson, B., Von Heijne, G., Sonnhammer, E.L.: Predicting transmembrane protein topology with a hidden Markov model: application to complete genomes. J. Mol. Biol. **305**(3), 567–580 (2001)
32. Möller, S., Croning, M.D., Apweiler, R.: Evaluation of methods for the prediction of membrane spanning regions. Bioinformatics **17**(7), 646–653 (2001)
33. Tusnady, G.E., Simon, I.: The HMMTOP transmembrane topology prediction server. Bioinformatics **17**(9), 849–850 (2001)
34. Webb, B., Sali, A.: Comparative protein structure modeling using MODELLER. Curr. Protocols Bioinf. **54**(1), 5–6 (2016)
35. Crasto, C.J.: Hydrophobicity profiles in G protein-coupled receptor transmembrane helical domains. J. Receptor Ligand Channel Res. **2010**(3), 123–133 (2010)
36. O'Boyle, N.M., Banck, M., James, C.A., Morley, C., Vandermeersch, T., Hutchison, G.R.: Open Babel: An open chemical toolbox. J. Cheminform. **3**, 33 (2011)
37. ArgusLab. http://www.arguslab.com/arguslab.com/ArgusLab.html. Accessed 20 oct 2024
38. The Good Scents Company. https://www.thegoodscentscompany.com/. Accessed 20 Oct 2024
39. Roberts, S.C., Havlíček, J., Schaal, B.: Human olfactory communication: current challenges and future prospects. Philos. Trans. R. Soc. Lond. B Biol. Sci. **375**(1800), 20190258 (2020)
40. Barwich, A.-S.: A sense so rate: Measuring olfactory experiences and making a case for a process perspective on sensory perception. Biol. Theory **9**(3), 258–268 (2014)

41. Auffarth, B.: Understanding smell–the olfactory stimulus problem. Neurosci. Biobehav. Rev. **37**(8), 1667–1679 (2013)
42. Crasto, C.J., Bhayani, H., Thomas, S. S., Thilakarathne, R., Perera, N.: Exploring the molecularity of specific olfactory and gustatory perceptive phenotypes—a computational approach. Biomed. J. Sci. Tech. Res. **53**(5), BJSTR. MS.ID.008459 (2023)
43. Bhayani, H.S., Thilakarathne, R., Perera, N., Crasto, C.J. Exploring the molecularity of odor and taste perceptions of "brown". In: A Computational Approach. International Conference on New Trends in Applied Sciences, Konya, Turkey pp. 98–102 (2023)

# Synergy Between Mechanistic Modelling and Ensemble Feature Selection Approaches to Explore Multiscale Biological Heterogeneity

Riccardo Aucello[1,2], Nicola Licheri[2,4], Elena Rosso[1,2,3], Giulio Ferrero[4], Sandro Gepiro Contaldo[1,2], Simone Pernice[1,2(✉)], Pietro Lió[5], Francesca Cordero[1,2], and Marco Beccuti[1,2]

[1] Computer Science Department, University of Turin, Turin, Italy
simone.pernice@unito.it
[2] Laboratorio InfoLife, Consorzio Interuniversitario Nazionale per l'Informatica (CINI), Rome, Italy
[3] Head and Neck Medical Oncology Department, Fondazione IRCCS Istituto Nazionale dei Tumori, Milan, Italy
[4] Biological and Clinical Sciences Department, University of Turin, Turin, Italy
[5] Computer Science and Technology Department, University of Cambridge, Cambridge, UK

**Abstract.** Biological phenomena are based on the precise and accurate cooperation of a non-random combination of molecules implicated in several pathways and networks. In the view of precision medicine, the plethora of omics data accrued sheds light on the comprehension of molecules cooperation. However, these data bring noise and redundancy that it is necessary to consider during the data analysis. A combination of omics data resources, integrated to parameterize mechanistic models, and multiphase Ensemble Feature Selection (EFS) is proposed. Through EFS, we characterized the metabolic heterogeneity of three distinct glycolysis-associated clusters (GACs) in colorectal cancer. Our study reveals that the EFS-derived genetic signatures associated with each GAC group also characterize three glycolysis profiles previously identified. GAC1 demonstrated unique separation, while GAC2 and GAC3 exhibited overlapping characteristics.

**Keywords:** Mechanistic Models · Machine Learning · Omics Data · Flux Balance Analysis

## 1 Introduction

In recent years, multi-omics datasets are becoming increasingly common in biological research leading to obtaining an overview of the complex interactions beyond the biological systems. Consequently, computational approaches for data


M. Vettoretti et al. (Eds.): CIBB 2023, LNBI 14513, pp. 117–130, 2025.
https://doi.org/10.1007/978-3-031-90714-2_9

integration become popular to extract hidden relationships insight from these several layers of information. We aim to deal with data analysis strategies to integrate diverse measures of biological information.

Typically, the three common data integration strategies are (i) early integration, (ii) middle integration, and (iii) late integration [2]. The early integration is a simple concatenation of features derived from each omics layer into one single data structure. Middle integration uses machine learning algorithms to consolidate data without concatenating features or merging results. Finally, the late integration independently merges the results obtained from modeling and analysis techniques performed at each omics layer. Among these strategies, the methodology should be chosen concerning the biological question and the computational context used to translate enormous amounts of data and information into clinical and biological understanding. One limitation of these strategies is the interpretability of the results obtained. Specifically, even though state-of-the-art algorithms are used to perform the data integration, a deep understanding from the biological functional point of view is not easy to obtain. To pursue this direction, we would like to integrate fluxomics and transcriptomics, providing a more comprehensive understanding of cellular processes under clinical conditions.

Indeed, fluxomics corresponds to the global analysis of fluxes in the metabolic network of a cell, which can measure and evaluate the rates of reactions (i.e., fluxes) for a metabolic network. It represents an innovative bio-omics research field, where the fluxome (the total set of fluxes in a metabolic network) depicts the integrative information on several cellular processes. Its integration with transcriptomics can help identify the genes responsible for changes in metabolic fluxes by Flux Balance Analysis (FBA). FBA is a mathematical modeling technique [22] that uses stoichiometric models of metabolic networks to predict metabolic fluxes (i.e., velocities at which metabolites are produced and consumed) under different environmental conditions by minimizing/maximizing an objective function (e.g., biomass growth). FBA can complement a novel workflow to integrate multi-omics data simultaneously or incorporate it with other FBA-derived tools, [10,17] and [15]. Furthermore, many studies have been trying to predict fluxes via transcriptomics and FBA or other constraint-based modeling techniques; recent approaches, such as [6,7] and [9] map from expression profile to constraints for metabolic fluxes adjusting fluxes based on gene expression levels

In this way, integrating fluxomics data simulated from transcriptomic data can provide a more comprehensive picture of cellular metabolism and help identify the genes and pathways distinctive for a given condition. This work proposes a new approach to characterize metabolic phenotypes from gene expression data. Specifically, we integrate the gene expression profiles in the FBA network. Then, the fluxes most influenced by the increase of cell biomass are identified using multiphase Ensemble Feature Selection (EFS) techniques.

Our approach reflects the extent to which different metabolic reactions in the network contribute to the phenotype. In a recent study [12], FS methods were used to identify top-contributing metabolic predictors based on a logistic regression model. In contrast, the EFS techniques applied in our work are part of an original multi-omics integration framework.

The case study presented in this paper is focused on the characterization of glycolysis-associated clusters (GACs) in the context of Colorectal cancer (CRC). The paper proposed by Wang et al. [24] identified GACs with unique clinical, genomic, and tumor micro-environment characteristics.

Our approach integrates functional data to unveil the metabolic heterogeneity of GACs. Specifically, we selected genes whose associated metabolic fluxes support the classification of cancer samples. These flux-derived genetic signatures aim to enhance our understanding of the diverse expression patterns within these clusters. We exploit the Cancer Genome Atlas (TCGA) COAD/READ dataset for the full analysis, while the GSE39582 dataset [16] as validation of the signatures.

## 2 Methods

In this section, we provide a concise overview of the components comprising our approach: (i) the FBA model,(ii) genetic algorithms, (iii) the feature selection technique and (iv) the characterization of flux-derived genetic signatures. Additionally, we describe the multi-omics data to which our approach has been applied.

### 2.1 Flux Balance Analysis

Beginning with a system of metabolic reactions, stoichiometric equations can be derived to describe the relationships among various products and biomass. Flux Balance Analysis (FBA) [25] leverages linear programming to determine the fluxes within these relationships. The primary goal of FBA models is to simulate genome-scale reconstructions of metabolic networks, which entail numerous reactions and metabolites. This is achieved by assuming the steady state of the system, meaning there is no surplus or deficit of any metabolite. Specifically, the model estimates the distribution of fluxes given specific lower and upper bounds for each flux, along with an objective function to maximize or minimize. This objective function serves as a surrogate for the most plausible physiological state within the system. Mathematically, modeling FBA with $m$ metabolites and $r$ reactions can be translated into a linear programming problem (LPP) as follows:

$$\begin{aligned} &\text{maximize} && f(v) \\ &\text{subject to} && S \cdot v = 0 \\ &\text{and} && v_i^{Low} \leq v_i \leq v_i^{Up} \qquad \forall i = 1, \dots, r \end{aligned} \tag{1}$$

where $v \in \mathbb{R}^r$ is the flux vector describing the activity of all the $r$ reactions, $S \in \mathbb{Z}^{r \times m}$ is the stochiometric matrix, $f : \mathbb{R}^m \rightarrow \mathbb{R}$ is the objective function to maximize, $v_i^{Low} \in (-\infty, \infty)$ and $v_i^{Up} \in (-\infty, \infty)$ are respectively the known upper and lower bound of the rates for the $i^{th}$ flux, with $v_i^{Low} < v_i^{Up}$.

Hereinafter, the integration of gene expression data can contribute to obtaining a biologically relevant and context-specific flux distribution so that the resulting model can predict active metabolic pathways from RNA-seq data and reconstruct a tissue-specific metabolic network. Specifically, RNA-Seq data were used to specify the activity of genes in each sample, and accordingly, reaction rates were quantitatively constrained to obtain flux distributions consistent with the available data and ascribe the impact of changes in gene expression to phenotypes according to [11]. This approach allows mapping the RNA-seq data for each reaction into a specific condition of the model. First of all, each model's reaction $i$ depends on the expression of a set of genes which can be represented by a logical formalism including gene names and boolean operators AND and OR. The formulation is then converted into expression level for a gene set of reaction $i$ from the expression of individual genes following two scenarios:

1. when two genes are connected with an AND operator, the gene set expression i for reaction i is the minimum of the individual gene making the gene set;
2. when two genes are connected by an OR operator, the gene set expression i for reaction i is the sum of the individual genes making the gene set.

This gene-reaction map provides compatibility with a boolean on-off approach, as it approaches zero when the expression level approaches zero. Moreover, we defined the multiplicative factor for the flux bounds given the gene set expression $GSE_i$ as follows:

$$v_i^{Low} h(GSE_i) \leq v_i \leq v_i^{Up} h(GSE_i), \tag{2}$$

where

$$h(GSE_i) = \begin{cases} (1 + |log(GSE_i)|)^{\frac{GSE_i - 1}{|GSE_i - 1|}} \text{if } GSE_i \in \mathbb{R}^+ \setminus \{1\} \\ 1 \text{ if } GSE_i = 1 \end{cases}$$

Based on the direct integration of the gene expression information into the flux bounds, we set fluxes to zero if the expression of their associated genes was low and the maximum allowable flux value as a function of measured gene expression [3]. Finally, the metabolic model fluxes are units of $mmol/gDW * h$, where $gDW$ is the dry weight of cell mass in grams, and $h$ is the reaction time in hours. Human metabolism reconstruction Recon3D can be queried and downloaded from http://bigg.ucsd.edu or http://vmh.life. Several models, including Recon3D, were initially deposited with default bounds set. Then, constraints were applied based on the recalculation of default bounds.

## 2.2 Genetic Algorithms

Genetic algorithms (GAs) are population-based metaheuristic optimization algorithms based on mechanisms inspired by Darwin's theory of evolution to iteratively refine a set of candidate solutions [26]. GAs are problem-independent approaches: the search space $\mathbb{G}$ (genome) differs from the problem space $\mathbb{X}$ (phenome), and a genotype-phenotype mapping is required. The search space is composed of chromosomes $g \in \mathbb{G}$, sequences of genes (e.g., bits, integers, real numbers) that are manipulated by the search operators: (i) the crossover that recombines the genes of two parent chromosomes, and (ii) the mutation, which applies small random changes to the chromosome. The population is iteratively evaluated through a fitness function in order to select the fittest chromosomes for reproduction and give light to a new generation, until convergence or another stop criteria.

### 2.3 Feature Selection in Machine Learning

In supervised learning, labeled datasets $D = (X, y)$ are used to learn mappings $f : X \rightarrow y$. Machine Learning frequently has to deal with high-dimensional data, then feature selection (FS) is a crucial step to identify a small subset of predictive features with respect to the target and remove the irrelevant and redundant features. FS methods are broadly categorized into filter, wrapper, and embedded methods [5,21,23]. It is challenging to choose the most suited FS method for a given task: a solution is provided by ensemble FS [1,19,20], namely to combine multiple selection methods in a single one, in order to get more predictive and stable results (i.e., more robust with respect to changes in the training set).

A two-phase FS approach has been implemented, which is composed of (i) a GA-driven dimensionality reduction, and (ii) a subsequent ensemble FS to perform the final selection from a small pool of features. Let us introduce the feature dependency graph (FDG), a graph formalism exploiting weighted vertices to represent the features' predictiveness with respect to the target and edges to indicate a redundancy between two features. The Genetic Algorithm (GA) approach utilizes integer chromosomes of length k, where each chromosome is represented as a vector of integers with k genes. This configuration easily allows for setting an upper bound on the number of selectable features. Each gene within the chromosome can either be coding, representing a selected feature, or non-coding, by not contributing to the final feature set. This setup allows for an upper bound on selectable features without specifying the exact number in advance. This flexibility helps the algorithm explore solutions of varying lengths, enhancing the search for an optimal feature subset. However, a chromosome with all genes set as non-coding would result in a solution with zero fitness, and such chromosomes are naturally excluded during the selection process.

The fitness function rewards predictive features (i.e. high vertex weights) and penalizes feature redundancies. The selection mechanism uses a k-tournament selection technique. The crossover operator is based on the one-point crossover, whereas the mutation operator can (i) insert and (ii) delete features, and (iii) replace features with its neighbors in the overall FDG.

The second phase performs an ensemble feature selection that combines two filters (ANOVA F-test and mutual information) and three embedded (logistic regression, random forest, and linear SVM) methods. K-fold cross-validation is exploited to build multiple samples of the training set, which are used to identify the most prominent features using the available selectors. The approach explores feature sets of different lengths in order to identify a proper trade-off between predictive accuracy and the complexity of the candidate solutions. The multiphase feature selection model has been implemented in Python 3, exploiting the scikit-learn [18] package.

### 2.4 Genetic Signature Characterization

A gene expression signature is a list of genes that can be employed to classify groups of samples in any independent dataset. The gene expression matrix contained gene expression levels reported as FPKM. The data were retrieved using the GDCquery function of the TCGAbiolinks R package setting "STAR - Counts" as the value of the *workflow.type argument*. To assess the informativeness of any groups of interesting genes both principal component analysis and hierarchical clustering were applied on normalized and log2 transformed gene expression matrix after outlier removal, i.e. removing any samples with the expression level of at least one gene falling below Q1 - 1.5 IQR or above the third quartile Q3 + 1.5 IQR, where Q1 and Q3 are the 25th and 75th percentile of the gene expression in the dataset, respectively and IQR represents the inter-quartile range given by Q3 - Q1.

Two clusters are considered and extrapolated from the dendrogram, defining a confusion matrix reporting the association between hierarchical clusters and GACs. The chi-squared test was performed on the confusion matrices when all the expected frequencies were greater than one and the 80% was greater than 5. Then, the empirical Bayes test in Limma package v3.54.2 was used to identify the differential expressed genes (DEGs) between the two clusters of samples corrected by age and sex. A gene was considered DEG when associated with a Benjamini-Hochberg adjusted p-value smaller than 0.05 and a log2 fold change greater than 0.5 in absolute value. Therefore, the subgroups of DEGs in any potential signature were identified. The R package bestglm v0.37.3 was used to identify the optimal subgroups of regressors in the signature. This was achieved through a 10-fold cross-validation process on logistic regression, which compared all possible combinations of the genes of interest. Finally both subgroups of gene, namely DE and glm, were analyzed as signature their self.

### 2.5 Multi-omics Data

Gene Expression profiles from 429 CRC tissues were obtained from COAD cohort of The Cancer Genome Atlas. Data were retrieved using the TCGAbiolinks R package v2.28.1 considering the STAR-counts as the GDC workflow type. We used the most comprehensive human Genome-scale metabolic model (GEM) to date, the community-curated Recon3D human metabolic reconstruction (8399 metabolites, 13543 reactions, 3698 genes) [4], to build tissue-specific models. For signature validation purposes, we exploited the GSE39582 dataset using the

GEOquery R package v2.66.0.b. The CRC metabolic subtypes for both datasets were obtained from [28].

## 3 Results

By combing the transcriptomics data and the FBA model, we offer a functional explanation of the three glycolysis profiles (GAC) identified in [24]. In detail, the glycolysis-associated clusters identified distinct clinical genomic and tumor environment properties. Here, we focus on expanding the description of the glycolysis profiles adding the metabolic layer. For each CRC sample, we generated the corresponding FBA model mapping the transcription profile on the set of reaction boundaries (Eq. 2), leading to a personalization of the metabolic model. Then, by maximizing biomass, we obtained a distribution of fluxes for each CRC sample.

These flux profiles were given as input to the EFS methodology. Specifically, three binary classification tasks have been designed: GAC1 vs. GAC2 (notated as *GAC12*), GAC1 vs. GAC3 (*GAC13*), and GAC2 vs. GAC3 (*GAC23*). The resulting datasets have been split into training (70% of samples) and test (30% of samples) sets. For each comparison group, the GA is run three times using three different feature score functions (i.e., Pearson correlation coefficient, ANOVA F-test, and mutual information), and 30 is imposed as the maximum number of selected features. Features sets selected by the GA are evaluated against the test set exploiting a leave-one-out setting and six classification algorithms. For each comparison group, the sets associated with the highest Area Under the Curve (AUC) values in at least five out of six classifiers are merged into a single set that became the feature pool for the ensemble features selection (FS): 57 features for *GAC12*, 84 for *GAC13*, and 87 *GAC23*. The ensemble FS is applied to identify the proper trade-off between the number of features and the achieved performance (Fig. 1). The resulting sets, which have no overlap among the features, contain 11 (*GAC12*), 7 (*GAC13*), and 9 (*GAC23*) GAC-related reactions that achieve an AUC performance on the test set of 0.98, 0.93, and 0.98, respectively.

These GAC-related reactions are linked to 9 *GAC12*, 8 *GAC13*, and 11 *GAC23*-related genes retrieved from the gene-reaction annotations incorporated by Recon3D. Through the KEGG pathway representation, we found pathways characterized by the network of enzyme-enzyme relations associated with the selected genes.

These pathways (Fig. 2A) were presented using AnnotationDbi, which provides an interface for connecting and querying the KEGG annotation database. Additionally, for reaction-related gene signatures, we found their involvement in various metabolic subsystems (Fig. 2B), where a metabolic subsystem corresponds to a set of reactions sharing a similar metabolic function.

GAC1 is attributed to pathways connected to nucleotide metabolism. Both annotation processes revealed this characteristic. GAC2 is associated with transport across the endoplasmic reticulum, the compartment reserved for the biogenesis of secretory proteins. Intermediate products of protein biogenesis, such

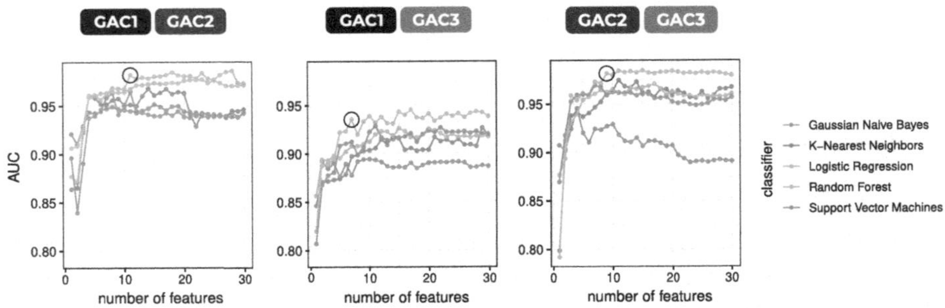


**Fig. 1.** Performances on the test set in terms of Area Under the Curve (AUC) using the lists of features identified by the ensemble FS. For each comparison group we show the performance of the five classifiers in terms of AUC values increasing the number of features. The selected sets are selected considering the minimum number of features associated with an $AUC_{set-selected}$ where $AUC_{max} - 0.01 < AUC_{set-selected} < AUC_{max}$.

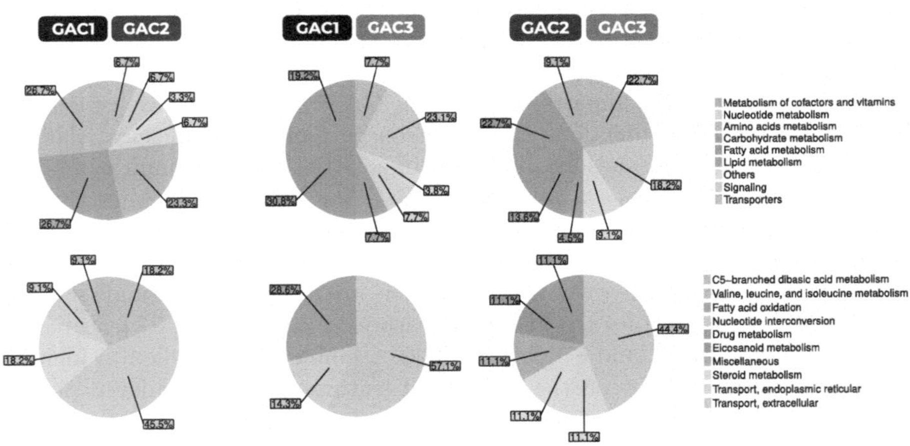


**Fig. 2.** Metabolic characterization of genes linked to the signatures identified by the ensemble FS within each comparison GACs groups (namely GAC1_GAC2 for GAC1 vs GAC2 and so on): (A) relative abundance of pathways annotated with KEGG modules, (B) relative abundance of reactions in various Recon3D metabolic subsystems.

as amino acids, are connected to GAC2. GAC3 tended to be enriched with pathways related to steroid hormones and fatty acid metabolism.

Finally, we analyzed the classification power of each reaction-related signature at the transcription level, inspecting PCA maps and hierarchical clustering. As shown in the top panel of Fig. 3, the first two main components built on the genes of each signature are nearly able to separate GAC groups. We computed the p-value using the chi-square test to evaluate the significance of the difference between the cluster solutions and we have to accept the dependence between hierarchical clusters and GAC labels in each group at the 0.05 level (pvalue $< 0.0001$ in *GAC12* and *GAC13*, pvalue $= 0.008$ in *GAC23*). Therefore, GAC1 demonstrates a distinct separation from GAC2 and GAC3, while the latter exhibited a closer association. However, to support the cluster quantitative separation with numerical data, additional metrics such as the silhouette score could be considered, which provide an measurement of cluster separation.

Then, we worked via a linear model in two ways. On the one hand, we discovered that three over nine genes in *GAC12*, three over eight in *GAC13*, and four over 11 in *GAC23* were differentially expressed. On the other hand, we work with a binomial generalized linear model utilizing 10-fold-cross validation to select the optimal subset of reaction-related genes for each comparison. We analyzed four expression matrices created as follow: i) removing samples that are outliers in at least one gene, ii) scaling the expression level, iii) both, and iv) none operations. In Fig. 3B, we can see for each comparison group the overlap of the DEGs genes (in red) and the four different optimal GLM models (in navy blue). For *GAC12* and *GAC13*, the optimum subset of genes is independent of the operations on the count matrix. However, for *GAC23*, the GLM lists have one out of 3 genes dependent on the outlier removal. Moreover, the gene found with this operation is the only gene detected by both GLM and DEGs. In Fig. 3C, we show PCA and hierarchical clustering for GLM signature (with outlier removal) in the TGCA cohort (pvalue $< 0.0001$ in each comparison).

Finally, we validated these findings by classifying CRC samples across another dataset. In Fig. 3D, we show the performance of the GLM signatures (p-value$<$ 0.0001 in each comparison). The GAC position recognized the three distinct GACs based on the EFS- and GLM-derived signatures, which aligned with previous TGCA results.

## 4 Discussion

Functional data integration could be a new breakout in the bioinformatics approaches.

This work represents a first step in demonstrating how combining mechanistic models integrating omic data with multiphase feature selection methodologies can decipher the complexity at the basis of the biological systems. As proof of concept, we used our approach to integrate transcriptomics into human genome-scale mechanistic models comparing glycolysis-related molecular subtypes and identify the different metabolic markers of GACs through feature selection of fluxes followed by annotation.

Our workflow aims to provide sample-specific maps of cancer metabolism for enhanced data analysis planning. Notably, the ensemble feature selection characterizes heterogeneity in the metabolism of GAC subtypes, returning non-overlapping feature sets and achieving an AUC performance on the test set of 0.98, 0.93, and 0.98 for GAC12, GAC13, and GAC23. The importance of the hybrid approach is highlighted through our metabolic process characterization, which revealed that different GAC groups are associated with different metabolic pathways.

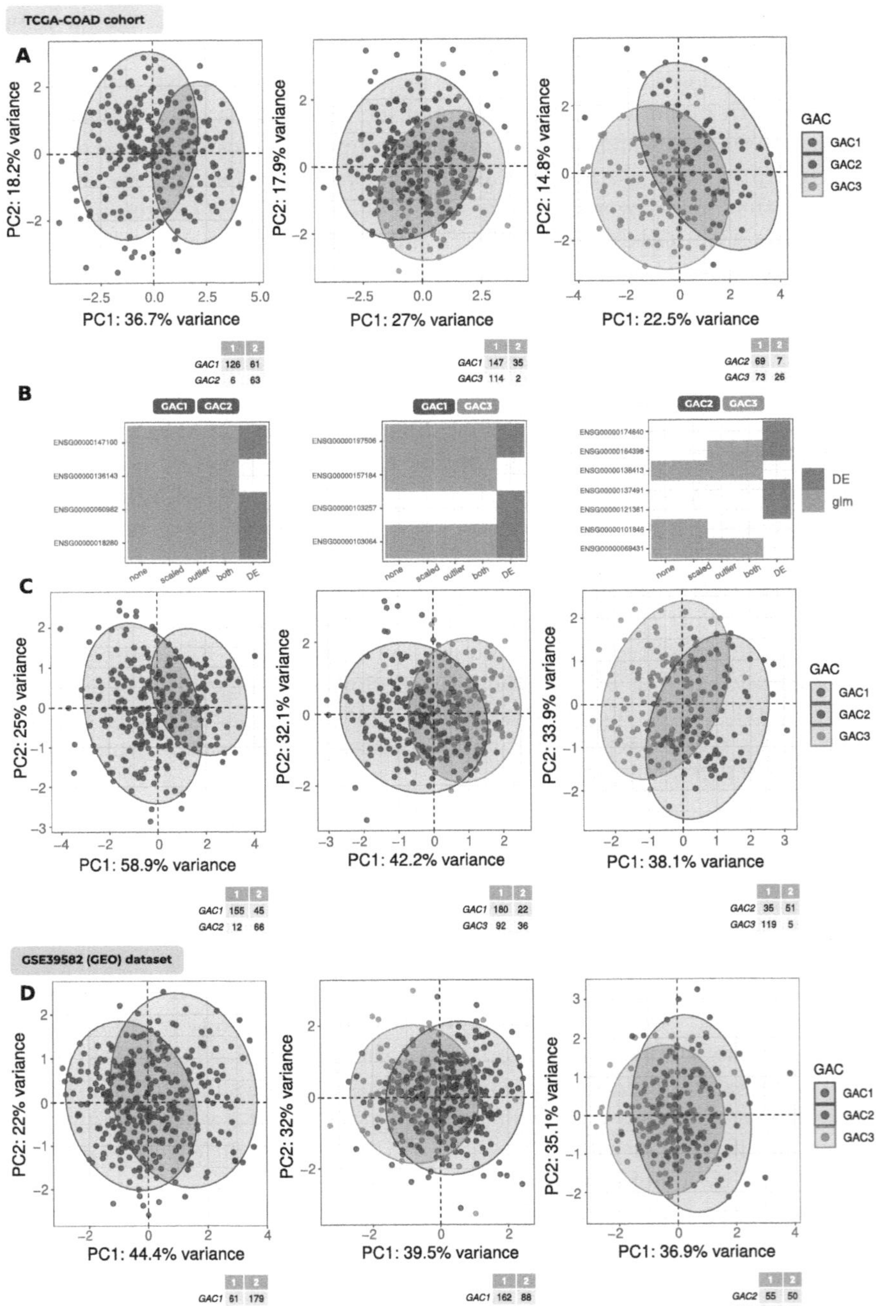


**Fig. 3.** Genetic signature characterization: A) PCA with reaction-related genes in the three comparison groups; B) genes involved in optimum GLM signatures with different operation on the starting matrix and DEGs in reaction-related signatures; C-D) PCA with optimum GLM genes. A-B-C involved the TGCA dataset, D the GEO one.

The results presented in this paper have limitations. First and foremost, while Recon3D is the most updated human metabolic reconstruction, it has default bounds of (-1000, 1000). These default bounds could be problematic for reactions not constrained by gene expression profile integration, such as constraints representing the availability of metabolic sources in the environment. Indeed, there is a type of reaction in metabolic modeling called boundary reactions, which usually describe the exchange of metabolites between the internal cell and the external environment.

Recon3D contains 13,543 reactions, 1,892 of which are devoid of gene-to-reaction associations because they are boundary reactions. If one uses default bounds to constrain reactions, there must be a check on the solution flux distribution to see if any of the fluxes are 1000 or -1000, and a warning should be given in such cases. Although the uniform model for metabolic sources in the environment was not as accurate, it could provide an initial prediction for metabolic characterization methods. Thus, the metabolic modeling framework was used as an initial prediction for metabolic fluxes, but more data are certainly needed to provide higher-quality data for feature selection.

We employed Ensemble Feature Selection (EFS) to characterize the metabolic heterogeneity of GAC subtypes. Meanwhile, by classifying the various expression patterns of glycolysis-related genes, we confirmed that the genes associated with the signatures identified by EFS within each comparative GAC group (GAC12, GAC23, and GAC13) also characterize, to some extent, the three glycolysis profiles (GACs) identified in [24]. Genetic signatures derived from three binary classification tasks were utilized to stratify CRC samples, revealing discernible patterns.

We employed dimensionality reduction and hierarchical clustering techniques to clearly visualize the differentiation among the GACs, thereby highlighting the heterogeneity within these clusters. This analysis classified three GACs, underscoring the heterogeneity within these clusters.

Notably, GAC1 demonstrated a distinct GAC2 and GAC3 separation. In contrast, the latter two clusters exhibited a closer association.

Based on the results, we aimed to ascertain whether these EFS-derived signature genes are differentially expressed genes (DEGs). We constructed a linear model that incorporated additional clinical variables, such as sex and age at diagnosis, as explanatory variables, in conjunction with the GAC subtypes. The DEGs identified among the various GAC subtypes were detected as partially overlapping genes associated with the signatures identified by the EFS.

While EFS is a potent tool for feature selection, we validated these findings using alternative feature selection methods to ensure robustness. We employed a Generalized Linear Model (GLM) for feature selection to identify the subset of genes that best predict the GAC subtypes. This was achieved by fitting a GLM to the data and utilizing cross-validation to select the optimal subset of features.

We constructed the GAC predictors using TGCA data to determine the GAC attribution of each CRC sample. The relationship between GAC displayed is analogous to that obtained using the EFS-identified signatures. We observed that

GAC2 and GAC3 predominantly overlap, while GAC1 samples exhibited greater heterogeneity. Similar observations were made in additional Gene Expression Omnibus (GEO) datasets. We classified CRC samples and examined how EFS- and GLM-derived signatures were positioned in GSE39582 achieved from GEO.

Finally, our analysis recognised three distinct Glycolysis-Associated Clusters (GACs) based on the expression of genes related to glycolysis. This differentiation can be attributed to several factors, including clinical, immune infiltration, biological processes, and genomic traits unique to each cluster.

From the perspective of biological pathways, GAC1 was characterized by a relative increase in WNT and MYC pathways. These pathways are known to play pivotal roles in cell cycle regulation and replication [8]. This could explain the unique metabolic profile of GAC1, as these pathways could promote a high rate of glycolysis to meet the energy and biosynthetic demands of rapidly proliferating cells.

The distinct separation of GAC1 from GAC2 and GAC3 also could be attributed to their unique immune landscapes. GAC1 is noticeably devoid of immune cell infiltration, setting it apart from GAC2 and GAC3, demonstrating significant immune infiltration properties. Despite this, GAC2 and GAC3 share overlapping characteristics, particularly in their active immune responses. However, the nature of these responses differs between the two. GAC2 tends to be immune-inflamed/excluded, characterized by a higher presence of immunosuppressive cells and stromal components. On the other hand, GAC3 is more likely to be an immune-activated type, marked by a high mutation rate and a greater number of active immune cells [24].

## 5 Conclusion

In conclusion, we have identified three feature sets that serve as GAC-related signatures. These signatures are derived from metabolic reactions leading to the retrieval of GAC-related genes from gene-reaction annotations. This approach has enabled our study to generate genetic signatures that enhance the characterization of molecular heterogeneity among colorectal cancer (CRC) subtypes associated with glycolysis. First, sample-specific FBA best accuracy can be achieved through omics integration into mechanistic models. Second, identifying genetic signatures through binary classification tasks provides a more nuanced understanding of the GACs.

Our approach reflects the extent to which different metabolic reactions in the network contribute to the phenotype. While our approach is innovative, it is part of a broader effort by the scientific community as reviewed by [27]. GAC1 demonstrated a distinct separation from both GAC2 and GAC3, suggesting variations perhaps due to differential metabolic processes and tumor microenvironments. These divergent immune microenvironments likely influence metabolic remodeling and necessitate distinctive glycolytic strategies between GAC1 and GAC2/GAC3. The closer association between GAC2 and GAC3 observed through analytical approaches may stem from shared stromal-focused and immune-active type despite harboring distinct mutational traits and pathway dependencies.

We developed genomes-scale mechanistic models to identify the different metabolic markers and then built predictive models capable of forecasting the GACs subtype for individual samples. In conclusion, the analysis supports the characterization of the heterogeneity within the GACs and further validates the initial findings.

However, given that GAC1 demonstrated a distinct separation from GAC2 and GAC3, further and more detailed investigation into the unique characteristics of GAC1 could be beneficial. FBA has significant potential for embracing increasingly available omics data [13]. Further research could explore the underlying mechanisms that contribute to the distinct metabolic profiles of these clusters. Meanwhile, advances in metabolomics and transcriptomics hold promise for synergistic outcomes by incorporating such data into the FBA framework. Our framework has a few limitations, such as focused solely on the biomass optimisation assumption. Parsimonious Flux Balance Analysis (pFBA) will form the basis of our future work. pFBA involves solving two sequential linear programs to achieve optimal growth rate while minimizing total flux [14]. Integrating pFBA can enhance the accuracy and reliability of flux predictions, ensuring our model is both optimal in biomass production and efficient in resource use.

**Acknowledgement.** We greatly appreciate the analytical data provided by the TCGA and GEO databases.

## References

1. Abeel, T., et al.: Robust biomarker identification for cancer diagnosis with ensemble feature selection methods. Bioinformatics **26**(3), 392–398 (2010)
2. Adossa, N., et al.: Computational strategies for single-cell multi-omics integration. Comput. Struct. Biotechnol. J. **19** (2021)
3. Angione, C., Conway, M., Lio, P.: Multiplex methods provide effective integration of multi-omic data in genome-scale models. BMC Bioinform. **17** (2016). https://doi.org/10.1186/s12859-016-0912-1
4. Brunk, E., et al.: Recon3D enables a three-dimensional view of gene variation in human metabolism. Nat. Biotechnol. **36**(3), 272–281 (2018)
5. Chandrashekar, G., Sahin, F.: A survey on feature selection methods. Comput. Electr. Eng. **40**(1), 16–28 (2014)
6. Damiani, C., et al.: Integration of single-cell RNA-seq data into population models to characterize cancer metabolism. PLOS Comput. Biol. **15**, e1006733 (2019). https://doi.org/10.1371/journal.pcbi.1006733
7. Di Filippo, M., et al.: INTEGRATE: model-based multi-omics data integration to characterize multi-level metabolic regulation (2021). https://doi.org/10.1101/2021.08.13.456220
8. El-Sahli, S., Xie, Y., Wang, L., Liu, S.: Wnt signaling in cancer metabolism and immunity. Cancers **11**, 904 (2019). https://doi.org/10.3390/cancers11070904
9. Galuzzi, B., Vanoni, M., Damiani, C.: Combining denoising of RNA-seq data and flux balance analysis for cluster analysis of single cells. BMC Bioinform. **23** (2022). https://doi.org/10.1186/s12859-022-04967-6

10. Jamialahmadi, O., Hashemi-Najafabadi, S., Motamedian, E., Romeo, S., Bagheri, F.: A benchmark-driven approach to reconstruct metabolic networks for studying cancer metabolism. PLOS Comput. Biol. **15** (2019). https://doi.org/10.1371/journal.pcbi.1006936
11. Kashaf, S.S., et al.: Making life difficult for clostridium difficile: augmenting the pathogen's metabolic model with transcriptomic and codon usage data for better therapeutic target characterization. BMC Syst. Biol. **11**(1), 1–13 (2017)
12. Kugler, A., Stensjö, K.: Machine learning predicts system-wide metabolic flux control in cyanobacteria (2023). https://doi.org/10.1101/2023.10.18.562898
13. Lee, J.M., et al.: Flux balance analysis in the era of metabolomics. Briefings Bioinform. **7**(2), 140–150 (2006). https://doi.org/10.1093/bib/bbl007
14. Lewis, N., et al.: Omic data from evolved E. coli are consistent with computed optimal growth from genome-scale mode. Mol. Syst. Biol. **6**, 390 (2010). https://doi.org/10.1038/msb.2010.47
15. Machado, D., Herrgård, M.: Systematic evaluation of methods for integration of transcriptomic data into constraint-based models of metabolism. PLoS Comput. Biol. **10**, e1003580 (2014). https://doi.org/10.1371/journal.pcbi.1003580
16. Marisa, L., et al.: Gene expression classification of colon cancer into molecular subtypes: characterization, validation, and prognostic value. PLoS Med. **10**, e1001453 (2013). https://doi.org/10.1371/journal.pmed.1001453
17. Opdam, S., Richelle, A., Kellman, B., Li, S., Zielinski, D., Lewis, N.: A systematic evaluation of methods for tailoring genome-scale metabolic models. Cell Syst. **4** (2017). https://doi.org/10.1016/j.cels.2017.01.010
18. Pedregosa, F., et al.: Scikit-learn: machine learning in Python. J. Mach. Learn. Res. **12**, 2825–2830 (2011)
19. Pes, B., et al.: Exploiting the ensemble paradigm for stable feature selection: a case study on high-dimensional genomic data. Inf. Fusion **35**, 132–147 (2017)
20. Rokach, L.: Taxonomy for characterizing ensemble methods in classification tasks: a review and annotated bibliography. Comput. Stat. Data Anal. **53**(12), 4046–4072 (2009)
21. Saeys, Y., et al.: A review of feature selection techniques in bioinformatics. Bioinformatics **23**(19), 2507–2517 (2007)
22. Schilling, C.H., et al.: Combining pathway analysis with flux balance analysis for the comprehensive study of metabolic systems. Biotechnol. Bioeng. **71**(4), 286–306 (2000)
23. Tang, J., et al.: Feature selection for classification: a review. Data classif. Algorithms Appl. 37 (2014)
24. Wang, Z., et al.: Machine learning-based glycolysis-associated molecular classification reveals differences in prognosis, TME, and immunotherapy for colorectal cancer patients. Front. Immunol. **14** (2023)
25. Watson, M.R.: Metabolic maps for the Apple II. Biochem. Soc. Trans. **12**(6), 1093–1094 (1984)
26. Weise, T.: Global optimization algorithms-theory and application. Self-Published Thomas Weise **361** (2009)
27. Zampieri, G., Vijayakumar, S., Yaneske, E., Angione, C.: Machine and deep learning meet genome-scale metabolic modeling. PLOS Comput. Biol. **15**, e1007084 (2019). https://doi.org/10.1371/journal.pcbi.1007084
28. Zhang, M., et al.: Metabolism-associated molecular classification of colorectal cancer. Front. Oncol. **10**, 602498 (2020)

# Homophily of Large Weighted Networks in a Data Streaming Setting

Nicola Apollonio[1], Paolo G. Franciosa[2](✉), and Daniele Santoni[3]

[1] Istituto per le Applicazioni del Calcolo "Mauro Picone", Consiglio Nazionale delle Ricerche, Via dei Taurini 19, 00185 Rome, Italy
nicola.apollonio@cnr.it

[2] Dipartimento di Scienze Statistiche, Università di Roma "La Sapienza", piazzale Aldo Moro 5, 00185 Rome, Italy
paolo.franciosa@uniroma1.it

[3] Istituto di Analisi dei Sistemi ed Informatica "Antonio Ruberti", Consiglio Nazionale delle Ricerche, Via dei Taurini 19, 00185 Rome, Italy
daniele.santoni@iasi.cnr.it

**Abstract.** We propose a rigorous and efficient method for evaluating homophily and heterophily in edge-weighted networks. In a network with nodes partitioned into classes, homophily (resp., heterophily) is defined as the tendency to have edges between nodes in the same class (resp., in different classes). Assuming a suitable null model, we provide a closed formula for the $z$-score of the total weight of homophilic/heterophilic edges for each class/pair of classes. The $z$-score directly measures how much this weight deviates from its expected value under the null model.

In addition, we also propose a global homophily measure, that gives a significant score of how the set of all classes at a glance tend to be homophilic.

The proposed statistics can be computed for very large networks since, as we show, they can be efficiently computed in a data streaming setting. For a network with $n$ nodes and $m$ edges, our algorithm only needs $O(n)$ internal memory space, optimal $O(m)$ worst case time, and a single scan of the $m$ input edges, in any order, is required. Experimental results are shown on ten Protein-Protein Interaction networks, reporting homophily w.r.t. protein functional classes.

**Keywords:** Network homophily · $z$-scores · Weighted networks · Data streaming algorithms

## 1 Introduction

Social or biological networks exhibit a typical behaviour, in which nodes preferentially interact if they show some similarity. This is observed when nodes are partitioned into classes, and network edges tend to have both endpoints in the same class. This behaviour is called *homophily*, and the corresponding edges are called *homophilic* edges, while by *heterophily* it is meant the tendency to have


M. Vettoretti et al. (Eds.): CIBB 2023, LNBI 14513, pp. 131–142, 2025.
https://doi.org/10.1007/978-3-031-90714-2_10

*heterophilic* edges, that is, edges joining nodes in different classes. Several statistical indices have been proposed in the literature to measure the homophily of each node class, and the heterophily of pairs of node classes, like Newman's *modularity* [13] and the 2-coloring model by Park and Barabasi [14]. For an up to date bibliography we refer the reader to [11,16].

More recently, a statistical model based on second order statistics has been proposed in [2]. The main idea consists in evaluating how much the number of homophilic/heterophilic edges deviates from its expected number under a null model, which we call the *random coloring model.* This deviation is exactly the $z$-score of the density of the homophilic/heterophilic edges. Positive larger values of the $z$-score denote higher homophily/heterophily. Since these statistics are based on counting the number of homophilic/heterophilic edges, this approach only applies to unweighted graphs. A Python 3 tool implementing these measures is available in [3].

We present an extension to the weighted case of the random coloring model introduced in [2], namely when each edge of the input graph carries a numeric weight. This extension, besides being interesting on its own, also appears to be particularly meaningful because it avoids choosing a weight threshold to define a significant unweighted graph. Indeed, real world networks usually come weighted, where the weight measures edge reliability, capacity, connection rate, or other features, and are selected by fixing a weight threshold. Fixing a threshold yields the unweighted subgraph consisting of the edges whose weight is above the given threshold. Getting rid of this threshold allows us to capture all the information in the input, and also avoids any possible bias due to the threshold choice. The homophily/heterophily measures introduced in [2] for unweighted graphs are well suited for assessing the homophily/heterophily of the graphs obtained by applying a cutting threshold as described above. It is clear that any reasonable homophily/heterophily measure depends on the chosen threshold. A variety of methods and a correspondingly large amount of literature have been produced to support the threshold definition, see [8]. To the best of our knowledge, this is the first attempt to measure homophily/heterophily in weighted networks. In addition to the significance of the proposed method, we show that our indices can be computed very efficiently and using a limited amount of storage.

After describing the new model in Sect. 2 for defining homophily and heterophily indices on single classes or pairs of classes, we show in Sect. 3 how a global homophily/heterophily index can be computed. Complexity issues and applications are dealt with in Sect. 4.

## 2 Statistical Setting and Methodology

### 2.1 The New Model

The model we propose for capturing the homophily of a weighted graph is an extension of what has been presented in [2], which we refer to for notation and terminology (see also [1] for combinatorial proofs). For a positive integer $s$, $[s]$ is the subset $\{1, 2, \ldots, s\}$ of $\mathbb{N}$. Also, for a graph $G$, $V(G)$ and $E(G)$

stand for the vertex-set and the edge-set of $G$, respectively. We are given a triple $(G, w, f)$, where $G$ is an undirected graph with $n$ vertices and $m$ edges, $w : E(G) \to \mathbb{R}_+$ is the weighting of the edges of $G$, i.e., $w(e)$ is the weight of edge $e$, and $f : V(G) \to [s]$ is a surijective map (i.e., a coloring with $s$ colors) which represents a partition of $V(G)$ into $s$-classes (*colors*). The *color profile* of $f$ is the $s$-tuple $c = (c_1, \ldots, c_s)$ in which $c_i$ is the cardinality of the $i$-th color class. For $i \in [s]$, edge $e = uv$ is said $i$-homophilic if $f(u) = f(v) = i$ . Analogously, for $i \neq j$, edge $e = uv$ is said $ij$-heterophilic if $\{f(u), f(v)\} = \{i, j\}$. Let

- $W = \sum_{e \in E(G)} w(e)$ be the total weight of the edges of $G$;
- $W^i = \sum\{e \in E(G) \mid e \text{ is } i\text{-homophilic}\}$, $i \in [s]$, be the total weight of the $i$-homophilic edges;
- $W^{i,j} = \sum\{e \in E(G) \mid e \text{ is } ij\text{-heterophilic}\}$, $i, j \in [s]$, $i \neq j$, be the total weight of the $ij$-heterophilic edges.

Notice that $W = \sum_i W^i + \sum_{i<j} W^{i,j}$. We now define the random coloring model in the weighted case. We consider the given coloring $f$ as sampled uniformly at random from the set of colorings of $V(G)$ having profile $c$. This amounts to equip the sample space of vertex colorings having profile $c$ with the uniform distribution $c_1!c_2! \cdots c_s!/n!$. Having the sample space, we consider the $\binom{s}{2} + s$ random variables $M^i$, $i = 1, \ldots, s$ and $M^{i,j}$, $i, j = 1, \ldots, s$, $i < j$, defined as the total weight of $i$-homophilic and $ij$-heterophilic edges in $G$, under a random coloring with the same profile as $f$. There is a handy representation of these random variables as linear combination of Bernoulli random variables as follows. Recall that the notation $X \sim P$ means that the random variable $X$ is distributed according to $P$. Also, for $a, b$ natural numbers with $a \geq b$, $a^{\underline{b}}$ denotes the falling power $a \cdot (a-1) \cdot \ldots \cdot (a-b+1)$. Consider the following Bernoulli random variables

- $X_v^i$, $i \in [s]$, $v \in V(G)$, the indicator of the event $(F(v) = i)$ with $F$ a random coloring. Recall from [2] that

$$X_v^i \sim \left(\frac{c_i}{n}\right)^{X_v^i} \left(1 - \frac{c_i}{n}\right)^{1-X_v^i};$$

- $X_e^i$, $i \in [s]$, $e \in E(G)$, $e = uv$, the indicator of the event $(F(u) = i, F(v) = i)$ with $F$ a random coloring. It holds that $X_e^i = X_u^i X_v^i$. Therefore

$$X_e^i \sim \left(\frac{c_i^{\underline{2}}}{n^{\underline{2}}}\right)^{X_e^i} \left(1 - \frac{c_i^{\underline{2}}}{n^{\underline{2}}}\right)^{1-X_e^i};$$

- $X_e^{i,j}$, $i, j \in [s]$, $i \neq j$, $e \in E(G)$, $e = uv$, the indicator of the event $(F(u) = i, F(v) = j) \vee (F(u) = j, F(v) = i)$ with $F$ a random coloring. It holds that $X_e^{i,j} = X_u^i X_v^j + X_u^j X_v^i$. Therefore

$$X_e^{i,j} \sim \left(2\frac{c_i c_j}{n^{\underline{2}}}\right)^{X_e^{i,j}} \left(1 - 2\frac{c_i c_j}{n^{\underline{2}}}\right)^{1-X_e^{i,j}}.$$

We are now in the position of defining the main random variables we want to infer, namely $M^i$ and $M^{i,j}$, defined for $i$, $j \in [s]$, $i \neq j$ as the total weight of the $i$-homophilic and $ij$-heterophilic edges, respectively, under a random coloring of profile $c$. These variables have the following representations

$$M^i = \sum_{e \in E(G)} w(e) X_e^i , \tag{1}$$

$$M^{i,j} = \sum_{e \in E(G)} w(e) X_e^{i,j} . \tag{2}$$

Note that $W^i$ and $W^{i,j}$ can now be thought of as the values of $M^i$ and $M^{i,j}$ observed after sampling the given coloring $f$. If we knew the distribution of $M^i$ and $M^{i,j}$, then we could immediately ascertain how significant the values $W^i$ and $W^{i,j}$ are by simply computing the tail probabilities

$$\alpha_i := \Pr\left\{M^i \geq W^i\right\}, \qquad \alpha_{i,j} := \Pr\left\{M^{i,j} \geq W^{i,j}\right\}$$

where, as usual in statistical inference theory, $\alpha_i$ and $\alpha_{i,j}$ are the significance levels (also called $p$-values) of $W^i$ and $W^{i,j}$, respectively, while their complements to 1 are the so-called confidence levels. Therefore, it is natural to take these confidence levels as measures of $i$-homophily and $ij$-heterophily, respectively. Unfortunately, it is shown in [4] that even in the unweighted case, the problem of deciding whether $\alpha_i$ and $\alpha_{i,j}$ are positive is not approximable in polynomial time (i.e., it is not in the APX class). However, after relying on Cantelli's inequality, we can bound the $p$-values of $W^i$ and $W^{i,j}$ from above and take these estimated $p$-values as homophily/heterophily measures. Let $\mathbb{E}(X)$ and $\text{Var}\{X\}$ denote the mean (expected value) and the variance, respectively, of the random variable $X$, recalling that $\text{Var}\{X\} = \mathbb{E}(X - \mathbb{E}(X))^2 = \mathbb{E}(X^2) - (\mathbb{E}(X))^2$. For a random variable $X$ with zero mean and unit variance, and for a positive number $t$, Cantelli's inequality (see [7]) reads as

$$\Pr\{X \geq t\} \leq \frac{1}{1+t^2}.$$

The random variables

$$Z_i = \frac{M^i - \mathbb{E}(M^i)}{\sqrt{\text{Var}\{M^i\}}} \quad \text{and} \quad Z_{i,j} = \frac{M^{i,j} - \mathbb{E}(M^{i,j})}{\sqrt{\text{Var}\{M^{i,j}\}}} \tag{3}$$

have zero mean and unit variance. Therefore, after setting

$$z_i = \frac{W^i - \mathbb{E}(M^i)}{\sqrt{\text{Var}\{M^i\}}} \quad \text{and} \quad z_{i,j} = \frac{W^{i,j} - \mathbb{E}(M^{i,j})}{\sqrt{\text{Var}\{M^{i,j}\}}}, \tag{4}$$

we can apply Cantelli's inequality to $Z_i$ and $z_i$ on the one hand, and to $Z_{i,j}$ and $z_{i,j}$ on the other hand to get

$$\alpha_i = \Pr\left\{M^i \geq W^i\right\} = \Pr\{Z_i \geq z_i\} \leq \frac{1}{1+z_i^2} := a_i$$

and

$$\alpha_{i,j} = \Pr\left\{M^{i,j} \geq W^{i,j}\right\} = \Pr\left\{Z_{i,j} \geq z_{i,j}\right\} \leq \frac{1}{1+z_{i,j}^2} := a_{i,j}.$$

Finally, we define the $i$-homophily index $h_i$ and the $ij$-heterophily index $h_{i,j}$ as

$$h_i = 1 - a_i = \frac{z_i^2}{1+z_i^2} \qquad \text{and} \qquad h_{i,j} = 1 - a_{i,j} = \frac{z_{i,j}^2}{1+z_{i,j}^2} \tag{5}$$

so that these indices have the meaning of estimated confidence in the hypothesis that $W^i$ and $W^{i,j}$ are large. In view of the inapproximability result, this is the best we can do unless $P = NP$. We call the numbers $z_i$'s and $z_{i,j}$'s the observed $z$-scores of $M^i$ and $M^{i,j}$. These $z$-scores thus measure how much the weight $W^i$ or $W^{i,j}$ deviates from its expected value by random chance: positive values of $z_i$-scores denote $i$-homophily (we observe a weight $W^i$ larger than expected by random chance), while positive values of $z_{i,j}$, $i \neq j$, denote $ij$-heterophily (we observe a weight $W^{i,j}$ larger than expected by random chance). In both cases $h_i$ and $h_{i,j}$ measure the intensity (or confidence) of such deviations. Moreover, the $h_i$'s and the $h_{i,j}$'s all range in the interval $[0, 1]$. So they have the same scale and are comparable. Given (4) and (5), to compute $i$-homophily and $ij$-heterophily we only need to compute the first two moments of $M^i$ and $M^{i,j}$ in the weighted case. The (1) and (2) representations make it possible to efficiently perform this task, as shown in the next section.

### 2.2 Weighted Homophily and Heterophily

Let $W = \sum_{e \in E(G)} w(e)$ be the total weight of the edges. The expectations of $M^i$ and $M^{i,j}$ straightforwardly follow by linearity:

$$\mathbb{E}\left(M^i\right) = \frac{c_i^2}{n^2} W \quad \text{and} \quad \mathbb{E}\left(M^{i,j}\right) = 2\frac{c_i c_j}{n^2} W.$$

We now give exact expressions for the variances of $M^i$ and $M^{i,j}$. Recall that the variance of a linear combination of finitely many random variables has a nice expression: if $Y = \sum_{j \in J} \lambda_j Y_j$, where $J$ is a finite set and $Y_j$ has finite first and second order moments $\forall j \in J$, then the variance $\operatorname{Var}\{Y\}$ of $Y$ has the expression

$$\operatorname{Var}\{Y\} = \sum_{j \in J} \lambda_j^2 \operatorname{Var}\{Y_j\} + \sum_{\substack{j,j' \in J \\ j \neq j'}} \lambda_i \lambda_{j'} \operatorname{Cov}\{Y_j, Y_{j'}\}.$$

Since $M^i$ and $M^{i,j}$ are linear combinations of the $X_e^i$'s and $X_e^{i,j}$'s, respectively, to compute the variances of the former it suffices to compute variances and covariances of the latter. Moreover, when the $Y_j$'s, $j \in J$, are Bernoulli random variables—and the $X_e^i$'s and $X_e^{i,j}$'s are such for $e \in E(G)$—, say $Y_j \sim p_j^{Y_j}(1-p_j)^{1-Y_j}$, $j \in J$, one has (i) $\operatorname{Var}\{Y_j\} = p_j(1-p_j)$ and (ii) $\operatorname{Cov}\{Y_j, Y_{j'}\} = \Pr\{Y_j = 1, Y_{j'} = 1\} - p_j p_{j'}$, where, clearly, $\mathbb{E}(Y_j Y_{j'}) =$

$\Pr\{Y_j = 1, Y_{j'} = 1\}$. Therefore, the variance of linear combination of $r$ Bernoulli random variables $Y_j$ with known probability of success, is completely determined by their $r(r-1)$ second order mixed moments $\mathbb{E}(Y_j Y_{j'})$. Hence, to compute $\mathrm{Var}\{M^i\}$ and $\mathrm{Var}\{M^{i,j}\}$ it suffices to compute

$\Pr\left\{X_e^i = 1, X_f^i = 1\right\}$ and $\Pr\left\{X_e^{i,j} = 1, X_f^{i,j} = 1\right\}$, respectively, for each pair of distinct edges $e$ and $f$ of $G$. This task has been accomplished in [2]. If we set $p = \frac{c^{\underline{2}}}{n^{\underline{2}}}$, $W^{\underline{2}} = \sum_{e \neq f} w(e)w(f)$, and define the weight of a path $P$ of length 2 as the product of the weights of its edges, then we arrive at the expression

$$\begin{aligned} \mathrm{Var}\{M^i\} =& \frac{c_i^{\underline{2}}}{n^{\underline{2}}} \sum_{e \in E(G)} w^2(e) - \left( \frac{c_i^{\underline{2}}}{n^{\underline{2}}} \sum_{e \in E(G)} w(e) \right)^2 \\ &+ 2 \left( \frac{c_i^{\underline{3}}}{n^{\underline{3}}} - \frac{c_i^{\underline{4}}}{n^{\underline{4}}} \right) \sum_{P \cong P_3} w(P) + \frac{c_i^{\underline{4}}}{n^{\underline{4}}} W^{\underline{2}}. \end{aligned} \tag{6}$$

Analogously,

$$\begin{aligned} \mathrm{Var}\{M^{i,j}\} =& 2\frac{c_i c_j}{n^{\underline{2}}} \sum_{e \in E(G)} w^2(e) - \left( 2\frac{c_i c_j}{n^{\underline{2}}} \sum_{e \in E(G)} w(e) \right)^2 \\ &+ 2 \left( \frac{c_i c_j^{\underline{2}} + c_i^{\underline{2}} c_j}{n^{\underline{3}}} - 4\frac{c_i^{\underline{2}} c_j^{\underline{2}}}{n^{\underline{4}}} \right) \sum_{P \cong P_3} w(P) + 4\frac{c_i^{\underline{2}} c_j^{\underline{2}}}{n^{\underline{4}}} W^{\underline{2}}. \end{aligned} \tag{7}$$

Remark that, if $w(e) = 1\ \forall e \in E(G)$, then (6) and (7) specialize to the corresponding variances of unweighted homophily and heterophily. All the moments in the above can be computed in linear time. Indeed, $W^{\underline{2}}$ and $\sum_{P \cong P_3} w(P)$ can be both computed in time $O(m)$ and space $O(n)$ by a single scan of edge weights. The former because $W^{\underline{2}} = W^2 - \sum_e w^2(e)$, the latter because, with $w^2(\delta_G(v)) = (\sum_{e \in \delta_G(v)} w(e))^2$, one has

$$2 \sum_{P \cong P_3} w(P) = \sum_{v \in V(G)} \left( w^2(\delta_G(v)) - \sum_{e \in \delta(v)} w^2(e) \right) \tag{8}$$

## 3 Assessing Global Homophily in Weighted Networks

The results of the previous section allow us, in the general weighted setting, to give a meaningful measure of the homophily of a given class or the heterophily between two given classes. However, we need a global notion if we want to assess whether the network as a whole is homophilic or not. A visual (albeit qualitative) perception of the global homophily of the network can be inferred by drawing (symmetric) heat maps on $[s] \times [s]$, where the temperature of the $(i, j)$ entry is $h_{i,j}$ if $i \neq j$, and $h_i$ if $i = j$. If the temperature ranges from blue (cold) to red (warm), then a heatmap whose diagonal tends to red, and freezes out of the diagonal, supports the hypothesis that the network is globally homophilic. This is the approach we have taken in [2] in the unweighted case, and can of course be reproduced here.

A precise quantitative approach to a global assessment of homophily, though still in the unweighted setting, is instead given in [4]. The idea is simple: whatever homophily means in a global sense, the generally accepted paradigm is to measure it by a monotone function of the number (sum of weights) of observed homophilic edges, possibly depending on certain graph parameters acting as constants. Each such function is then considered to be a homophily score. One of the simplest of such scores is the *homophily ratio* [17] (*edge-inside fraction* in [9]), simply defined as the fraction of observed homophilic edges (in the unweighted setting) and as the ratio $\frac{\sum_i W^i}{W}$, in the weighted case. Since

$$W = \sum_i W^i + \sum_{i<j} W^{i,j}$$

it is clear that if the homophily ratio is high, then the network is not globally heterophilic and conversely. It is also clear that

$$W = \sum_i M^i + \sum_{i<j} M^{i,j}. \tag{9}$$

Thus, after setting $M = \sum_i M^i$ (total weight of the homophilic edges under a random coloring) and $N = \sum_{i<j} M^{i,j}$ (total weight of the hetherophilic edges under a random coloring), it follows that $M$ and $N$ determine the distribution of each other. As shown in [4], in the unweighted setting, Newman's celebrated modularity index can also be seen as a homophily score in the previously defined sense. Within the random coloring model, any such score is viewed as the observed value of a corresponding random variable. Accordingly (and depending on the chosen score), we measure homophily in a global sense by the (estimated) statistical significance of the observed score rather than by its value. The advantages of this approach are obvious:

- our measures always lie in the interval $[0, 1]$. In contrast, the problem of locating the range $[a, b]$ of a homophily measure (e.g., homophily ratio, Newman's modularity) for a specific graph is typically difficult, because determining $a$ and $b$ requires solving NP-hard optimization problems;
- our measures are interpreted directly as significance levels;
- our measures can be computed efficiently, as we show in the next section.

In a sense, compared to existing methods, our method can be seen as a nonparametric one, and as such, in addition to those listed above, it has the typical advantages and (of course) disadvantages of this class of methods. See [4] for an annotated technical discussion of these issues. Let us illustrate the method by choosing as homophily score the arithmetic mean

$$\bar{z} = \frac{1}{s} \sum_i z_i. \tag{10}$$

In light of the definition of the $z$-scores (4), $\bar{z}$ is a homophily score: the higher is $\bar{z}$, the higher is the overall weight of homophilic edges. Within the random coloring model $\bar{z}$ is the observed value of the random variable $\bar{Z} = \frac{1}{s}\sum_i Z_i$, where the $Z_i$'s are defined in (3). The random variable $\bar{Z}$ has zero mean and variance equal to $S(\Gamma)/s^2$, where $S(\Gamma)$ is the sum of the entries of the covariance matrix $\Gamma = \mathbb{E}(ZZ')$ of the $s$-dimensional random vector $Z = (Z_1, \ldots, Z_s)'$ (the apex denotes transposition). Actually, $\Gamma$ is the correlation matrix of $M$. By Cantelli's inequality

$$\Pr\left\{\bar{Z} \geq \bar{z}\right\} \leq \frac{S(\Gamma)}{S(\Gamma) + (s\bar{z})^2}.$$

Therefore

$$h = \frac{(s\bar{z})^2}{S(\Gamma) + (s\bar{z})^2}$$

measures the (weighted) homophily of $G$ in a global sense in terms of the confidence that $G$ is globally homophilic according to the score $\bar{z}$. It remains to compute $\Gamma$ to show that $h$ can be computed efficiently. This is easily seen, however, because by exploiting the results in [4] it can be shown that the diagonal entries of $\Gamma$ are all equal to 1, while for $i \neq j$ one has

$$\Gamma_{i,j} = \frac{c_i^{\underline{2}} c_j^{\underline{2}} \left[\left(\frac{W^{\underline{2}} - 2\sum_{P \simeq P_3} w(P)}{n^{\underline{4}}}\right) - \left(\frac{W}{n^{\underline{2}}}\right)^2\right]}{\sqrt{\mathrm{Var}\{M^i\}}\sqrt{\mathrm{Var}\{M^j\}}}$$

## 4 Computational Complexity Issues and Results

The homophily/heterophily $z$-scores derived from expressions (6) and (7) can be efficiently evaluated exploiting (8), since they can be computed by an algorithm whose worst case time complexity is linear in the size of the graph. If we consider a network with $n$ nodes and $m$ edges, computing $z$-scores of homophily/heterophily for all pairs of classes requires $O(n + m + s^2)$ worst case time and space, where $s$ is the number of classes. We assume from now on that $s$ is a small constant, so the space and time complexity of a straightforward implementation is $O(m + n)$.

When processing a very large network, a *data streaming* approach might be beneficial. In the *data streaming* model of computation [10,12] it is assumed that the algorithm may only read/write in a limited memory space, smaller than the input size, and input may only be sequentially read one or more times, and possibly re-written in output data streams for successive input scans. Thus, the algorithm is not allowed to store and randomly access input data, but has to rely on some "compact" data representation that is updated in memory while scanning the input. In a more restrictive model, input data cannot even be reordered between input scans, but can only be read in the original order. The complexity of a data streaming algorithm is measured in terms of the number of passes over the input stream, the amount of internal storage used, and the time needed to process the input.

An additional advantage in adopting the data streaming model, even in the case when large main memory is available, comes from the fact that this model forces the algorithm to adopt a *local* memory access pattern, and it is widely confirmed that sequential scans of input data are much faster in practice than random accesses.

The data streaming model has been successfully applied to graph algorithms [5,6], where a *semi-streaming* model is often considered. In a semi-streaming model the algorithm is only allowed to store $O(n)$ data in memory, where $n$ is the number of nodes in the graph, while sequentially scanning the $O(m)$ graph edges. Consider that even in a sparse large network, $m$ can be some orders of magnitude larger than $n$, for example a sparse network with 1 billion nodes may easily contain thousands of billion edges.

Expression (6) and (7) can indeed be evaluated by a semi-streaming algorithm, starting from two input files containing the description of nodes (node labels with their classes) and edges (pairs of node labels with edge weight). Nodes and edges are not required to appear in any special order. It is only required to know in advance whether each edge appears in the input file as a single entry, or the two symmetric copies are given, i.e., edge $(x, y)$ appears both as $(x, y)$ and as $(y, x)$. Our algorithm requires a single input scan on node and edge files, $O(n + m)$ internal memory operations, and only $O(n)$ memory space, assuming that the number of classes is small with respect to the number of nodes, or in any case $s$ is $O(\sqrt{n})$. A Python 3 semi-streaming prototype for computing $z$-scores according to (6) and (7) is available. The implementation is straightforward, but special attention has been given to avoid propagation of numerical errors. While scanning the input edge file, information is maintained for each node about its degree, the total weight of incident edges and the sum of weight squares for incident edges. Due to expression (8), this is sufficient to evaluate homophily and hetherophily scores without explicitly considering the whole set of $P_3$, as required by expressions (6) and (7).

Theoretical complexity analysis is confirmed by preliminary experiments performed on Terastat2 HPC cluster[1] on large network datasets.

### 4.1 Data Used in the Experiment

We applied the model described above to ten weighted Protein-Protein Interaction (PPI) networks obtained from STRING database [15], mostly derived by bacterial organisms, in order to evaluate homophily/heterophily with respect to functional protein classes. Functional classes of proteins of the considered ten organisms were obtained from NCBI database[2]. Proteins were partitioned into 18 different functional classes. As reported in [2], homophily and heterophily for those PPI networks was evaluated and analysed in the case of unweighted graph. The unweighted networks were derived setting a high confidence weight cut-off

[1] a description of cluster Terastat2 is available at: https://www.dss.uniroma1.it/en/node/8271.

[2] ftp://ftp.ncbi.nih.gov/pub/COG/COG/.

of 700 on the edge weights, this threshold value is usually applied in order to work on reliable interactions. Anyway, the choice of a weight cut-off could deeply impact on the performed analysis and moreover it implies a loose of information, even if the less reliable interactions should be removed.

### 4.2 Homophily of Weighted Vs. Unweighted PPI Networks

The extension of the model to the weighted case allows us to work on the weighted graph without any apriori pre-processing of the network to build the unweighed network. Homophily $z$-score values of functional classes were computed through the new model for the ten weighted PPI networks. In the same way homophily $z$-score values of functional classes were computed for the same PPI networks, taking into account 16 different weight cut-off values 150, 200, 250, .., 900, so that for each of the ten organism we obtained 16 different unweighted networks.

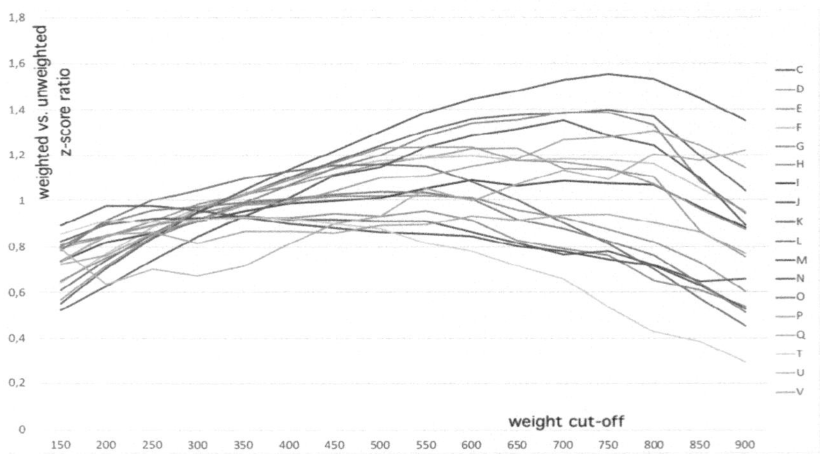


**Fig. 1.** Average ratio ($y$-axis) between weighted and unweighted homophily $z$-scores values reported for each functional class (in different colors) as a function of the weight cut-off ($x$-axis), ranging from 150 to 900.

In Fig. 1 the average ratio between weighted and unweighted homophily $z$-scores values is reported for each functional class (in different colors) as a function of the weight cut-off, ranging from 150 to 900. For each functional class and cut-off value, ratioes were averaged on the ten organisms. As can be observed, all the plots show a common pattern, even if with remarkable differences. An average ratio value ($y$-axis) higher (smaller) than 1 indicates that on average the weighted $z$-score value for the given functional class (color plot) is higher (smaller) than the unweighted one for the given weight cut-off ($x$-axis). Interestingly, most functional classes (in particular J, C, G, P and I) show an ascending slope reaching a peak around 700–800 and then going down in the last weight cut-off value.

## 5 Conclusion

Evaluation of network homophily/heterophily has been proved to be a powerful and effective instrument to shed light on the nature of many complex systems. In the present work we propose an efficient method, implemented in a data streaming setting, that extends a previously presented model to the more general case of edge-weighted networks, thus providing more reliability to homophily/heterophily evaluation.

**Acknowledgment.** Daniele Santoni is member of the Gruppo Nazionale Calcolo Scientifico - Istituto Nazionale di Alta Matematica (GNCS-INdAM).
Nicola Apollonio is a member of Gruppo Nazionale per le Strutture Algebriche, Geometriche e le loro Applicazioni - Istituto Nazionale di Alta Matematica (GNSAGA-INdAM)

**Conflict of Interests.** All the authors declare that there are no conflicts of interests.

**Availability of Data and Software Code.** An $\alpha$ version of a Python3 tool is available at:

http://www.statistica.uniroma1.it/users/pfrancio/weightedHomophilyStreaming.py

For help, type `python3 weightedHomophilyStreaming.py --help`
where `python3` is the Python interpreter.

## References

1. Apollonio, N.: Second-order moments of the size of randomly induced subgraphs of given order. Discret. Appl. Math. **355**, 46–56 (2024). https://doi.org/10.1016/j.dam.2024.04.025
2. Apollonio, N., Franciosa, P.G., Santoni, D.: A novel method for assessing and measuring homophily in networks through second-order statistics. Sci. Rep. **12**, 9757 (2022). https://doi.org/10.1038/s41598-022-12710-7
3. Apollonio N., Blankenberg D., Cumbo F., Franciosa P.G., Santoni D.: Evaluating homophily in networks via HONTO (HOmophily Network TOol): a case study of chromosomal interactions in human PPI networks. Bioinformatics **39**(1) (2023) https://doi.org/10.1093/bioinformatics/btac763
4. Apollonio N., Franciosa P.G., Santoni D.: Network homophily via tail inequalities. Phys. Rev. E **108**(5) (2023). https://doi.org/10.1103/PhysRevE.108.054130. ISSN 2470-0045
5. Ausiello, G., Demetrescu, C., Franciosa, P.G., Italiano, G.F., Ribichini, A.: Graph spanners in the streaming model: an experimental study. Algorithmica **55**, 346–374 (2009)
6. Ausiello, G., Franciosa, P.G., Italiano, G.F.: Small stretch $(\alpha, \beta)$-spanners in the streaming model. Theor. Comput. Sci. **410**, 3406–3413 (2009)
7. Boucheron, S., Lugosi, G., Massart, P.: Concentration Inequalities-A Nonasymptotic Theory of Independence. Oxford University Press (2013). ISBN 978-0-19-953525-5

8. Bozhilova, L.V., Whitmore, A.V., Wray, J., Reinert, G., Deane, C.M.: Measuring rank robustness in scored protein interaction networks. BMC Bioinform. **20**, 446 (2019)
9. Chakraborty, T., Dalmia, A., Mukherjee, A., Ganguly, N.: Metrics for community analysis: a survey. ACM Comput. Surv. **50**(4), 1–37 (2017). https://doi.org/10.1145/3091106. Article 54
10. Feigenbaum, J., Kannan, S., McGregor, A., Suri, S., Zhang, J.: Graph distances in the streaming model: the value of space. In: Proceedings of the 16th Annual ACM-SIAM Symposium on Discrete Algorithms (SODA 2005), pp. 745–754 (2005)
11. Lancichinetti, A., Kivelä, A., Saramäki, J., Fortunato, S.: Characterizing the community structure of complex networks. PLoS ONE **5**(8), e11976 (2010)
12. Munro, I., Paterson, M.: Selection and sorting with limited storage. Theor. Comput. Sci. **12**, 315–323 (1980)
13. Newman, M.: Modularity and community structure in networks. Proc. Nat. Acad. Sci. USA **103**(23), 8577–8582 (2006)
14. Park, J., Barabasi, A.L.: Distribution of node characteristics in complex networks. Proc. Natl. Acad. Sci. U.S.A. **104**(46), 17916–17920 (2007)
15. Szklarczyk, D., et al.: STRING v11: protein-protein association networks with increased coverage, supporting functional discovery in genome-wide experimental datasets. Nucleic Acids Res. **47**, D607-613 (2019)
16. Yang, J., Leskove, J.: Defining and evaluating network communities based on ground-truth. Knowl. Inf. Syst. **42**, 181–213 (2015)
17. Zhu, J., Yan, Y., Zhao, L., Heimann, M., Akoglu, L., Koutra, D.: Adv. Neural. Inf. Process. Syst. **33**, 7793 (2020)

# Living Along COVID-19: Assessing Contention Policies Through Agent-Based Models

Daniele Baccega[1,2(✉)], Simone Pernice[1,2], Paolo Castagno[1], Matteo Sereno[1,3], and Marco Beccuti[1,2,3]

[1] Computer Science Department, Universitá di Torino, 10149 Turin, Italy
daniele.baccega@unito.it
[2] Laboratorio InfoLife, Consorzio Interuniversitario Nazionale per l'Informatica (CINI), Rome, Italy
[3] Competence Centre for Scientific Computing C3S, Universitá di Torino, 10124 Turin, Italy

**Abstract.** SARS-CoV-2 has become an endemic disease, and we will have to face the continuous rise of new variants. Designing and evaluating the effects of new containment policies is of primary importance to keep social activities going as safely as possible according to the different stages of the pandemic. Therefore, we propose an Agent-Based Model to study the evolution of SARS-CoV-2 spread in a well-defined environment (of small/medium size, like a shop, a restaurant, an office, a school, with fewer than a hundred or a few hundred people) to assess the efficacy of different non-pharmaceutical interventions and vaccination strategies. Specifically, we focused on schools, given that the COVID-19 quarantine has resulted in substantial disruptions to education, leading to a transition to remote learning and worsening educational inequalities. We consider using face masks and several real-world testing protocols combined with quarantine policies. All protocols/policies have been evaluated at various stages of the pandemic evolution. Results show that testing campaigns are effective as far as the testing process is faster than the virus diffusion. Also, vaccination campaigns covering less than 40% of the susceptible population provide poor results in protecting from a major outbreak, even if the circulation of the virus is low. Our empirical findings reveal a decreasing effectiveness of various policy implementations as viral circulation intensifies. Additionally, our analysis shows a consistent reduction in infection rates due to the continued use of facial protective measures, regardless of the level of viral circulation.

**Keywords:** Agent-Based Simulation · Non-Pharmaceutical Interventions · COVID-19 · School

## 1 Introduction

The COVID-19 pandemic highlighted the difficulties in managing a world-wide emergency. One of the primary challenges has been keeping people safe


M. Vettoretti et al. (Eds.): CIBB 2023, LNBI 14513, pp. 143–161, 2025.
https://doi.org/10.1007/978-3-031-90714-2_11

while doing their normal activities, such as going to school. Different non-pharmaceutical interventions (NPIs)—such as distancing requirements, masks, rooms' air ventilation, screening, and quarantine—were used to control the spread and to avoid the collapse of often fragile and overloaded healthcare systems. In addition, the onset of new SARS-CoV-2 variants, such as the Omicron variant, exhibited a high degree of immune evasion, leading to increased infection rates worldwide [1]. The advent of vaccines has significantly contributed to reducing the number of deaths [2]. However, in the school setting, children were not initially covered by the vaccine, and even after that, their vaccine coverage grew slowly.

The efficacy of NPIs was studied using different types of modeling [3]: Susceptible-Exposed-Infected-Recovered-Susceptible (SEIRS) compartmental model [4], aerosol transmission model [5–7], Agent-Based Models [8–14]. In this work, we propose an extension of the Agent-Based Model developed in [15] to study the spread of SARS-CoV-2 infections within an explanatory scholastic environment (see Fig. 1) and to analyze different intervention strategies to provide policymakers valuable insight to face future emergencies.

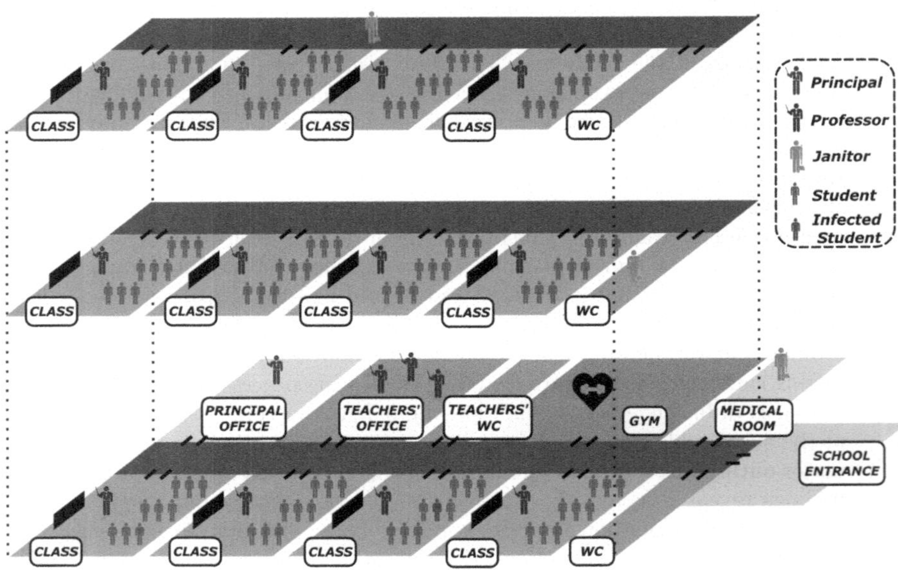


**Fig. 1.** Graphical representation of a generic school featuring 12 classrooms distributed across three floors, along with common areas such as the entrance, corridors, bathrooms, stairs, gym, medical room, teachers' office, and principal's office. Each classroom includes twenty-four desks, evenly spaced apart, along with a teacher's desk and a blackboard at the front. The school population consists of a total of 290 individuals, including 240 students, 49 teachers, and a principal [15].

## 2 Data and Methods

Agent-Based Models (ABMs) simulate complex systems by representing individual entities, or agents, each with specific rules and behaviors. These agents interact with one another and their environment, leading to the emergence of system-level patterns and dynamics. ABMs allow modelers for a nuanced exploration of the intricate relationships and behaviors that shape complex systems.

This work proposes an ABM, developed in the multi-agent programmable modeling environment NetLogo version 6.1.1 [16], that compounds *i)* a detailed description of the environment and the population dynamics within such environment—including both how agents move within the physical spaces and social contact structure of the population—, *ii)* the COVID-19 progression model—modeled as a Susceptible-Exposed-Infected-Recovered (SEIR) compartmental model—, and *iii)* the non-pharmaceutical interventions (NPIs) available to contain the infection spreading.

As an example scenario, graphically depicted in Fig. 1, we considered a school with 12 classrooms (arranged on three floors) with shared spaces (i.e., entrance, corridors, bathrooms, stairs, gym, medical room, teachers' office, and principal's office). Each floor consisted of one corridor connecting all the rooms and one bathroom only for students. The school entrance, the gym, the medical room, the teacher's office, the principal's office, and the teachers' bathroom are all on the first floor. Each classroom is characterized by twenty-four desks (equally distanced from each other), one teacher's desk, and a blackboard. Students and personnel—overall 290 people divided into 240 students, 49 teachers, and a principal—move across the school according to their role and schedule (e.g., students stay in classrooms and at most they can go to the bathroom during classes, and teachers are allowed to move across the school during class changes). Students' schedules foresee 30 classes of 50 min each, from Monday to Friday. Students arrive at school from 7:35 a.m., with the first class beginning at 8:10 a.m. A school day consists of six classes of 50 min each and two breaks. The first break is between the second and the third classes, while the second one is after the fourth class. At the beginning of each simulation students are allocated to classrooms (i.e., the first twenty agents are allocated to one classroom, and so on), and this association lasts for the whole simulation. Teachers move from one classroom to another according to their schedules.

The COVID-19 progression model is a Susceptible-Exposed-Infected-Recovered (SEIR) model (showed in Fig. 2) where the population is divided into four compartments: *i)* the Susceptible (S) population consists of healthy individuals who have not yet been exposed to the disease, *ii)* the Exposed (E) population includes those who have been exposed and are infected, but are not yet infectious, *iii)* the Infected (I) population comprises individuals who can transmit the disease to others, and *iv)* the Recovered (R) population includes those who have recovered from the disease. Each agent is labeled with a specific disease progression status, whose parameters are reported in Table 2. In particular, we labeled each agent with one of the four statuses of the disease progression, updating it as follows: *i)* from the susceptible status to the exposed one—to

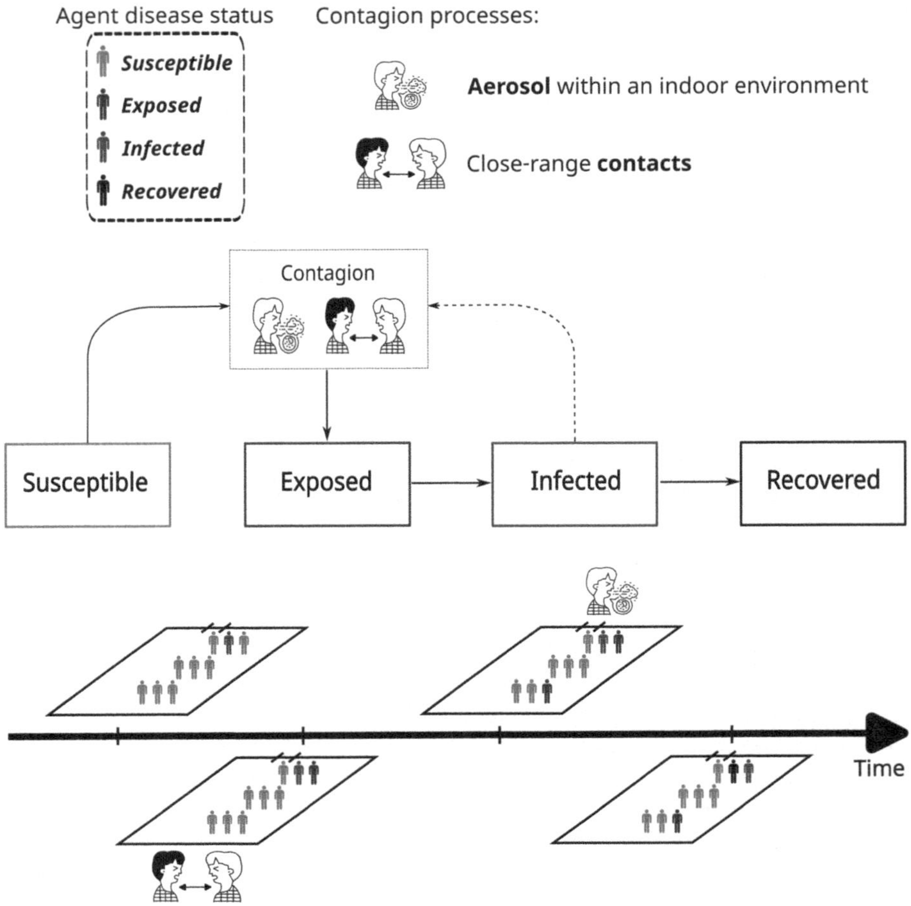


**Fig. 2.** The COVID-19 progression model (SEIR)—the dashed line indicates that contagion occurs via close-range contacts with an infected individual [17] or through aerosol transmission [7]—with an example of its evolution. We considered four time points: $t_1$, $t_2$, $t_3$, and $t_4$ ($t_1 < t_2 < t_3 < t_4$). At time $t_1$, all agents are susceptible except one infected. By time $t_2$, a nearby agent becomes exposed due to close-range transmission. At time $t_3$, a distant agent becomes exposed via aerosol transmission. Finally, at time $t_4$, the initially infected agent recovers, the close-range contact exposed agent becomes infected, and the aerosol-exposed agent remains exposed.

model the virus exposure—, *ii)* from the exposed status to the infected one—when the incubation period ends, and agents become contagious—, *iii)* from the infected status to the recovered one—when the infectious period ends.. The virus exposure is modeled through two different drivers: either close-range contacts [17] between susceptible and infected agents or through aerosol [7] within an indoor environment.

In the former case, the probability of successful contagious contact is associated with each susceptible agent ($S_i$) that remains within an area ($\mathbf{A}$) of 4.41 $m^2$ surrounding an infected agent. This probability is determined by the following function [17]:

$$P_{S_i} = c_r \frac{C_{S_i}}{\mathbf{A}} \tag{1}$$

where $c_r$ represents the *contamination risk*, and $C_{S_i}$ denotes the total time (in minutes) that the $S_i$ agent remains within the vicinity of an infected agent.

In contrast, the spread of the pathogen via aerosol is calculated based on the *quanta* concentration in each room. A quanta is defined as the dose of airborne droplet nuclei necessary to infect 63% of susceptible individuals [18]. Following the approach in [7], we determine the number of quanta inhaled by each agent in a given room (e.g., classrooms, bathrooms, etc.) to compute the probability of infection without direct contact with an infected agent. This probability is given by the following function:

$$P_{S_i} = 1 - e^{-\frac{N_{\text{virus}}}{k_p}} \tag{2}$$

where $k_p$ represents the reciprocal of the probability that a single pathogen particle will trigger an infection [7,19], and $N_{virus}$ denotes the number of quanta inhaled by each susceptible agent during a time interval $\Delta t$. This value is determined by solving the following differential equation:

$$dN_{virus} = (1 - \eta_{mask} f_{\text{mask}})\ C(t) Q_{inh}\ dt \tag{3}$$

where $\eta_{mask}$ is the efficacy of the masks, $f_{mask}$ represents the proportion of agents correctly wearing a mask, $C(t)$ denotes the quanta concentration at time $t$, and $Q_{inh}$ is the inhalation rate. Observe that quanta concentration at time $t$ mainly depends on the number of people within a room, the number of infections, and the room's ventilation.

In Fig. 2, we illustrate an example of the transitions between different statuses over four time points: $t_1$, $t_2$, $t_3$, and $t_4$ ($t_1 < t_2 < t_3 < t_4$). At time $t_1$, all agents in the room are susceptible, except for one infected. At time $t_2$, an agent close to the infected agent becomes exposed due to close-range transmission [17]. At time $t_3$, another agent, located further away from the initially infected individual, becomes exposed through the aerosol transmission [7]. Finally, at time $t_4$, the initially infected agent has recovered, the agent previously exposed at close-range transmission has become infected, and the agent exposed through aerosol transmission remains exposed.

Eventually, several NPI policies can be implemented to prevent the infection from spreading within the school. A contention policy might require the application of one or more NPIs, compounding both actions aimed at reducing contact between students and their exposure to the infection. In particular, the considered NPIs are distancing requirements, rooms' air ventilation, masks, quarantine, and testing campaigns. As a baseline for all the experiments, we assumed that desks are equally distanced from each other, and if masks are required, students use them correctly and for all the time they are at school. Likewise, the ventilation process to replace contaminated with clean air is carried out constantly at 3 Air Changes per Hour (ACH).

Several modifications to the model presented in [15] were introduced to study the efficacy of different contention policies. First, the arising of new variants required to take into account in the model the presence of several pathogens, with different infection characterizations—namely Alpha, Beta, Delta, Omicron BA.1, and Omicron BA.2 variants were introduced. In particular, COVID-19 variants are versions of the SARS-CoV-2 virus with specific genetic mutations. Some variants may exhibit changes in transmissibility, severity, and resistance to immunity. We used the results of [20] to adjust the transmissibility of the selected variant. Furthermore, to account for interactions with the environment outside the school, we considered external screening campaigns and infections due to personal interactions in other environments, such as at home. In particular, two different probabilities of external screening are applied to students every day: the probability of testing a student due to sporting activities that involve screening campaigns (e.g., playing soccer) and the probability of testing a student outside the school due to the presence of some symptoms.

The efficacy of a given contention policy strongly depends on the virus circulation. Modeling the prevalence of the infection in the general population and reflecting it on the spreading dynamics within the school required further modifications. Specifically, the model must account for the fact that any student will likely catch the infection outside the school. Therefore, we introduced the infection outside the school as a stochastic arrival process, so that the intensity of people getting the infection outside the school reflects the infection's prevalence in the population. In particular, a probability is applied to each susceptible agent every day, representing the possibility of infection outside the school (see Table 2).

## 3 Results

For studying the efficacy of NPIs and vaccination policies in schools, we considered both active and passive interventions: the former requires actions from the healthcare system, while the latter only requires the students to behave according to some specific indications. The active interventions consist of vaccinations and testing campaigns. While we considered as passive interventions social distancing as a common baseline behavior, rooms' air ventilation, and using masks—surgical or FFP2—and quarantine as possible contention policies.

In all scenarios, the principal and all teachers are vaccinated. We considered two screening policies, three quarantine policies, three mask policies, four levels of vaccine efficacy (100%, 70%, 50%, and 30%), four levels of student vaccination coverage (70%, 40%, 10%, and 0%), and three levels of virus circulation (low, medium, and high). Table 1 summarizes the analyzed scenarios, with 32 scenarios for each combination of students' vaccination coverage and virus circulation levels, resulting in a total of 384 scenarios. Specifically, we define vaccine efficacy as the fraction of the vaccinated population that is effectively immune to infection, while the remaining portion remains susceptible despite having been vaccinated. To react to the circulation of infection within the school, we considered two alternative screening policies: *i) none (NN)* where no screening policy is performed, and *ii) round robin (RR)* where the students in each classroom are divided into groups, ensuring that over a month, every student is tested, with one group being tested each week [21].

Once the infection has been discovered, three actions were considered: *no quarantine (NQ)*, *class quarantine (CQ)*, or *student quarantine (SQ)*. Specifically, NQ does not trigger any intervention when new infections are discovered, while CQ sparks a class quarantine only when the virus circulation within the classroom is over a threshold $n_q$, and $SQ$ sparks the quarantine of single infected students when we discover them. CQ and SQ are the quarantine policies used in the Piedmont Region in November/December 2021 [22] and January/February 2022, respectively. Specifically, an infected student in quarantine does not contribute to the infection spreading. Thus, using the CQ policy, if one or more infections are detected in a given classroom, the entire class is tested. The entire classroom goes into quarantine if the total number of infections reaches or exceeds $n_q$. If the threshold is not met, only the infected students are quarantined, while the remaining students are retested after $d_{test}$ days, following the same quarantine procedure described earlier. Eventually, we considered three cases for the use of masks: *i) no masks (NM)* where masks are not required, *ii) FFP2 when required (FWR)* and *iii) always on (AO)* requiring to wear no masks and surgical masks, respectively. If $n_m$ infections are detected in a classroom, these two policies mandate the use of FFP2 masks by all students and teachers in that classroom for $d_{ffp2}$ days.

Specifically, we considered eight clusters of results for each combination of vaccine efficacy, students vaccination coverage, and virus circulation levels (see Table 1): *i)* scenarios 1 to 4 represent our baseline, modeling a policy in which NPIs countermeasures are not implemented, *ii)* scenarios 5 to 8 account for screening policy RR and quarantine policy CQ, *iii)* scenarios 9 to 12 for quarantine policy CQ alone, *iv)* scenarios 13 to 16 for quarantine policy SQ alone, *v)* scenarios 17 to 20 for mask policy FWR and quarantine policy CQ, *vi)* scenarios 21 to 24 for mask policy FWR and quarantine policy SQ, *vii)* scenarios 25 to 28 for mask policy AO and quarantine policy CQ, and *viii)* scenarios 29 to 32 for mask policy AO and quarantine policy SQ.

Figures 3, 4 and 5 show the cumulative distribution function (CDF) of the total number of infected agents at the final time for different vaccine efficacy—decreasing moving from left to right—, several levels of students vaccination coverage—decreasing moving from top to bottom—and different NPIs combinations.

**Table 1.** Summary of the scenarios simulated for each students vaccination coverage (70%, 40%, 10% and 0%) and circulation level (low, medium and high).

| | Quarantine Policy | | | Screening Policy | | Mask policy | | | Vaccination Effectiveness | | | |
|---|---|---|---|---|---|---|---|---|---|---|---|---|
| | *NQ* | *CQ* | *SQ* | *NN* | *RR* | *NM* | *FWR* | *AO* | 30% | 50% | 70% | 100% |
| 1 | × | | | × | | × | | | × | | | |
| 2 | × | | | × | | × | | | | × | | |
| 3 | × | | | × | | × | | | | | × | |
| 4 | × | | | × | | × | | | | | | × |
| 5 | | × | | | × | × | | | × | | | |
| 6 | | × | | | × | × | | | | × | | |
| 7 | | × | | | × | × | | | | | × | |
| 8 | | × | | | × | × | | | | | | × |
| 9 | | × | | × | | × | | | × | | | |
| 10 | | × | | × | | × | | | | × | | |
| 11 | | × | | × | | × | | | | | × | |
| 12 | | × | | × | | × | | | | | | × |
| 13 | | | × | × | | × | | | × | | | |
| 14 | | | × | × | | × | | | | × | | |
| 15 | | | × | × | | × | | | | | × | |
| 16 | | | × | × | | × | | | | | | × |
| 17 | | × | | × | | | × | | × | | | |
| 18 | | × | | × | | | × | | | × | | |
| 19 | | × | | × | | | × | | | | × | |
| 20 | | × | | × | | | × | | | | | × |
| 21 | | | × | × | | | × | | × | | | |
| 22 | | | × | × | | | × | | | × | | |
| 23 | | | × | × | | | × | | | | × | |
| 24 | | | × | × | | | × | | | | | × |
| 25 | | × | | × | | | | × | × | | | |
| 26 | | × | | × | | | | × | | × | | |
| 27 | | × | | × | | | | × | | | × | |
| 28 | | × | | × | | | | × | | | | × |
| 29 | | | × | × | | | | × | × | | | |
| 30 | | | × | × | | | | × | | × | | |
| 31 | | | × | × | | | | × | | | × | |
| 32 | | | × | × | | | | × | | | | × |

Figure 3 shows the results obtained under a low virus circulation level. The first setting shows the baseline (quarantine NQ, screening NN, mask policy NM) and allows us to understand the effect of the vaccination alone. In the case of 100% vaccine efficacy, increasing the fraction of vaccinated students reduces the probability of encountering a high number of infected agents after 60 days. As expected, decreasing the vaccine efficacy increases the probability of a higher number of infected agents. Comparing this initial setting with the second one, we observe that the introduction of screening and quarantine policies (quarantine CQ, screening RR, mask policy NM) leads to a reduction in the probability of having a higher number of infections after 60 days. Also, it can be noted that vaccination campaigns covering less than 40% of the population provide poor results in terms of number of infections. It is important to notice that the most relevant results are obtained with higher vaccine efficacy: the testing process gets less effective as infections become more likely to happen, meaning that the virus spreads faster than the pace at which testing is performed.

**Table 2.** Model parameters.

| Parameter | Value |
|---|---|
| Number of classrooms | 12 |
| Students per classroom | 20 |
| Initial infected | 0 |
| Vaccine efficacy | 100 / 70 / 50 / 30% |
| Students vaccination coverage | 70 / 40 / 10 / 0% |
| Average number of days of incubation [23,24] | 3 |
| Average number of days of infection [25] | 7 |
| Duration of quarantine in days [22] | 10 |
| Virus variant | Omicron BA.1 |
| Ventilation | 3 ACH |
| Mask efficacy [20] | 59% (surgical) / 90% (FFP2) |
| Mask policy | NM / FWR / AO |
| Quarantine policy [22] | NQ / CQ / SQ |
| Screening adhesion | 100% |
| Screening policy [21] | NN / RR |
| Fraction of population wearing mask | 100% |
| Probability of going to the bathroom (per minute) | $8e^{-4}$ |

| Parameter | Value |
|---|---|
| Probability of going to the blackboard (per minute) | $6e^{-4}$ |
| Probability of going to the principal office (per minute) | $1.4e^{-4}$ |
| Probability of going into the hall during interval | 0.532 |
| First probability of external screening (per day) | 0.0075 |
| Second probability of external screening (per day) | 0.03 |
| Probability of outside contagion | 0.0004 (low circulation) / 0.008 (medium circulation) / 0.02 (high circulation) |
| Contamination risk ($c_r$) [17] | 0.024 |
| $d_{test}$ [22] (Section 3) | 5 |
| $d_{ffp2}$ (Section 3) | 10 |
| $n_q$ [22] (Section 3) | 5 |
| $n_m$ (Section 3) | 1 (FWR) / 5 (AO) |
| Duration of the lessons in minutes | 50 |
| Simulated days for each run | 60 |
| Number of runs for each scenario | 500 |

The last six settings illustrate the effects of the NM, FWR, and AO mask policies in combination with CQ and SQ quarantine policies, along with NN screening. Comparing the second setting to the third one (quarantine CQ, screening NN, mask policy NM) is clear that the screening RR has a negligible impact for the used quarantine policy. Focusing the attention on quarantine policies, we can state that CQ works much better than SQ, especially in the settings with FWR and AO mask policies. Regarding mask policies, we can conclude that the AO mask policy gives the best results (especially combined with the CQ quarantine policy). This is because the AO mask policy provides a boost in the protection of about 50% by forcing the use of FFP2 masks—see Table 2. In addition, opposite to what happens in the case of testing, the impact of the AO mask policy increases as the vaccine efficacy decreases. Indeed, despite how fast the infection spreads, masks always provide the same level of protection.

Figures 4 and 5 show the same eight settings under a medium and high virus circulation level. Once again, we find that increasing the fraction of vaccinated students reduces the probability of observing a high number of infected agents after 60 days while decreasing vaccine efficacy increases this probability. These results indicate that containing the spread is more challenging under medium and high virus circulation levels compared to low virus circulation level. However, a combination of masks and quarantine policies can effectively reduce the final number of infected agents. Moreover, Fig. 6 displays the area under the CDF (normalized between 0 and 1) for each scenario, making it easier to compare the effects across different levels of virus circulation.

Figures 7, 8 and 9 illustrate the distribution of infections over time and the average number of infections. Representing the evolution of the infection distribution allows us to understand how interventions contribute to containing the

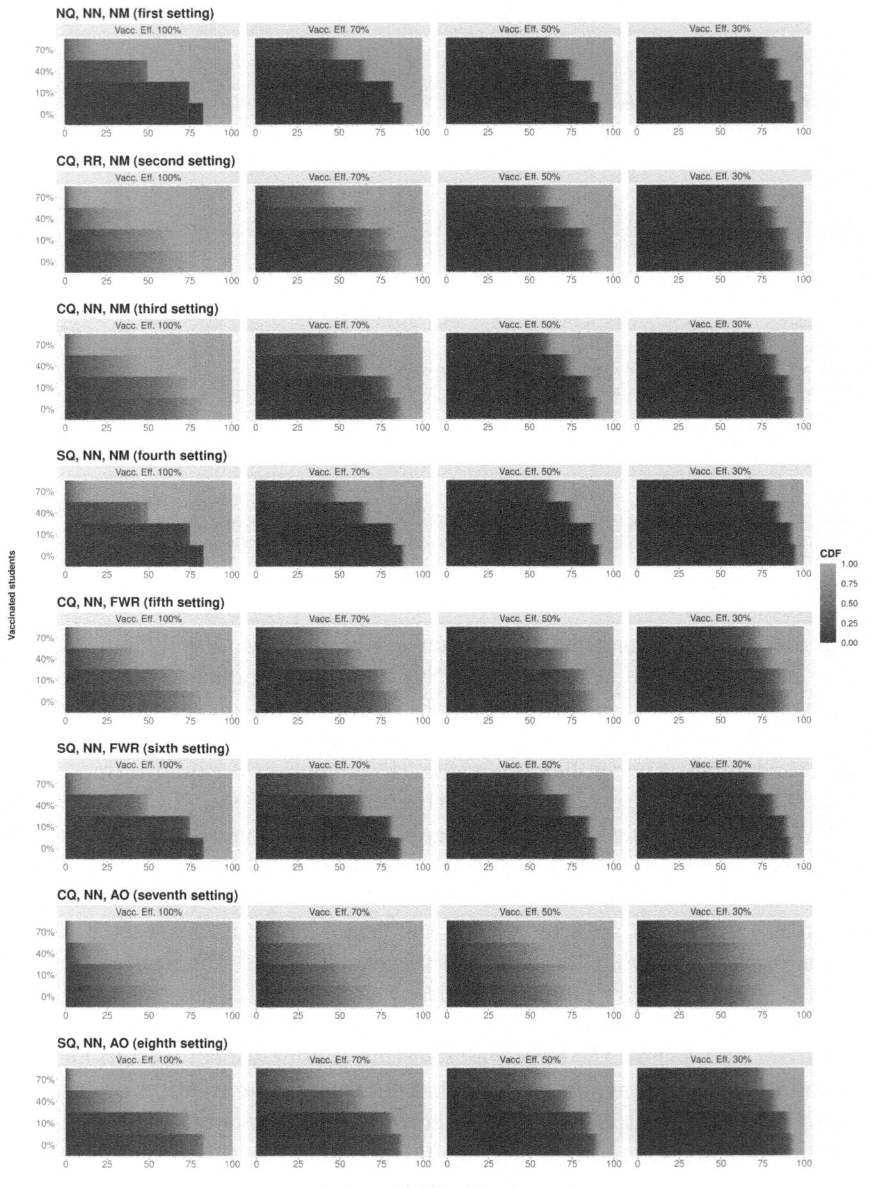


**Fig. 3.** Cumulative distribution function (CDF) of the total number of infected agents at the final time for each considered scenario under a low virus circulation level. For each scenario—identified by the setting (on the rows), the vaccine efficacy (on the columns), and the fraction of vaccinated students (on the rows of each subplot)—we run 500 simulations—see Table 2. Lighter CDFs indicate more effective countermeasures. The first, second, and seventh settings are the most noteworthy. The first setting (NQ, NN, NM) shows that higher infections are associated with lower vaccine efficacies and lower fractions of vaccinated students. The second setting (CQ, RR, NM) shows that a fraction of vaccinated students $< 40\%$ provides poor results in terms of levels of protection—the testing process gets less effective as infections become more likely to happen. Finally, the seventh setting (CQ, NN, AO) shows that masks are more effective with lower vaccine efficacies—masks always provide the same level of protection.

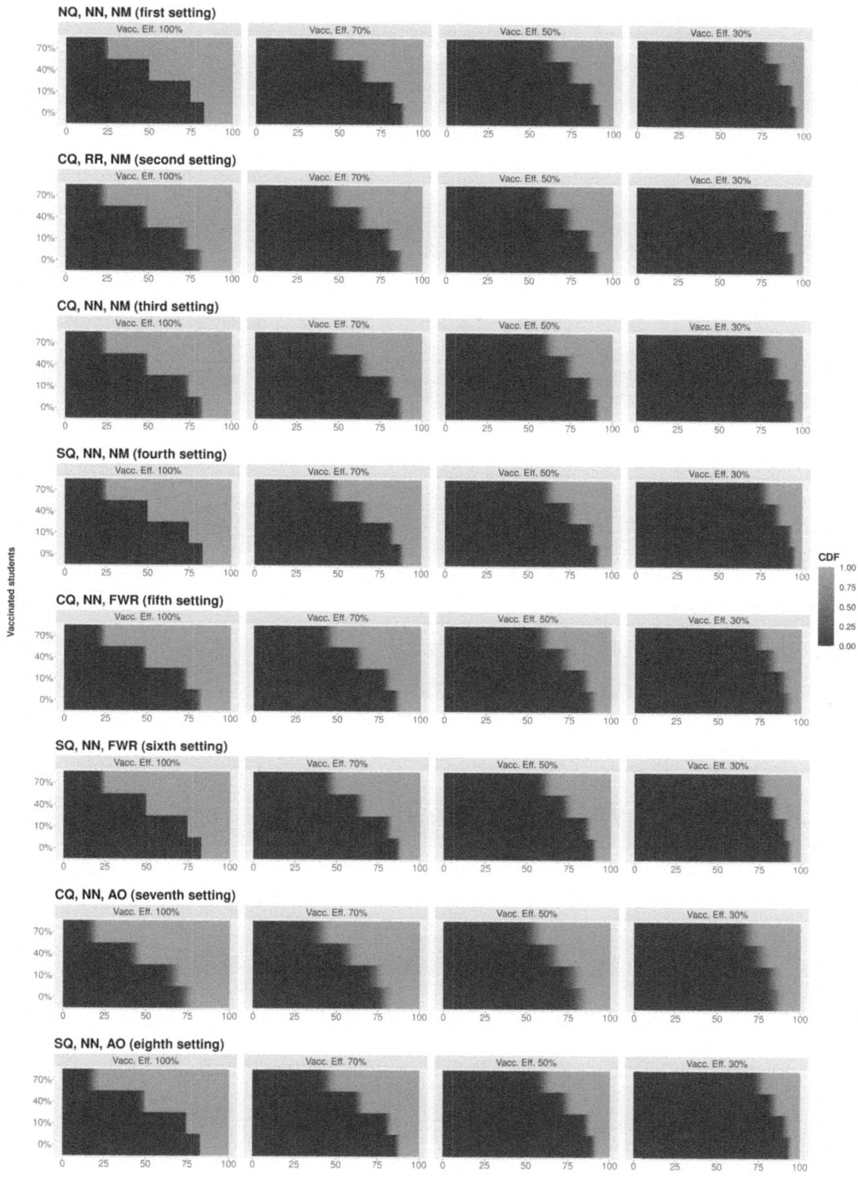


**Fig. 4.** Cumulative distribution function (CDF) of the total number of infected agents at the final time for each considered scenario under a medium virus circulation level. For each scenario—identified by the setting (on the rows), the vaccine efficacy (on the columns), and the fraction of vaccinated students (on the rows of each subplot)—we run 500 simulations—see Table 2. Lighter CDFs indicate more effective countermeasures. We can state that higher infections are associated with lower vaccine efficacies and lower fractions of vaccinated students and a combination of masks and quarantine policies allows to reduce the final number of infected agents.

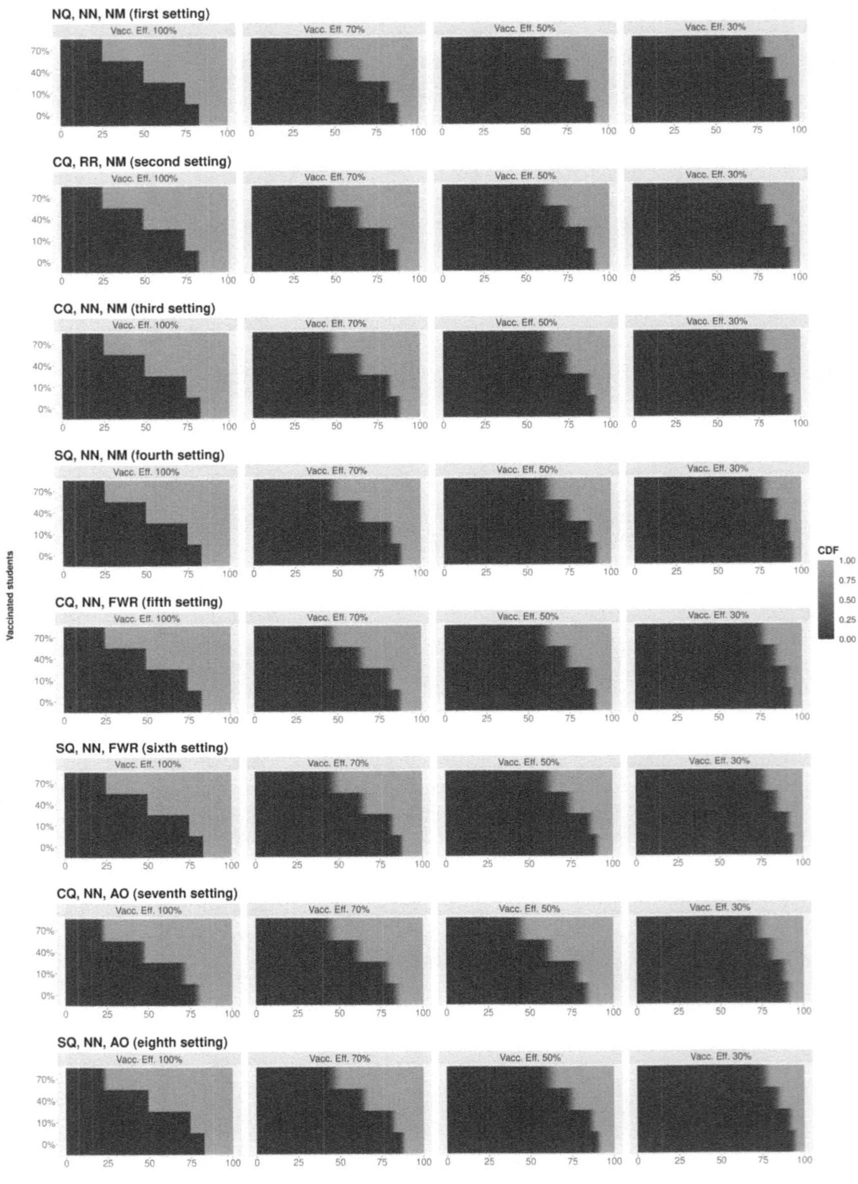


**Fig. 5.** Cumulative distribution function (CDF) of the total number of infected agents at the final time for each considered scenario under a high virus circulation level. For each scenario—identified by the setting (on the rows), the vaccine efficacy (on the columns), and the fraction of vaccinated students (on the rows of each subplot)—we run 500 simulations—see Table 2. Lighter CDFs indicate more effective countermeasures. We can state that higher infections are associated with lower vaccine efficacies and lower fractions of vaccinated students and a combination of masks and quarantine policies allows to reduce the final number of infected agents.

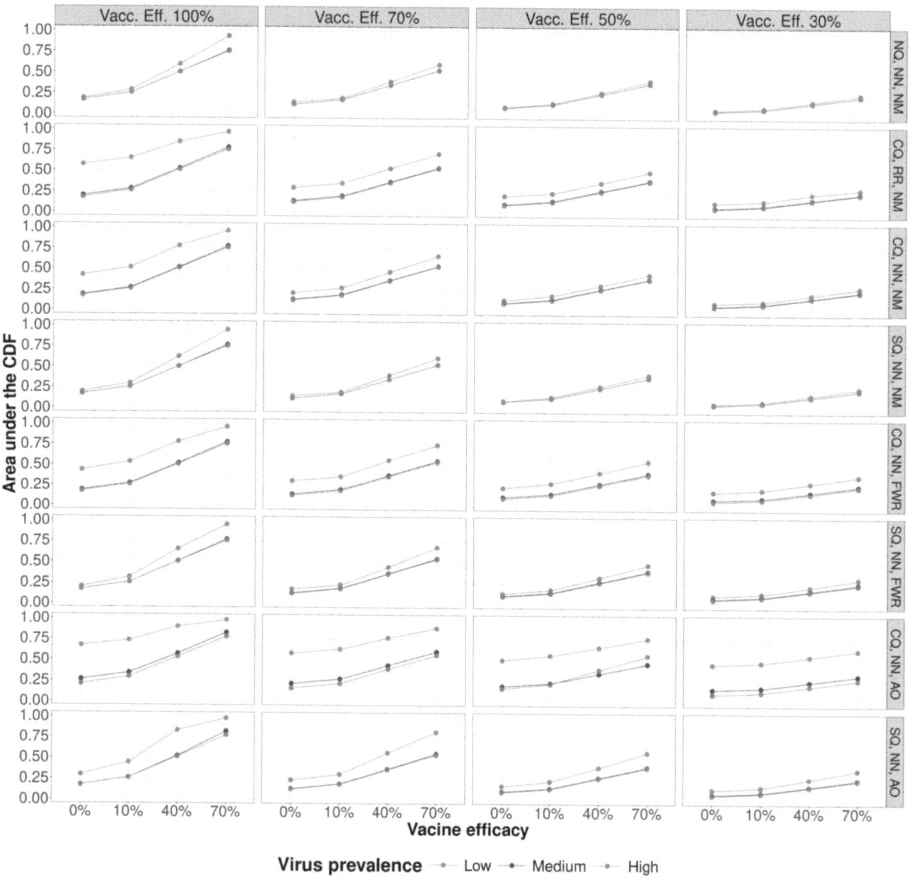


**Fig. 6.** Area under the CDFs—normalized between 0 and 1—showed in Fig. 3—low virus circulation, in red—, Fig. 4—medium virus circulation, in green—, and Fig. 5—high virus circulation, in blue. For each scenario—identified by the setting (on the rows), the vaccine efficacy (on the columns), the fraction of vaccinated students (on the columns of each subplot), and the virus circulation level (distinguished by the colors)—, the values represent the effectiveness of the implemented countermeasures in the considered scenario. Higher values indicate greater effectiveness of the countermeasures. Virus prevalence appears to influence the area under the CDF differently across the scenarios, with low prevalence often having higher values compared to medium and high prevalence, particularly in scenarios with higher vaccine efficacy.

spread of the infection and, more significantly, how they shift the probability of a negative outcome towards a more convenient one in the presence of different probability masses. In particular, Fig. 7a clearly shows two main outcomes under a low virus circulation level: the infection fades or explodes in a major outbreak. Such behavior is easily observed in the violin plot, where the probability mass gets either at zero or between the susceptible population size and half of the sus-

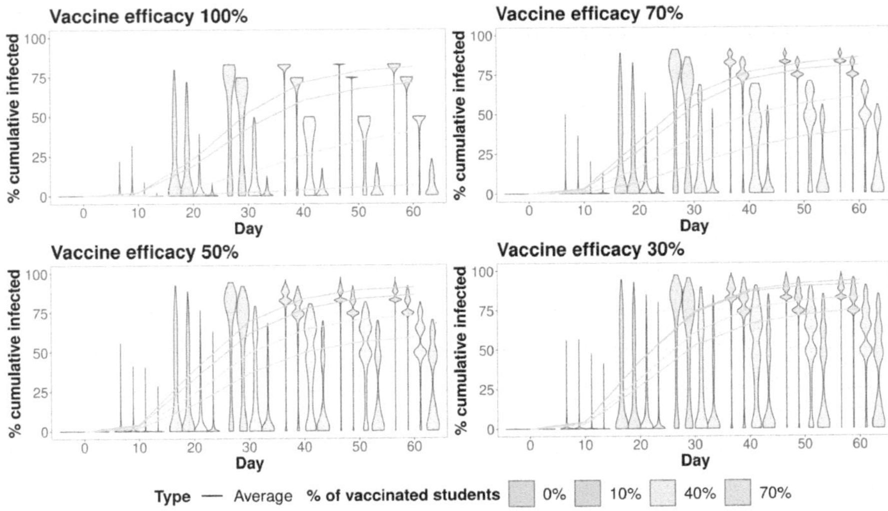


(a) First setting (NQ, NN, NM). Various probability masses emerge with two main outcomes: the infection fades or explodes in a major outbreak.

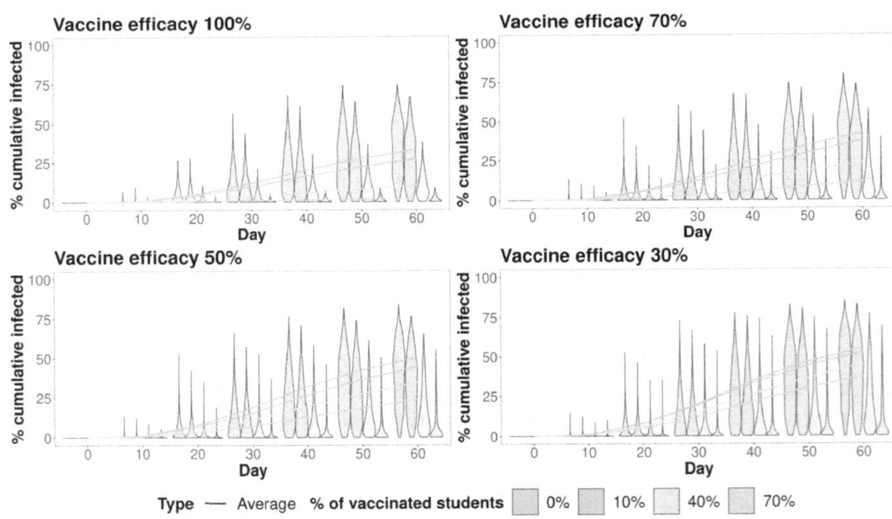


(b) Seventh setting (CQ, NN, AO). The outcome leading to a major outbreak is no longer present: masks and quarantine slow down the infection and contain the risk of a major outbreak.

**Fig. 7.** Distributions of infections over time and the average number of infections for each considered scenario in the first and seventh settings under a low virus circulation level. For each scenario—the sub-figures illustrate two distinct settings, with each subplot representing a different vaccine efficacy level and the varying colors in each subplot correspond to different fractions of vaccinated students—we run 500 simulations—see Table 2.

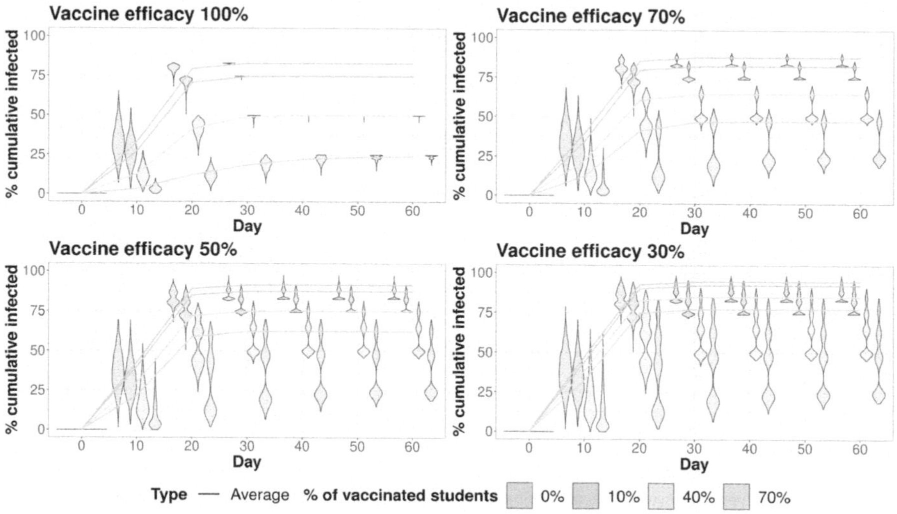


(a) First setting (NQ, NN, NM).

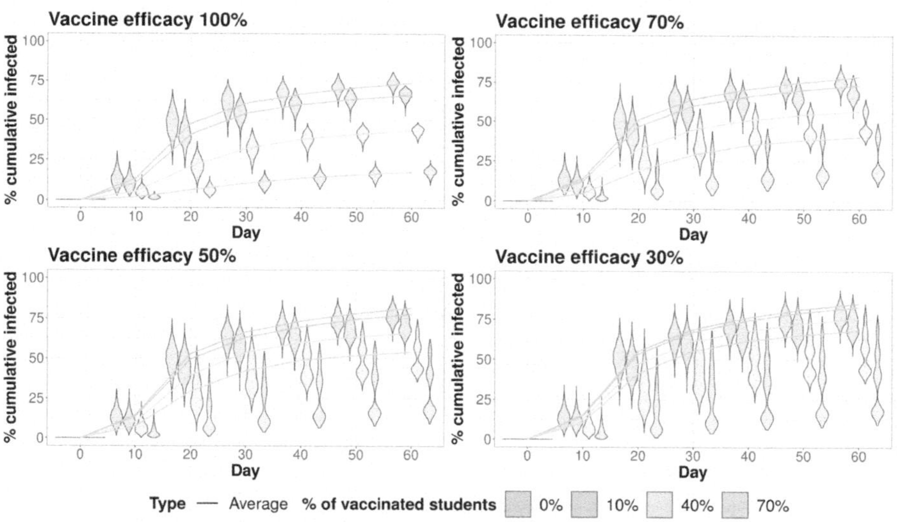


(b) Seventh setting (CQ, NN, AO).

**Fig. 8.** Distributions of infections over time and the average number of infections for each considered scenario in the first and seventh settings under a medium virus circulation level. For each scenario—the sub-figures illustrate two distinct settings, with each subplot representing a different vaccine efficacy level and the varying colors in each subplot correspond to different fractions of vaccinated students—we run 500 simulations—see Table 2. The vanishing of the infection is no longer present and different probability masses emerged. The most favorable outcome is represented by the probability mass with the lowest mean. A combination of masks and quarantine policies allows to reduce the final number of infections.

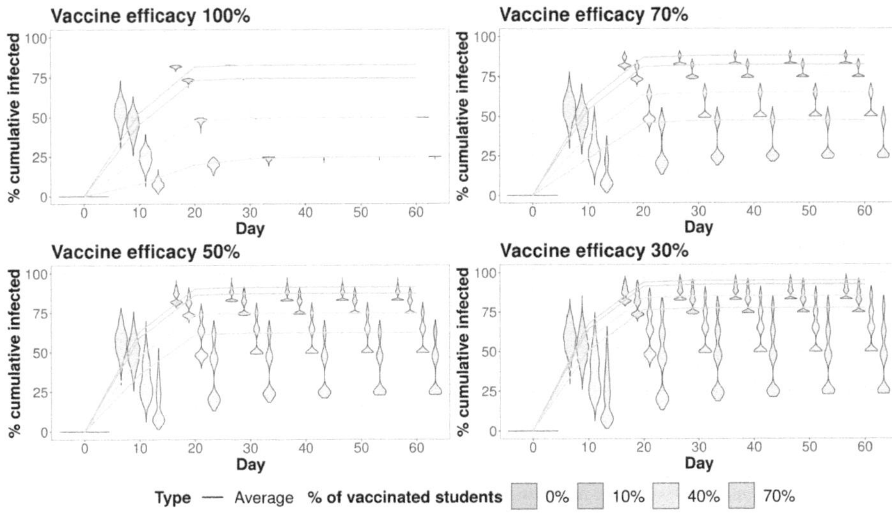


(a) First setting (NQ, NN, NM).

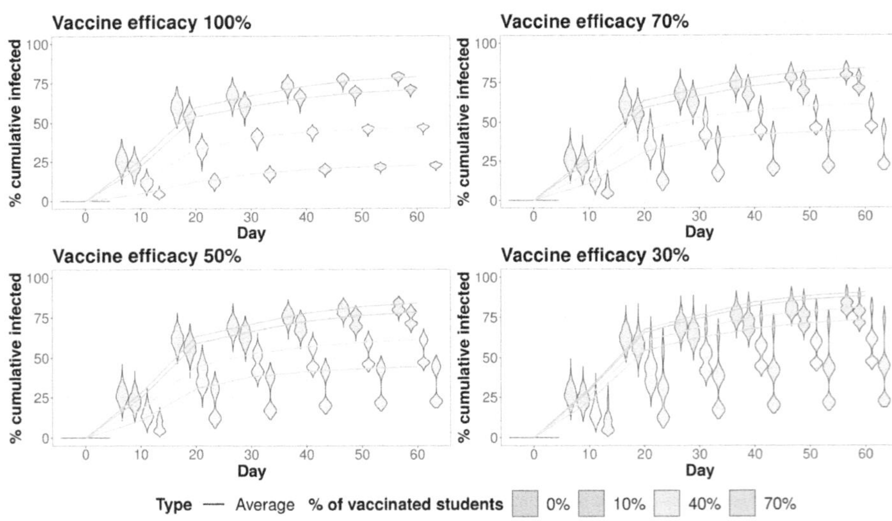


(b) Seventh setting (CQ, NN, AO).

**Fig. 9.** Distributions of infections over time and the average number of infections for each considered scenario in the first and seventh settings under a high virus circulation level. For each scenario—the sub-figures illustrate two distinct settings, with each subplot representing a different vaccine efficacy level and the varying colors in each subplot correspond to different fractions of vaccinated students—we run 500 simulations—see Table 2. The vanishing of the infection is no longer present and different probability masses emerged. The most favorable outcome is represented by the probability mass with the lowest mean. A combination of masks and quarantine policies allows to reduce the final number of infections.

ceptible population size—intended as the number of agents not covered by the vaccine immunity. The impact of introducing the AO mask policy and the CQ quarantine policy is clear when we consider Fig. 7b. Indeed, here the probability mass is shifted towards a lower number of infections at each time point. Also, the outcome leading to a major outbreak is not present anymore. This result clearly shows the efficacy of the NPIs countermeasure: although masks and quarantine are not enough to stop the spread of the infection they allow to slow it down and contain, if not avoid, the risk of a major outbreak under a low virus circulation level.

Figures 8a, 8b, 9a and 9b show the possible outcomes under a medium and high virus circulation level. In particular, it is straightforward that the best outcome, i.e., the vanishing of the infection, is no longer present. With a vaccine efficacy less or equal to 70% different probability masses emerge from the results. In particular, the best outcome is represented by the probability mass with the lowest mean. From these figures, it is clear that a combination of masks and quarantine policies allow to reduce the final number of infections. Specifically, the CQ quarantine policy and the AO mask policy represent the best outcome also under a medium and high virus circulation level. More generally, even if we no longer have the extinction of the infection, the use of NPIs makes it possible to slow down its spread. We can conclude by observing that NPIs are effective when they manage to outrun the infection, which is why increasing virus circulation decreases effectiveness.

## 4 Conclusion

We developed an ABM [15] to study the evolution of SARS-CoV-2 spread in a school. We used this model to showcase different NPI approaches to contain the spreading of the virus. Specifically, we considered a baseline setting where no interventions are implemented. That allows us to better understand what happens when NPIs are introduced. In particular, we compared the effects of wearing masks all the time and the implementation of a testing protocol, using two quarantine policies used in the Piedmont region of Italy in 2021 and 2022.

The computational modeling allowed us to pinpoint several useful takeaways for policymakers. First of all, vaccination campaigns covering less than 40% of the susceptible population provide poor results in protecting from a major outbreak. Furthermore, we showed that testing campaigns are effective as far as the testing process is faster than the virus diffusion. On the other hand, wearing masks properly is an effective countermeasure also when the infection speeds up.

Eventually, the introduction of an infection external to the school allows us to consider different settings of infection prevalence—where in this work we consider three different levels of infection's prevalence in the general population. In particular, in a setting with a medium or high circulation of the virus, the results show that it is more challenging to contain the spread for a low virus circulation level, but a combination of masks and quarantine policies allows to reduce the final number of infections.

NPIs are useful for infection, especially in the early stages and can help to keep activities open even in scenarios with higher circulation, but they must be supported by other public health instruments.

**Acknowledgments.** DB is a Ph.D. student enrolled in the National Ph.D. in Artificial Intelligence, XXXVII cycle, course on health and life sciences, organized by Università Campus Bio-Medico di Roma.

**Funding Information.** This work was supported by grants from "Creation of a computational framework to model and study West Nile Disease" (Cod. ROL 67410, PI Marco Beccuti), and "Ripresa delle attività socio-economiche e delle scuole: modelli per la progettazione e supporto di linee guida per la convivenza con il Covid-19" (Cod. ROL 73459, 2020, PI Matteo Sereno), projects funded by CRT foundation. This work is part of the project TrustAlert which has received funding from the Fondazione Compagnia San Paolo and Fondazione CDP under the "Artificial Intelligence" call.
**Conflict of interests.** All authors declare that they have no conflicts of interest.

**Availability of data and software code.** The ABM developed for the analysis presented in this work and the configurations presented above are available at https://github.com/qBioTurin/epischool-abm/tree/CIBB.

## References

1. Eales, O., et al.: Dynamics of competing SARS-CoV-2 variants during the Omicron epidemic in England. Nat. Commun. **13**(1), 4375 (2022)
2. Watson, O.J., Barnsley, G., Toor, J., Hogan, A.B., Winskill, P., Ghani, A.C.: Global impact of the first year of COVID-19 vaccination: a mathematical modelling study. Lancet Infect. Dis. **22**(9), 1293–1302 (2022)
3. Perra, N.: Non-pharmaceutical interventions during the COVID-19 pandemic: a review. Phys. Rep. **913**, 1–52 (2021)
4. Pernice, S., et al.: Impacts of reopening strategies for COVID-19 epidemic: a modeling study in Piedmont region. BMC Infect. Dis. **20**, 1–9 (2020)
5. Wells, C.R., et al.: Optimal COVID-19 quarantine and testing strategies. Nat. Commun. **12**(1), 356 (2021)
6. Villers, J., et al.: SARS-CoV-2 aerosol transmission in schools: the effectiveness of different interventions. Swiss Med. Wkly, **152**(2122), w30178 (2022)
7. Gkantonas, S., Zabotti, D., Mesquita, L.C.C., Mastorakos, E., de Oliveira, P.M.: airborne. cam: a risk calculator of SARS-CoV-2 aerosol transmission under well-mixed ventilation conditions (2021)
8. Kerr, C.C., et al.: Covasim: an agent-based model of COVID-19 dynamics and interventions. PLOS Comput. Biol. **17**(7), e1009149 (2021)
9. Woodhouse, M.J., Aspinall, W.P., Sparks, R., Brooks-Pollock, E., Relton, C.: Alternative Covid-19 mitigation measures in school classrooms: analysis using an agent-based model of SARS-CoV-2 transmission. Roy. Soc. Open Sci. **9**(8), 211985 (2022)
10. Fouad, A.A., Antably, A.E.: Agent-based modeling and simulation of pandemic propagation in a school environment. Int. J. Architectural Comput. **21**(1), 120–135 (2023)

11. Faucher, B., et al.: Agent-based modelling of reactive vaccination of workplaces and schools against COVID-19. Nat. Commun. **13**(1), 1414 (2022)
12. Lasser, J., Sorger, J., Richter, L., Thurner, S., Schmid, D., Klimek, P.: Assessing the impact of SARS-CoV-2 prevention measures in Austrian schools using agent-based simulations and cluster tracing data. Nat. Commun. **13**(1), 554 (2022)
13. Lorig, F., Johansson, E., Davidsson, P.: Agent-based social simulation of the Covid-19 pandemic: a systematic review. J. Artif. Soc. Soc. Simul. **24**(3), 5 (2021)
14. Tatapudi, H., Das, T.K.: Impact of school reopening on pandemic spread: a case study using an agent-based model for COVID-19. Infect. Dis. Model. **6**, 839–847 (2021)
15. Baccega, D., et al.: An agent-based model to support infection control strategies at school. J. Artif. Soc. Soc. Simul. **25**(3), 2 (2022)
16. Uri Wilensky. NetLogo (1999). http://ccl.northwestern.edu/netlogo/
17. Hoertel, N., et al.: A stochastic agent-based model of the SARS-CoV-2 epidemic in France. Nat. Med. **26**(9), 1417–1421 (2020)
18. Buonanno, G., Stabile, L., Morawska, L.: Estimation of airborne viral emission: quanta emission rate of SARS-CoV-2 for infection risk assessment. Environ. Int. **141**, 105794 (2020)
19. Watanabe, T., Bartrand, T.A., Weir, M.H., Omura, T., Haas, C.N.: Development of a dose-response model for SARS coronavirus. Risk Anal. Int. J. **30**(7), 1129–1138 (2010)
20. Peng, Z., et al.: Practical indicators for risk of airborne transmission in shared indoor environments and their application to COVID-19 outbreaks. Environ. Sci. Technol. **56**(2), 1125–1137 (2022)
21. Baccini, M., Cereda, G.: Screening plans for SARS-CoV-2 based on sampling and rotation: an example in a European school setting. PLOS ONE **16**(9), 1–13 (2021)
22. Piemonte, R.: Scuola: operative entro il 17 novembre le nuove regole per le quarantene. https://www.regione.piemonte.it/web/pinforma/notizie/scuola-operative-entro-17-novembre-nuove-regole-per-quarantene
23. Zaki, N., Mohamed, E.A.: The estimations of the COVID-19 incubation period: a scoping reviews of the literature. J. Infect. Public Health **14**(5), 638–646 (2021)
24. World Health Organization. Coronavirus disease 2019 (COVID-19): situation report, 73. Technical documents (2020)
25. Byrne, A.W., et al.: Inferred duration of infectious period of SARS-CoV-2: rapid scoping review and analysis of available evidence for asymptomatic and symptomatic COVID-19 cases. BMJ Open, **10**(8) (2020)

# Stochastic Modeling and Dosage Optimization of a Cancer Vaccine Exploiting the EpiMod Framework

Marco Beccuti[1], Giuliana Franceschinis[2,3], Marzio Pennisi[2,3](✉), Simone Pernice[2], and Irene Terrone[1]

[1] Department of Computer Science, University of Turin, Turin, Italy
[2] Computer Science Institute, DiSIT, University of Eastern Piedmont, Alessandria, Italy
[3] Laboratorio Integrato di Intelligenza Artificiale e Informatica in Medicina DAIRI, AO AL - DISIT, UPO, Alessandria, Italy
marzio.pennisi@uniupo.it

**Abstract.** Ordinary Differential Equations (ODEs) and Agent-Based Models (ABMs) represent nowadays the two main approaches for Immune System (IS) modeling. While the former approach does not allow for representing aleatory variations, the latter lacks a clear well-defined semantics, entailing possible biases on simulation results. We present here the application of our modeling pipeline, that has been designed to cope with these shortcomings, to a case-study about the competition between cancer and IS under the administration of a pre-clinical vaccine in transgenic mice. The pipeline involves the use of Extended Stochastic Symmetric Nets (ESSN) for a formal definition of the conceptual model, and allows to study the domain problem from a macro-perspective by means of the Stochastic Simulation Algorithm (SSA) or from a micro-perspective through an Agent Based Model with a clear defined semantics. The numerical results obtained in this study using SSA are presented and global sensitivity analysis is performed using Latin Hypercube Sampling - Partial Rank Correlation Coefficients (LHS-PRCC) to analyze and improve vaccine dosages and timings.

**Keywords:** Cancer · Immune System · Modeling · Stochastic Simulation · Petri Net · LHS-PRCC

## 1 Introduction

Mechanistic modeling and simulation have been applied in recent years for the simulation of complex systems. Their use is highly recommended in many fields, including immunology, where the scarcity of data coming from reduced sets of experiments can be compensated for by basic knowledge and laboratory observations that describe how the system works. In this scenario, a mechanistic description of the immunological landscape can usually be coupled with some quantitative information to create a knowledge-driven simulation model capable


M. Vettoretti et al. (Eds.): CIBB 2023, LNBI 14513, pp. 162–176, 2025.
https://doi.org/10.1007/978-3-031-90714-2_12

of representing the temporal evolution of the real scenario. These knowledge-driven models can be used to verify or confute hypotheses about how the system works, to verify *in silico* the effects of novel treatments, and to optimize drug dosages and timing.

For years, systems of ordinary and partial differential equations (ODE and PDE) have been the main modeling techniques for these immunological models, but more recently the need to move towards stochastic approaches has become increasingly evident. Despite the need to describe the difference among individuals that can be however encoded by means of a vector of features to uniquely characterize specific patients, the presence of aleatory uncertainty can be represented by stochastic simulations only. Deterministic models work, in fact, by providing mean behaviors, and thus any possibly significant deviation from the mean value is lost. Furthermore, ODE and PDE based models are not easily understandable by clinicians and involve a description from a macroscopic point-of-view (global behavior or the population) with some approximations that are not always representative of the real biological scenario. Stochastic models based for example on ABMs or stochastic Petri nets can instead be used to represent such stochastic variations by executing multiple simulations with different random seeds, thus allowing to capture possible deviations from the mean value. In particular, ABMs resulted in being particularly suitable for IS related problems, thanks to their ability to give a detailed bottom-up description of the immunological events and interactions that may occur, and also account for stochastic uncertainty.

However, in general, ABMs lack clear well-defined semantics, so the use of different approaches, techniques, and frameworks may lead to different results, e.g. Anylogic [7] adopts a continuous-time dynamics approach, while most NetLogo models [19] are implemented with a discrete-time approach. Furthermore, ABM models may be affected by unconscious biases introduced by the programmers and the modelers at the time of definition of the models, e.g., on the order of actions and/or interactions that the agents may carry on. Just to give an example, in [9] it was presented that, starting from the same conceptual model for the Cancer-Immune System competition, an ABM implemented in AnyLogic leads to different results with respect to classical Ordinary Differential Equations (ODE) or Stochastic Simulation Algorithm (SSA). We showed in [3] that the same model implemented with an ABM with clear and well-defined semantics leads instead to identical results.

Finally, as in agent-based approaches, all agents are generally followed and stored individually, and their application to large-scale simulations may be computationally challenging.

In this work, we re-analyse the effects of a cancer immunoprevention vaccine against mammary carcinoma by using a modeling pipeline that starts with the formal definition of the conceptual model through the use of a Petri Net extension named Extended Stochastic Symmetric Nets (ESSN). From such a well-defined model, it is possible to analyze the results from a macro (top-down) or a micro (bottom-up) perspective exploiting, upon necessity, the use of SSA (presented here) and/or an Agent Based Model (as future work) but with a clearly defined semantics as presented in [3,5]. The proposed modeling pipeline offers several advantages over previous approaches present in the literature that utilize ODE models [4,6,12,14] and agent-based models (ABM) [15,16]. Specifically, it minimizes the potential for errors during the conceptual model definition, facilitates communication with domain experts by employing graphical formalism rather than equations, and allows for the simulation of system evolution at both macroscale (using ODE or SSA) and microscale (using ABM), depending on the requirements.

# 2 Methods

## 2.1 Clinical Background

The pre-clinical *in vivo* experiment has been designed on BALB-neuT virgin female mice which over-express the activated rat HER-2/neu oncogene. Such mice start to produce cancer cells starting from 6 weeks of age. This causes the onset of carcinoma-in-situ (CIS) formation (a conglomerate of cancer cells that duplicate in the same location) that will eventually lead to the formation of solid tumors shortly after. To avoid CIS formation an immunopreventive vaccine named Triplex [11] has been developed. The vaccine takes into account three different stimuli i.e., p185neu oncoantigen that characterizes the tumor and is expressed on cancer cells' surface, H-2q Major Histocompatibility Complex (MHC) molecules that favor recognition of such vaccine cells from Immune System (IS) cells, and interleukin-12 to boost IS functions. For testing the efficacy of the Triplex Vaccine, the pre-clinical experiment contemplated various mice groups with diverse administration protocols. One group was treated with a Placebo solution (the control untreated group); the other groups with the Triplex vaccine, but with different administration schedules. Among these, we recall here the Early vaccination protocol, that is composed of twice-weekly intra-peritoneum vaccinations repeated for 2 weeks followed by 2 weeks of rest. This regimen is repeated for 3 times. The Early protocol demonstrated unable to prevent CIS formation at the end of the experiment, even if a significant delay in respect to the untreated protocol was observed. The Chronic vaccination protocol that uses the same regimen as the Early protocol, but repeated until the end of the experiment, demonstrated the ability to prevent the onset of mammary carcinoma. Other tested protocols demonstrated much less effective, with little or no gain in efficacy when compared to untreated control mice [11].

## 2.2 Modeling Pipeline

The first step of our modeling pipeline is represented by the drawing of the conceptual model exploiting Extended Stochastic Symmetric Nets (ESSN) [18]. We will briefly sketch the main concepts of ESSN, leaving the reader the possibility to explore the topic in detail in [18]. As an extension of Petri Nets (PNs), ESSNs are a graphical formalism composed of bipartite graphs whose nodes are *places* (represented by circles) and *transitions* (represented by rectangles). Places are used to describe a local state or aspect of a system or an entity; transitions represent events that modify the current state of the system. Places are connected to transitions by directed annotated *arcs*. While places, transitions and arcs give a qualitative description of the net structure, *tokens* contained into places give a quantitative description of the PN (its *marking*). Arcs may connect a place to a transition to indicate that the event associated with the transition consumes some tokens from an input place, or a transition to a place to describe that the event associated with the transition produces some tokens into the output place. In general, a transition is enabled if the number of required tokens of the transition is lower than or equal to the number of tokens in all the input places, and thus may occur (fire) according to a given rate and law associated with the transition. If we suppose to use places to represent entity types and/or sub-types, tokens to represent individual entities, and transitions to represent interactions among entities, the previous statement can be seen in our case as follows: if the constraints on the number of involved entities (i.e., tokens) required for a given interaction (i.e., transition) are satisfied, the interaction may occur (i.e., the transition is enabled), entering in competition with other enabled interactions (each one with a given associated firing rate). Clearly, only one interaction can occur (fire) at each time, so only one interaction will be randomly selected with a probability that is proportional to the transition rate (i.e., the higher is the rate, the higher will be the probability of being chosen).

Furthermore, in ESSN there are colors classes, i.e. elementary types that represent finite and disjoint sets that are used to define specific properties that tokens may have. A color domain, defined as the Cartesian product of *color classes*, is associated with each place, to specify that it can contain tokens belonging to a given color domain. Similarly, a color domain is also associated with transitions, defining which types of tokens can be consumed and/or produced. It is also possible to define constraints on transitions by using *Guards*, i.e. logical expressions defined on the color domain of transitions. Finally, for each transition it is possible to define a rate representing the parameter of the exponential distribution used for modeling its random firing time. ESSN also allow defining semantic rates using general real functions. The use of stochastic firing delays sampled from negative exponential distributions allows us to automatically derive the underlying stochastic process defined as a Continuous-Time Markov Chain (CTMC) [1]. The CTMC provides a structured way to capture the probabilistic behavior of the system by defining how likely it is for the system to move from one marking of the ESSN to another at any given time. This derivation leads to the related Master Equations, which describe how the probabilities of being in each marking evolve over time. These equations are crucial for quantitatively evaluating the system's behavior. However, for complex systems, the Master Equations are often computationally intractable, thus the evaluation of the system relies on other techniques, such as the Stochastic Simulation Algorithm (SSA) [10]. which simulates the exact sequence of events in the system based on the probabilistic rules defined by the CTMC. This methodology is particularly useful for large scale simulations because allows us to observe the system from a macroscopic perspective. This kind of models can be designed exploiting GreatSPN 3 [2] and Epimod framework [8,17].

Another option we will not explore here is given by the automatic translation (implemented into GreatSPN 3) of a properly annotated ESSN into an ABM with precisely defined semantics [3,5]. The translated model is in NetLogo ABM format [19] and can be more convenient to observe the system from a microscopic scale allowing to define, even at later times and directly inside the NetLogo code, more complex agent behaviors, functions, and/or individual-centered measures. The two approaches can be in principle combined to achieve multi-scale descriptions of the systems under study, using the micro-scale ABM representation to accurately simulate how the system evolves in small (and possibly highly heterogeneous) scenarios, and simulating by means of SSA (mostly homogeneous) large-scale systems. We note here that within GreatMod also ODE generation and deterministic simulation are available. In Fig. 1 we summarize the proposed modeling pipeline.

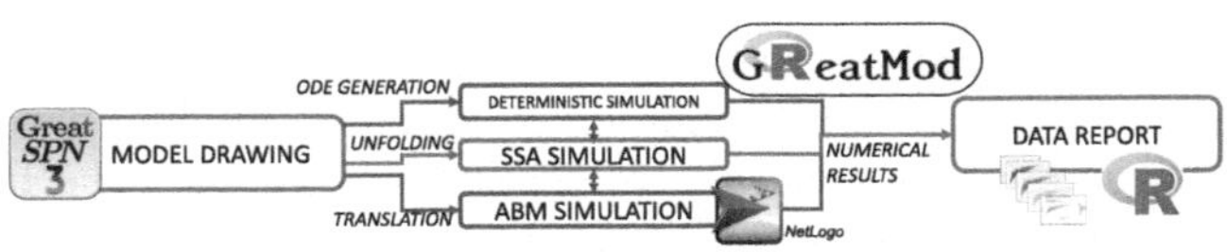


**Fig. 1.** From the definition of the model using ESSN (GreatSPN 3) it is possible to either move towards SSA or ODE simulation with GreatMod, or ABM simulation with Netlogo. Data report and analysis is obtained thanks to GreatMod R modules. In particular, for the study we used the data visualization and LHS-PRCC analysis modules. In red the path followed in the present article. (Color figure online)

## 2.3 Model Description

The ESSN model we present in Fig. 2 is inspired by the equations in [6]. In the following, we briefly describe the presented net.

The model is composed of various sub-nets, each one identifying a different entity type: Vaccine Cells (VC), Cancer Cells (CC), Antibodies (AB), Plasma B cells (B), Tumor Associated Antigens (TAA), Antigen-Presenting Cells (APC), Cytotoxic TC cells (TC), Interleukin-2 (IL2), Interleukin-12 (IL-12) and Helper T cells (TH).

**VCsubnet.** VC are injected by following a pre-defined vaccination protocol (e.g., Chronic or Early) at given times (Injection) and disappear from the system due to natural death (DieVC), killing by TC (IntVC_TC) or AB (IntVC_AB). Because of their death they release into the system a given quantity of TAA and IL12 (transitions DieVC, IntTC_VC, IntVC_AB).

**TAAsubnet.** TAA are introduced as a consequence of the death of VC (transitions DieVC, IntVC_TC, IntVC_AB) or CC (DieCC, IntTC_CC, IntCC_AB), can disappear for natural degradation (DieTAA), for binding with AB (TAA_AB), or for being phagocyted by APC (TAA_APC).

**APCsubnet.** APCs identify a broad class of cells including B cells, dendritic cells and macrophages that can capture antigens, being active and present such antigens to TH. We do not explicitly model this activation process, but we consider the introduction of activated APC by considering TAA presence with transition APC_TAA. Furthermore, APC may die for a natural death (DieAPC), may consume TAA (TAA_APC), may activate TH (transition TH_APC).

**ABsubnet.** ABs are produced by plasma B cells (IntAB_B), can die for natural degradation (DieAB), or formation of immunocomplexes with TAA (TAA_AB), thus they can bind to and eventually kill VC and CC cells (IntVC_AB,IntCC_AB, IntAB_CC), disappearing from the system. Note here that IntAB_CC IntCC_AB are separate events, the first one representing the binding of an AB to a CC that however remains alive, and the second one representing the killing of a (possibly already partially bound) CC.

**CCsubnet:** CC can die for a natural death (DieCC) or because they are killed by TC or AB (IntTC_CC, IntCC_AB). During their death they release a given quantity of TAA. CC can also duplicate (GrowCC) and appear due to the transgenic nature of BALB-neuT mice (GrowCC2).

**Bsubnet.** B cells can produce antibodies (IntAB_B), die (DieB) and absorb IL2 to eventually boost their duplication (IntIL2_B and IntB_IL2). Also here IntIL2_B and IntB_IL2 represent two asymmetric events, one for accounting of consumption only, and the other for cell duplication, just because more IL2 may be needed for B duplication. Finally, they can be introduced into the system according to the presence of TH cells that activate naive B cells (intTH_B).

**TCsubnet.** Active TC are introduced on the basis of VC thanks to the presence of their allogenic MHC molecules (intTC_VC). Once activated they can kill VC (intVC_TC). TC can die for natural death (DieTC), can absorb and eventually duplicate thanks to IL2 (IntIL2_TC and IntTC_IL2), can kill CC cells (IntCC_TC) and absorb IL12 molecules for remaining active (IntIL12_TC). Also here IntIL2_TC and IntTC_IL2 represent two asymmetric events.

**THsubnet.** Active TH are introduced based on APC (intTH_APC). Once activated they consume IL12 and use it to eventually duplicate (IntIL12_TH, IntTH_IL12, respectively). Moreover, they release IL2 and use it to self-promote their duplication (IntIL2_TH and IntTH_IL2), they die for natural death (DieTH) and promote the introduction of activated B cells (IntTH_B).

**IL2.** IL2 is produced by TH (IntIL2_TH) and consumed by TC (IntIL2_TC and IntTC_IL2) and B (IntIL2_B and IntB_IL2) cells to eventually promote their duplication. IL2 may eventually disappear from the system due to natural degradation (DieIL2).

**IL12.** IL12 is introduced as a consequence of the death of VC (DieVC, IntVC_TC, IntVC_AB), it is consumed and eventually boosts TH duplication (IntIL2_TH and IntTH_IL2), is consumed by TC for remaining active (IntIL12_TC) and disappears due to natural degradation (DieIL12).

Generally, the model is characterized by ten places modeling a specific entity and 33 transitions representing the events that may occur in the system. For instance, the VC place models the number of Vaccine Cells which increases due to the *Injection* transition and decreases due to the *DieVC*, *IntVC_AB* and *IntVC_TC* transitions. Hence, the *DieVC* transition is enabled and can fire only if the number of tokens in the VC place is greater than one, while the *IntVC_TC* transition when both the TCcells and the VC places have more than one token. Observe that the transition *IntVC_VC* does not change the marking of the place given the presence of two arcs connecting the place with the transition, which, at the same time, removes and adds a token at every transition firing. Thus, the ODE describing the Vaccine Cells (i.e. VC place) can be derived as follows:

$$\frac{dx_{VC}}{dt} = +r_{Injection} - r_{DieVC} \cdot x_{VC} - r_{IntVC\_AB} \cdot x_{VC} \cdot x_{ABcell} - r_{IntVC\_TC} \cdot x_{TCcells} \cdot x_{VC} \tag{1}$$

where the $x_{\cdot}$ represents the number of tokens in the corresponding place "$\cdot$", $r_{Int\cdot}$ the rate associated with the exponential distribution of the corresponding transition "$Int\cdot$", and the product $r_{IntVC\_AB} \cdot x_{VC} \cdot x_{ABcell}$ represents the intensity of the transition defined by the Mass Action law (i.e., the product of the number of tokens of each input place with the transition rate).

Most transitions follow a Mass Action law, excluding transitions that represent duplication processes such as IntTH_IL2, IntTC_IL2, IntTH_IL12 and IntB_IL2 that follow a Michaelis-Menten kinetics, and GrowthCC is modeled using a Gompertz growth as in [6]. As the model was designed to take into account populations that become active upon vaccine administrations, the initial conditions are the same for all entities with 0 tokens in every place at time 0.

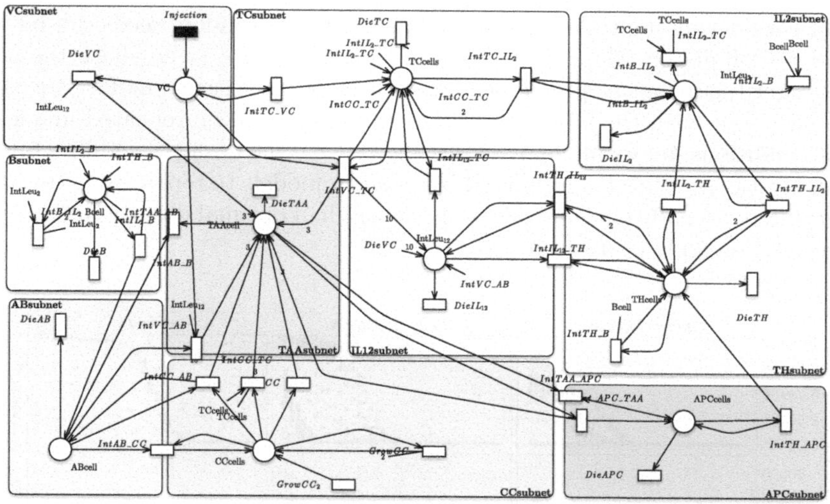


**Fig. 2.** The ESSN conceptual model. Some arcs, color domains and variables are hidden for readability.

# 3 Results

## 3.1 Simulation Results

In this subsection, we show the simulation results obtained by executing over 1000 simulations for the Untreated, Early and Chronic protocols (1 year). In Fig. 3 we show CC, AB and TC (final actors of tumoral and cellular-mediated immune responses) behaviors. The boxplots report median values, interquartile ranges, and outliers for different time points. The Early protocol significantly delays cancer growth with respect to the untreated case, but only the Chronic protocol is capable of ensuring complete protection against tumor growth (excluding some outliers). Having a deeper look at the Early protocol plot, we observe that even if the median values remain limited, for multiple cases the cancer growth becomes exponential just after day 200. For these cases, it is possible, in principle, to look at the behaviors of the other entities to understand the mechanisms responsible for the lack of efficacy. Note here that, over a very high number of CC (arbitrarily set to $10^5$ in [6]), Carcinoma-in-Situ is formed, and, as biological conditions are different, the model results should no longer be trusted. Furthermore, it is worth nothing here that even if both the Early and Chronic vaccination protocols clearly entitle a similar immune system response at the initial stages (as shown in the AB and TC plots), the lack of later administrations in the Early protocol leads, in general, to a decline in the number of activated immune entities that thus become unable to cope with the reappearance of tumor cells.

The presented results, which are in line with *in vivo* observations [11], underline the importance of including stochastic variability in simulations. The process of considering multiple simulations with different random seeds can be thus considered similar to the process of using multiple mice (who share the same genetic heritage) during pre-clinical studies in order to account for stochastic variability, but with the advantage that simulations can be repeated under different conditions, for example by increasing the vaccine dosage or adding further administrations. The chance offered by *in silico* models to repeat the same tests under different conditions may allow the research of optimal dosages and timings for population-wide protection.

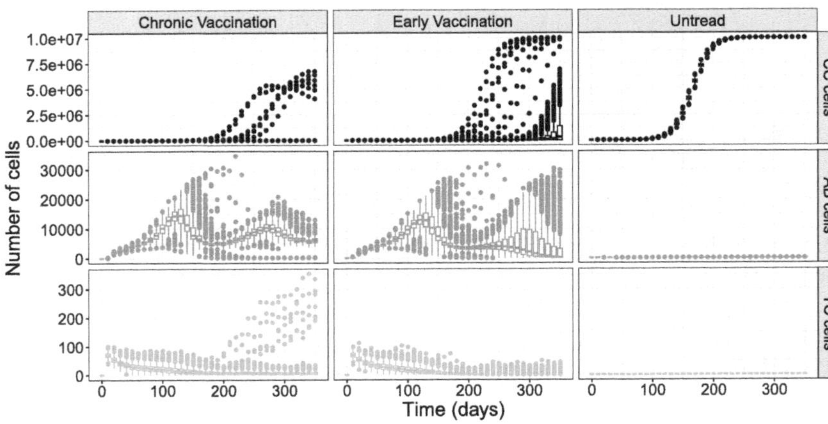


**Fig. 3.** Boxplots for median, interquartile ranges, min, max, and outliers for the time-evolution of CC, AB and TC.

### 3.2 LHS-PRCC Analysis for Dosage Optimization

LHS-PRCC (Latin Hypercube Sampling - Partial Rank Correlation Coefficients) is a global sensitivity analysis technique that can be used to statistically determine the effects of parameters' variation on the model outputs [13]. In practice, the LHS method, renowned for its efficient stratified sampling approach, is first used to generate samples of the model's input variables. Subsequently, the model is executed different times, each time using a different combination of the generated input variable samples within a specified interval. This exhaustive exploration allows for a comprehensive assessment of how variations in input parameters influence the model output. Following the model executions, PRCC are calculated between the generated input variables and the resulting model outputs across the chosen interval. PRCC analysis provides information on the strength and direction of the relationships between inputs and outputs. A PRCC value close to 1 indicates a positive monotone relationship, while a value close to −1 suggests a negative monotone relationship. In this context, we leveraged the sensitivity functionality offered by GreatMod. In contrast to a previous investigation in which the impact of model parameters on the initial ODE model was explored [6], our study examines how the variation in the amount of VC injected affects the model output. This aspect remained unexplored in previous analyses. Specifically, we show the effects of vaccine dosage variation considering TC, AB, and CC as the targeted variables.

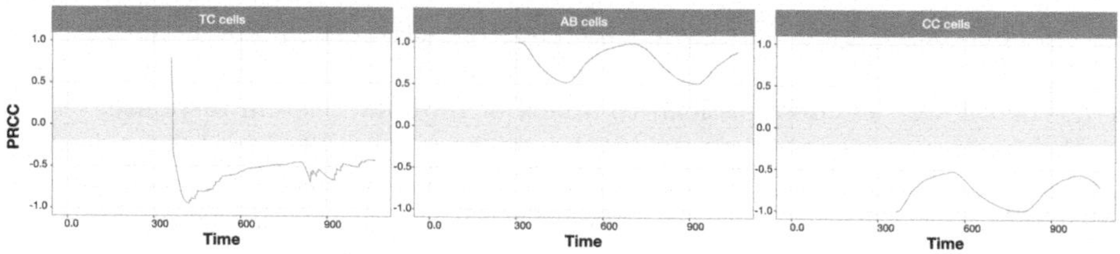


**Fig. 4.** PRCC analysis generated exploiting GreatMod with TC, AB, and CC as the targeted variables.

In this context, our objective is twofold: firstly, to employ the LHS-PRCC methodology to assess the sensitivity of the model to its parameters, particularly focusing on variations in vaccine dosage. Secondly, we aim to uncover insights that could guide the optimization of vaccination dosages and schedules. From a translational standpoint, we must consider the possibility of reducing the frequency of vaccine administrations.

It is important to note that the Chronic protocol has been the only protocol capable of consistently demonstrate its efficacy in ensuring lifelong protection against the development of mammary carcinoma. Therefore, it served as the baseline for our analysis.

As first step we analysed the effect of dosage variation on the Chronic protocol. In Fig. 4 we show the LHS-PRCC values vs time for CC, AB, TC. No correlation is revealed up to approximately time 300, as all simulations return the same value (see also, for example, plot A) on Figs. 5, 6 and 7). After that, it is possible to observe that a negative correlation between VC quantity and CC cells holds, with values that show an oscillatory behavior between -0.5 up to -1.0. A positive correlation is instead present with the total number of AB, showing a pattern that somewhat mirrors the pattern observed for CC. Interestingly, VC seem to have a, although low, negative correlation with TC cells, principal actors of the cellular-mediated immune system responses.

In Figs. 5, 6 and 7, panel A), left plot, we show the effects of the Chronic protocol considering 1000 average simulations (each trajectory represents the average trajectory of the SSA simulations with a fixed VC quantity) obtained at different values of VC cells in the interval 20–60. Different colors (from dark purple to yellow) are used to represent increasing VC quantities. Injections times are here represented by dashed vertical lines.

Always in panel A), right plot, each dot is used to represent the final average value reached in the corresponding place (CC, AB, or TC) for each simulation (with a different VC quantity). It is interesting to notice that a threshold on the number of VC (around 35 injected VC) can be observed for the CC cells plot (Fig. 5). This threshold leads to bifurcation on the behavior of the model. More specifically, under the threshold we observe the lack of treatment efficacy with CC that grow exponentially, while complete protection is instead achieved for every simulation that adopts dosages over such a threshold. This relevant finding suggests us that it is possible to optimize the vaccine dosage by lowering the number of injected VC right over the threshold. It is worth noting that similar scenarios are also present for the AB and TC plots (Figs. 6 and 7, panel A)), where lower VC dosages lead to a lack of Immune System response.

We further analyzed Fig. 4 trying to catch up possible insights on how to optimize the vaccination schedule. By looking again at the PRCC value on the CC cells plot, we argued that it is possible lower the number of injections at the times in which the PRCC is smaller, i.e. when it is close to -0.5. At these times the negative correlation is weaker, and thus the injections may be less important. We then tried, as shown in panels B) and C), left plots, to remove 8 or 16 injections, respectively, at these less critical times (see the missing dashed vertical lines in such plots). The Figs. 5, 6 and 7, panel B) and C), right plots, confirm again the persistence of a dosage threshold (around 35 VC) that assures cancer depletion for all the dosages above it, even with the fewer injections.

To better understand in how many simulations the vaccine is able to prevent carcinoma formation, we considered in Figs. 5, 6 and 7, panel D), the estimation of the distribution (namely density distributions) of the involved quantities (CC, AB and TC) at the final simulation time, obtained always varying the VC cells quantity in the range 20–60. Furthermore, the percentages of the final quantity values are shown, distinguishing between the values before and after the vertical black dashed line that separates the two peaks characterizing the distributions. By looking at the CC cells density plot, it is possible to observe that all the curves show a very similar pattern, i.e. a bimodal distribution with two peaks, one near 0 (for dosages over the threshold that almost assure protection) and the other one toward the right-hand side of the plot (for lower dosages that lead to exponential CC growth). This also holds for the AB and TC density plots, indicating how the number activated IS cells tends to zero (peaks on the left-hand side of the plot) when the cancer has been depleted, and instead is higher (peaks on the right-hand side of the plot) when cancer cells are still up and growing. Furthermore, it should be noted that the number of simulations for the first (left) peak in the CC plot is superimposable for the Chronic protocol and the protocol with 8 removed injections, while the protocol with 16 removed injections leads to a slight lower peak, suggesting that it is in general less effective at similar dosages in respect to the other two. The situation is clearly reversed for the AB and TC plots, as the two more effective protocols may entitle slightly higher residuals of IS active entities even at the end of the experiment in respect to the protocol with 16 removed injections.

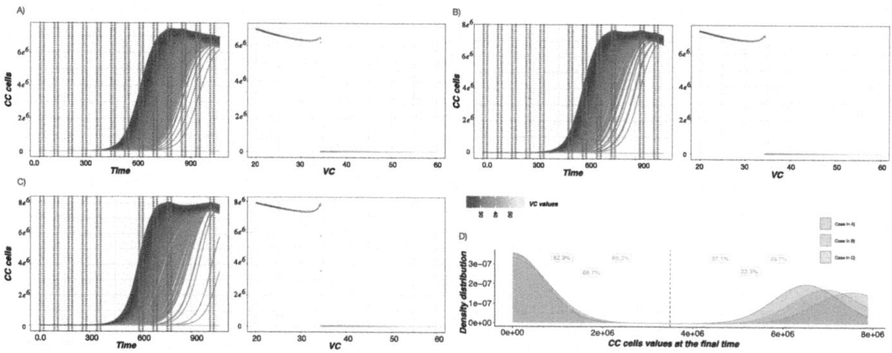


**Fig. 5. CC Cells Dynamics.** We examine the dynamics of CC cells under different values of VC. In panel A), we depict the effects on the Chronic protocol. Panel B) illustrates the scenario where 8 injections are removed, while panel C) shows the effects of removing 16 injections. The left plots in panels A), B), and C) showcase the temporal evolution of CC cells. Dashed vertical lines in these plots denote the timing of injections. On the right side of these plots, we display how variations in VC values (on the x-axis) influence the final value of CC cells (on the y-axis), considering the simulation obtained with the corresponding VC value. In panel D), we present the three density distributions at the final simulation time. The colored percentages represent the number of values at the final simulation time lower and greater than the vertical black dashed line.

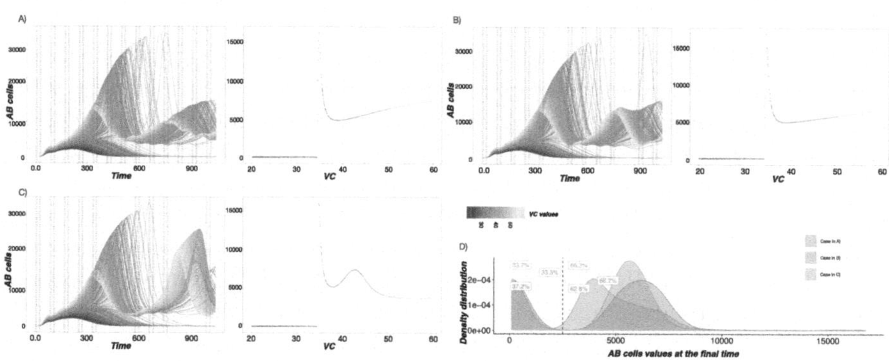


**Fig. 6. AB Cells Dynamics.** We examine the dynamics of AB cells under different values of VC. In panel A), we depict the effects on the Chronic protocol. Panel B) illustrates the scenario where 8 injections are removed, while panel C) shows the effects of removing 16 injections. The left plots in panels A), B), and C) showcase the temporal evolution of AB cells. Dashed vertical lines in these plots denote the timing of injections. On the right side of these plots, we display how variations in VC values (on the x-axis) influence the final value of AB cells (on the y-axis), considering the simulation obtained with the corresponding VC value. In panel D), we present the three density distributions at the final simulation time. The colored percentages represent the number of values at the final simulation time lower and greater than the vertical black dashed line.

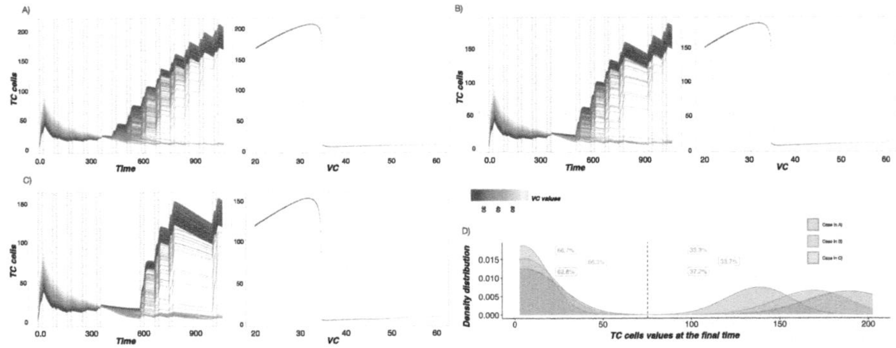


**Fig. 7. TC Cells Dynamics.** We examine the dynamics of TC cells under different values of VC. In panel A), we depict the effects on the Chronic protocol. Panel B) illustrates the scenario where 8 injections are removed, while panel C) shows the effects of removing 16 injections. The left plots in panels A), B), and C) showcase the temporal evolution of TC cells. Dashed vertical lines in these plots denote the timing of injections. On the right side of these plots, we display how variations in VC values (on the x-axis) influence the final value of TC cells (on the y-axis), considering the simulation obtained with the corresponding VC value. In panel D), we present the three density distributions at the final simulation time. The colored percentages represent the number of values at the final simulation time lower and greater than the vertical black dashed line.

## 4 Conclusion

In this work, we applied our modeling pipeline for reproducing the effects of a cancer immunoprevention vaccine on mammary carcinoma in a pre-clinical setup. The proposed pipeline uses ESSN for drawing the initial conceptual model and then exploits SSA for obtaining, *in silico*, the IS response stimulated by the use of the Triplex vaccine against mammary carcinoma. Results show a good agreement with observations from the *in vivo* experiment and underline the importance of stochastic simulation in immunology. While this problem has been previously examined by means of ODE [6] and ABMs [16], the use of the proposed modeling pipeline has various advantages. The ESSN representation allows minimizing the introduction of errors during the definition of the conceptual model, making also easier communication with domain experts thanks to the use of a standardized graphical formalism instead of a set of equations. The produced ESSN model can be then used to simulate the evolution of the system by using either a macro-scale (using SSA or ODE) or a micro-scale simulation approach (using ABM) upon necessity. In this study we used SSA for a macro-scale analysis of the system that also includes stochastic variability and, exploiting a global sensitivity analysis technique (LHS-PRCC), we analysed the effects of vaccine dosage variation on the model outputs. The analysis allowed us to make some important considerations that may be fruitful for further dosage optimization.

First of all, it was possible to obtain a clear minimal value on the number of VC injections that should be injected to assure protection against carcinoma formation. Then, by considering the times where PRCC was smaller, it was possible to obtain some vaccination schedules with fewer injections that can assure similar protection as the Chronic protocol. From the biological point of view these results may be considered relevant, as in a translational perspective dosage and schedule optimization is advisable. From the methodological point of view, it is worth noting that the approach adopted here about the application of LHS-PRCC, not only to perform the sensitivity analysis of the model but also to optimize some of its input variables, lead to some interesting results that suggest how such an approach can be in principle adopted in other studies and domains to gather up the best strategies for optimizing model results.

**Acknowledgments.** This research was supported by the scientific activities of the CINI InfoLife Laboratory.

**Data Availibility Statement.** The two tools: GreatMod https://qbioturin.github.io/epimod/ and GreatSPN 3 https://github.com/greatspn/SOURCES. The model, the parameters configuration and the file necessary to reproduce the results are available at https://github.com/qBioTurin/CarcinomaModel.

**Conflict of interests.** The authors declare no conflicts of interest.

## References

1. Ajmone Marsan, M., Balbo, G., Conte, G., Donatelli, S., Franceschinis, G.: Modelling with Generalized Stochastic Petri Nets. Wiley, Hoboken (1995)
2. Amparore, E.G., Balbo, G., Beccuti, M., Donatelli, S., Franceschinis, G.: 30 years of GreatSPN. In: Principles of Performance and Reliability Modeling and Evaluation, pp. 227–254. Springer (2016)
3. Amparore, E.G., Beccuti, M., Castagno, P., Franceschinis, G., Pennisi, M., Pernice, S.: Multiformalism modeling and simulation of immune system mechanisms. In: Proceedings - 2021 IEEE International Conference on Bioinformatics and Biomedicine, BIBM 2021, pp. 3259–3266 (2021)
4. Atitey, K., Anchang, B.: Mathematical modeling of proliferative immune response initiated by interactions between classical antigen-presenting cells under joint antagonistic il-2 and il-4 signaling. Front. Mol. Biosci. **9** (2022). https://doi.org/10.3389/fmolb.2022.777390, https://www.frontiersin.org/journals/molecular-biosciences/articles/10.3389/fmolb.2022.777390
5. Beccuti, M., Castagno, P., Franceschinis, G., Pennisi, M., Pernice, S.: A Petri net formalism to study systems at different scales exploiting agent-based and stochastic simulations. In: Proceedings of 17th European Workshop, EPEW 2021, and 26th International Conference, ASMTA. LNCS, vol. 13104. Springer (2022)
6. Bianca, C., Chiacchio, F., Pappalardo, F., Pennisi, M.: Mathematical modeling of the immune system recognition to mammary carcinoma antigen. BMC Bioinform. **13 Suppl 17** (2012). https://doi.org/10.1186/1471-2105-13-S17-S21, https://pubmed.ncbi.nlm.nih.gov/23281916/
7. Borshchev, A.: XJ technologies: Anylogic 6. In: Proceedings of the 37th Winter Simulation Conference, Orlando, FL, USA, December 4-7, 2005, p. 82. IEEE Computer Society (2005). http://dl.acm.org/citation.cfm?id=1163276

8. Castagno, P.: A computational framework for modeling and studying pertussis epidemiology and vaccination. BMC Bioinform. **21**, 344 (2020)
9. Figueredo, G.P., Siebers, P.O., Owen, M.R., Reps, J., Aickelin, U.: Comparing stochastic differential equations and agent-based modelling and simulation for early-stage cancer. PLoS ONE **9**(4), e95150 (2014)
10. Gillespie, D.T.: Exact stochastic simulation of coupled chemical reactions. J. Phys. Chem. **81**(25), 2340–2361 (1977)
11. Giovanni, C.D., et al.: Immunoprevention of her-2/neu transgenic mammary carcinoma through an interleukin 12-engineered allogeneic cell vaccine. Cancer Res. **64**, 4001–4009 (2004). https://doi.org/10.1158/0008-5472.CAN-03-2984
12. Lima, E.A., Wyde, R.A., Sorace, A.G., Yankeelov, T.E.: Optimizing combination therapy in a murine model of her2+ breast cancer. Comput. Methods Appl. Mech. Eng. **402**, 115484 (2022). https://doi.org/10.1016/j.cma.2022.115484, https://www.sciencedirect.com/science/article/pii/S0045782522005060. a Special Issue in Honor of the Lifetime Achievements of J. Tinsley Oden
13. Marino, S., Hogue, I.B., Ray, C.J., Kirschner, D.E.: A methodology for performing global uncertainty and sensitivity analysis in systems biology. J. Theor. Biol. **254**(1), 178–196 (2008)
14. Nave, O., Elbaz, M., Bunimovich-Mendrazitsky, S.: Analysis of a breast cancer mathematical model by a new method to find an optimal protocol for her2-positive cancer. Biosystems **197**, 104191 (2020). https://doi.org/10.1016/j.biosystems.2020.104191, https://www.sciencedirect.com/science/article/pii/S0303264720300861
15. Pappalardo, F., Lollini, P.L., Castiglione, F., Motta, S.: Modeling and simulation of cancer immunoprevention vaccine. Bioinformatics **21**, 2891–2897 (2005). https://doi.org/10.1093/BIOINFORMATICS/BTI426, https://academic.oup.com/bioinformatics/article/21/12/2891/268025
16. Pennisi, M., Catanuto, R., Pappalardo, F., Motta, S.: Optimal vaccination schedules using simulated annealing. Bioinformatics **24**, 1740–1742 (2008). https://doi.org/10.1093/BIOINFORMATICS/BTN260, https://academic.oup.com/bioinformatics/article/24/15/1740/264927
17. Pernice, S., et al.: Computational modeling of the immune response in multiple sclerosis using epimod framework. BMC Bioinform. **21**(17), 1–20 (2020)
18. Pernice, S., et al.: A computational approach based on the colored Petri net formalism for studying multiple sclerosis. BMC Bioinform. **20**, 1–17 (2019)
19. Wilensky, U.: NetLogo (1999). http://ccl.northwestern.edu/netlogo/

# Extension of the GreatMod Modeling Framework to Simulate Non-Markovian Processes with General-Distributed Events

Irene Terrone[1,2], Daniela Volpatto[1,2,3], Simone Pernice[1,2](✉), Elvio Amparore[1], Roberta Sirovich[3], Francesca Cordero[1,2], and Marco Beccuti[1,2]

[1] Department of Computer Science, University of Turin, Turin, Italy
[2] Laboratorio InfoLife, Consorzio Interuniversitario Nazionale per l'Informatica (CINI), Rome, Italy
simone.pernice@unito.it
[3] Department of Mathematics G. Peano, University of Turin, Turin, Italy

**Abstract.** This paper introduces significant advancements in the GreatMod modeling framework, enhancing its capacity to simulate systems characterized by non-Markovian dynamics accurately. These enhancements include the definition of a novel graphical formalism tailored to represent such complex models, alongside an extension of the Stochastic Simulation Algorithm to accommodate their simulation efficiently. Moreover, we validate the robustness of these improvements through two case studies: the Susceptible-Infected-Recovered model and the Parallel-Producer-Consumer model.

**Keywords:** Non Markovian models · Stochastic simulation · SIR

## 1 Introduction

The analysis and study of complex systems can help scientists better understand the real world in all its implications. Among the various techniques that can be successfully exploited to deal with this challenge, **computational modeling** is one of the most powerful since capable of describing and studying a complex real system in a simple but meaningful way through computer simulations and mathematical models. In literature, several modeling and simulation techniques have been proposed, depending on the system under study and the research questions motivating the analysis. For quantitative analyses of the temporal evolution of systems, two simulation approaches are mainly utilized: (i) **the stochastic simulation** of the Continuous Time Markov Chain (CTMC) underlying the system behavior, and (ii) the computation of the deterministic solution based on mean-field approximation [15] of the CTMC, where the evolution is described as a system of Ordinary Differential Equations (ODEs). In detail, the former can capture relevant random features like bi-modality and provides information that cannot be represented using a deterministic setting such as tail behavior and variance, yet it requires a higher computational cost than the solution of the ODEs system. Let us observe that the main assumption behind these approaches


M. Vettoretti et al. (Eds.): CIBB 2023, LNBI 14513, pp. 177–191, 2025.
https://doi.org/10.1007/978-3-031-90714-2_13

is the Markov property, i.e. changes from one state to another are independent of the past, given the present state, and every inter-event's time is exponentially distributed. However, even if CTMC is largely exploited in computational modeling, it is known that the behavior of some biological and epidemiological systems may not be represented accurately by these models. For instance, in [2] the authors showed how the non-Markov epidemic spreading model captures the typical patterns of COVID-19 pandemics.

Several significant efforts have been made to combine multiple modeling formalisms into a single tool, notably SHARPE [23], SMART [7], and MÖBIUS [9]. These frameworks represent some of the earliest attempts to create versatile modeling tools that allow users to build, simulate, and analyze performance, reliability, and dependability models using a range of formalisms, including Petri nets, Markov, semi-Markov, and non-Markov models. However, some components of these tools, such as the user interface and documentation, are outdated, which can deter new users or slow down the modeling process. Furthermore, these tools provide discrete-event simulators without considering the possibility of exponential distributions and their properties, and so the SSA proposed by Gillespie. More recently, in the context of continuous-time multi-agent models where agents interact according to network topology, the RED (Renewal Equation Dynamics) algorithm [13] was proposed for generating statistically correct trajectories of non-Markovian spreading models on networks. This algorithm was compared with the state-of-the-art methods for simulating non-Markovian dynamics on networks, the non-Markovian Gillespie algorithm (nMGA) developed by Boguná et al. [3], and also with the Laplace-Gillespie algorithm (LGA) [17] which is an improvement of the nMGA algorithm. In 2022, the authors of [21] (unpublished work) introduced a Python-based library to address the existing limitations of the LGA and nMGA algorithms.

In this paper, we describe how we extended the user-friendly general modeling framework **GreatMod** [5], to support the simulation of non Markovian models in which events may be characterized by eventually non-exponential inter-time distributions. In details, this has required, firstly, the definition of a new generalization of the graphical Petri Net formalism [1], called Generally Distributed Extended Stochastic Petri Net and, secondly, the extension of the exact stochastic algorithm proposed in 1976 by Daniel Gillespie, called Stochastic Simulation Algorithm (SSA) [12], for efficiently simulating non Markovian models.

## 2 Background

In this section, we first introduce the modeling framework GreatMod, and secondly we reconstruct the theoretical structure within which the devised simulation algorithm is situated.

### 2.1 The GreatMod Framework

It is a general modeling framework that provides an intuitive graphical interface for model construction based on Petri Nets (PNs) and extensions, and automatic derivation of the low-level mathematical processes characterizing the system dynamics.

In particular, GreatMod's novelties and strengths can be summarized as follows: (i) the use of a graphical formalism based on PNs and extensions, which allow the encoding of events whose dynamics are characterized through complex functions, to simplify model construction and to provide an intuitive description of system behavior; (ii) the containerization (into Docker images) of all the implemented analysis techniques to improve the framework portability and to ensure the reproducibility of the derived results; (iii) an R library *Epimod* [5], which provides a user-friendly interface for easy access to the essential modeling techniques, from the automatic derivation of the underlying mathematical processes of the PN model to the sensitivity and calibration analysis of the unknown parameters. Finally, GreatMOD provides different algorithms for the numerical solution of ODEs systems and the simulation of the stochastic process (e.g., SSA and $\tau$-leap).

## 2.2 Stochastic Petri Nets

PNs and their extensions are widely recognized to be a powerful tool for modeling and studying complex systems, thanks to their ability to represent those systems in a natural graphical manner and to analyze their behavior from both qualitative and quantitative points of view. PNs are defined as bipartite directed graphs with two types of nodes, namely *places* and *transitions*. The former ones correspond to state variables of the system and are graphically represented as circles. The latter ones correspond to the events that can generate a state change and are graphically represented as boxes. Nodes of different types are connected by *arcs*, which express the relation between states and event occurrences. Places can contain *tokens* drawn as black dots, and the *marking* $m$ of the PN is the number of tokens in each place. A specific cardinality (multiplicity) is associated with each arc, and it describes the number of tokens removed from (or added to) the corresponding place upon the firing of the transition the arc is connected to. In particular, the firing of an enabled transition removes from the input places (i.e., those places connected to the transition with an arc whose direction is from the place to the transition) as many tokens as the multiplicity of the corresponding arches, and adds in the output places (i.e., places with arc whose direction is from the transition to the place) the number of tokens described by the multiplicity of the corresponding arches. A transition is enabled if each input place contains a number of tokens equal to or greater than a given threshold defined by the multiplicity of the corresponding input arc. Thus, the evolution of the system (i.e., how the marking of the system changes) is given by the firing of enabled transitions.

A possible way to include time in such a framework is to assign to each transition a holding time and therefore assuming that the next transition to be firing at any given time corresponds to the minimum among such holding times (*race execution policy*). Whenever the holding times are not deterministically assigned, but rather sampled according to probability distributions, the PN is called Stochastic Petri Net (SPN) and it is possible to depict an isomorphism between each SPN (equipped with an initial marking) and a stochastic process, with state space representing the marking.

In particular, if exponentially distributed independent random variables are used as holding times, the stochastic process has been proved to satisfy the Markov property, therefore many studies have revolved around this hypothesis. Indeed, thanks to their properties, exponential random variables allow to overcome several issues: first of all, being the remaining time to fire distribution still exponentially distributed with the same mean, thanks to the *memoryless* property of such distributions, there is no need to keep track of the history of the process nor to define a so-called *memory policy*, as it is possible to resample every holding time each time a transition fires; then, since the minimum among independent exponential random variables is an exponential random variable itself, with proper parameter, there is a natural way to move from a given semantic (i.e. *single-server*, *infinite-server*, *mass-action*,...) to the other without changing the structure of the underlying process.

In this work, we define an SPN generalization capable of dealing with generally distributed times associated with transitions.

## 3 Methods

We now describe the new PN formalism, called Generally Distributed Extended Stochastic Petri Net (GDESPN) and how its simulation has been encoded in GreatMod.

### 3.1 Generally Distributed Extended Stochastic Petri Net

We defined GDESPNs as an extension of SPNs in which waiting times associated with transitions are allowed to be generally distributed. In the PNs framework, they fall within the following categories:

1. race execution policy;
2. enabling memory policy, i.e. the value of the timers of the transitions that are disabled are discarded and the value of the timer of the transitions that are still enabled are kept;
3. infinite-server semantic[1] (possibly but not necessarily with *mass action* law), i.e. when the enabling degree of a transition is larger than one, it is assumed that the service represented by the transition is executed in parallel for each client, hence when the enabling degree of a transition is incremented a new timer is initialized.

[1] Notice that any infinite server transition can be transformed into a single server one with the addiction of an input place with a single token removed and replaced back at each firing.

From a mathematical point of view, it is not possible, under such a general hypothesis, to characterize the stochastic process associated with the reachability graph, unless for some special cases. In particular, in [10] conditions on the transitions are drawn for the general distributed PN associated process to be a Semi-Markov, while [6] explores the regenerative case.

In GDESPN formalism the set of transitions $T$ is split into two subsets: the set of transitions with exponentially distributed firing time, $T_{exp}$, and the set of transitions whose firing time is defined by a generic probability distribution, $T_{\neg exp}$. Since the SSA general idea relies on the exponential distribution properties to generate a stochastic trajectory of the system, inspired from [4], we modified it to simulate the process underlying the GDESPN model, without changing the general idea of the two-steps process behind the SSA.

### 3.2 Extended SSA Algorithm for Non Markovian Processes

Let $\theta_i$ be the random variable giving the waiting time to fire of transition $t_i \in T$, with probability distribution defined as $Exp(a_i(x))$ if $t_i \in T_{exp}$, or defined as $\mathcal{D}_i \in \mathbf{D}$ if $t_i \in T_{\neg exp}$. The parameter $a_i(x)$ is the propensity function, also known as the rate function, associated with the exponential distribution which might depend on the state $x$ of the system. To obtain the time spent in the current state, we have to calculate the minimum time required for the first event occurrence which is a random variable defined as

$$\min_{t_i \in T}(\theta_i) = \min\left(\min_{t_i \in T_{exp}}(\theta_i), \min_{t_j \in T_{\neg exp}}(\theta_j)\right) = \min\left(\theta_0, \min_{t_j \in T_{\neg exp}}(\theta_j)\right),$$

where $\theta_0 \sim Exp\left(\sum_{i=1}^{n} a_i(x)\right)$ (as for the SSA), and assuming that $\{\theta_i\}_{t_i \in T}$ are independent. So, let $\tau$ be the realization of the random variable $\min_{t_i \in T}(\theta_i)$, then we can summarise the extended SSA algorithm as in **Algorithm** 1

**Algorithm 1** SSA algorithm extended to deal with semi-Markov processes

Set time $time = time_0$, state $x = x_0$ and $FEL = Null$
**while** ($time < time_{final}$) **do**
  Compute $\tau \sim \theta_0$ as for the SSA.
  Compute $\forall t \in T_{\neg exp}$ its enabling degree $en_t$ in $x$ and updating $FEL$ adding or removing events according with $en_t$.
  **if** ($\tau > FEL.top().time$) **then**
    The corresponding fired event is $e_i = FEL.pop()$
    $\tau = e_i.time$
  **else**
    $e_i$ is defined as in the SSA.
  **end if**
  Updating $x$ firing $e_i$
  $time = time + \tau$
**end while**

where *FEL* represents the future event list implemented as a *Binary Heap* [18] data structure.
A Binary Heap is a complete binary tree, typically represented as an array.

The representation is done as for any element in position $i$ the indexes of other nodes for the $i^{th}$ node can be calculated as follows:

- the root element of the tree will be at $Arr[0]$,
- $Arr[(i-1)/2]$ returns the parent node,
- $Arr[(2*i)+1]$ returns the left child node,
- $Arr[(2*i)+2]$ returns the right child node.

Typically, a Heap does not perform well for arbitrary searches, requiring $\Theta(N)$ comparisons, with $N$ being the length of the heap. However the extraction of the minimum element is immediate, $O(1)$, since it is always in the first position, and the addition of a new element and the deletion occurs in time $O(\log N)$. This allows us to reduce the impact of managing events into the FEL on the whole simulation time. Let us observe that the number of events stored in *FEL* at time $\nu$ is defined as the enabled degree of all transitions $t \in T_{\neg exp}$, i.e., the minimum number of tokens required in the input places of $t$ to be enabled. Thus, for each event an inter-time value is sampled from the generic distributions $\mathcal{D}_i$ of the corresponding transitions $t_i \in T_{\neg exp}$.

Furthermore, the criterion to keep the memory of the past history of the process for the $t_i \in T_{\neg exp}$ is an *enabling memory policy*: the time to fire is re-sampled only when the transition is disabled. Inside the *FEL* there are as many events as the number of times the transition can fire according to the marking in the input places. When a transition fires it moves tokens from all its input places to all its output places, and this can modify the marking of input places to other transitions. So, when the enabling degree of a transition is modified, events will be added, sampling new waiting times, or randomly deleted, from the *FEL*. Instead, all events associated with persistent transitions are kept in memory with their waiting time reduced by the sojourn time in the previous marking.

## 4 Results

In this section, we delve into the evaluation of the new GreatMod functionality, focusing on its accuracy and its adaptability to model various types of PNs with different enabling policies. For our analysis, we chose two distinct case studies: (i) the Susceptible-Infected-Recovered (SIR) model and (ii) the Parallel Producer-Consumer (PPC) model.

The SIR model, despite its simple architecture, serves as an exemplary instance of a complex system that could benefit from the utilization of general distributed holding times. This is particularly relevant since the recovery in many diseases has been proved not to follow an exponential low. By incorporating non-exponential recovery distributions, the mathematical framework of the SIR model embodies a stochastic process that defies classification as a

Semi-Markov process, as it encompasses concurrent non-exponential transitions for certain markings. Additionally, it diverges from being a regenerative process as well. Notably, the fields of Semi-Markov and Markov-regenerative processes have undergone extensive study, yielding various theoretical insights and results [6,10].

Conversely, the PPC model serves as a prototypical example of a regenerative process. To thoroughly assess the capabilities of our extended framework, we conducted a comparative analysis with a state-of-the-art simulation tool. This comparison allowed us to gauge the effectiveness and efficiency of GreatMod in accurately modeling and simulating regenerative processes.

### 4.1 Susceptible-Infected-Recovered Model

Figure 1 shows how to encode the SIR model using GDESPN formalism when all the transitions are associated with exponential distributions, see panel A, or when *Recovery* is associated with a generic probability distribution $\mathcal{D}$, see panel B. This model consists of three places with their respective labels: **S**, the susceptible individuals, accounts for the subjects who could potentially become ill if exposed, **I**, the infected individuals, accounts for the currently infected subjects, and **R**, the recovered individuals who were successfully recovered by the infection. Only two possible events might occur: the infection of a susceptible individual after contact with an infected one, modeled by the transition *Infection*, and the recovery from the disease, represented by the transition *Recovery*.

The enabling policy chosen for the simulation has been the *mass action* as it is reasonable to assume that infections are directly proportional to the number of infected and susceptible individuals.

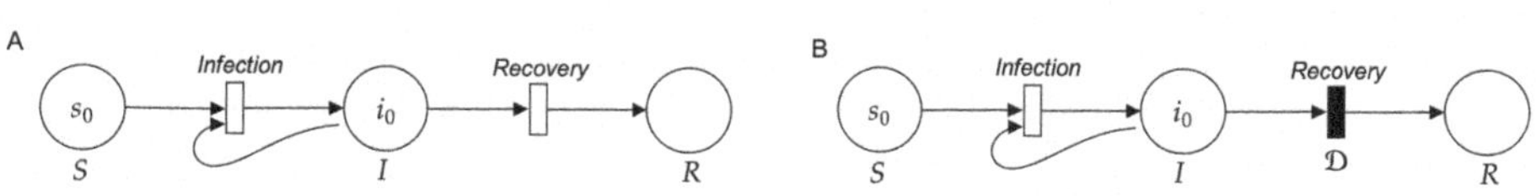


**Fig. 1.** Two SIR models described through GDESPN formalism in which all the transitions are associated with exponential distributions (A) or when *Recovery* is associated with a generic probability distribution $\mathcal{D}$ (B). [11].

*SIR with Gamma Distribution.* As mentioned before the probabilistic law governing a general GDESPN is unknown, moreover the SIR model is neither a Semi-Markov, nor a regenerative process. Though, being the SIR a wide-spread model in literature, there have been many studies capable of deriving theoretical results about it without using any probabilistic known framework (such as properties acquired from being Markov, semi-Markov,...). Hence, to evaluate the GreatMOD new functionality we started from the results presented in [8] and we derived a theoretical equation to estimate the expected number of susceptible individuals surviving the epidemics as follows: let $S(0) = s_0$ and $I(0) = i_0$ be the starting number of susceptible and infected individuals[2], respectively, if infections happen at an exponential rate $\beta$ and recoveries follow an almost surely finite distribution $Y$ such that $\phi(x) := \mathbb{E}(e^{-xY})$, then calling $T = \inf\{t \geq 0 : I(t) = 0\}$ the time at which the epidemics terminate,

$$\mathbb{E}\left(S(T)\right)) = \sum_{n=0}^{s_0} \frac{s_0!}{(s_0 - n)!}\phi(n\beta)^{s_0+i_0-n}G_n \tag{1}$$

where $G_0, G_1, \ldots, G_{s_0}$ are Abel-Gontcharoff pseudopolynomials defined recursively by the system of equations

$$\sum_{m=0}^{n} \frac{G_m}{(n-m)}\phi(n\beta)^{n-m} = \frac{n}{n!},\ n = 0, 1, \ldots, N$$

Then, we tested the SSA extended algorithm with the theoretical results estimated by the above equation. In particular, we considered a SIR model having exponential inter-times for *Infection* transition and $\Gamma(k, \theta)$-distributed *Recovery* inter-times ($\mathcal{D}$ in Fig. 1B).

In this example Eq. (1) holds with $\phi(x) = \frac{1}{(1+\theta x)^k}$, and hence it is possible to estimate the theoretical expected number of susceptible that never undergo infection and compare it with the mean observed value (i.e. the average marking of place $S$ once reached the absorbing state $I = 0$). We performed several experiments letting the parameters $k$ and $\theta$ varying across $\{2, 3, 4, 5\}$ and $\{0.01, 0.025, 0.5, 1\}$ respectively, as well as the initial settings $s_0$ selected in $\{10, 50, 100, 150\}$ and $i_0$ in $\{1, 10, 25, 50\}$. Among these, we considered all the 127 4-tuples of parameters having the expected number of survivors in an acceptable range according to the simulation tolerance, that is expectation computed with Eq. (1) larger than 0.001. For each of them, we run 10.000 simulations stopping when no more infected are left and saving the number of susceptibles.

These values have been used to perform a t-test on the mean to test for the null hypothesis "the mean number of survived susceptibles corresponds to the theoretical expected value given by Eq. (1)" with significance level $\alpha = 0.05$. Results are illustrated in Fig. 2A, where it is possible to notice that the percentage of rejected tests is very close to the significance level, in agreement with the definition of $\alpha$ being the probability that a true null hypothesis is rejected.

To further check for the validity of the results, for each tuple we have divided its 10.000 runs into 50 group of 200 and on each group we have performed the very same t-test to confirm the stability of the result. In Fig. 2B it can be seen a graphical representation of the proportion of tests in each group rejecting the null hypothesis; in Fig. 2C a boxplot illustrates more clearly the percentage of groups that have rejected the t-test.

*SIR with Exponential Distribution.* Afterwards, we took into account another experiment using a SIR model having both exponential inter-times for *Infection*

[2] We are assuming the number of initial recovered individuals equal to zero.

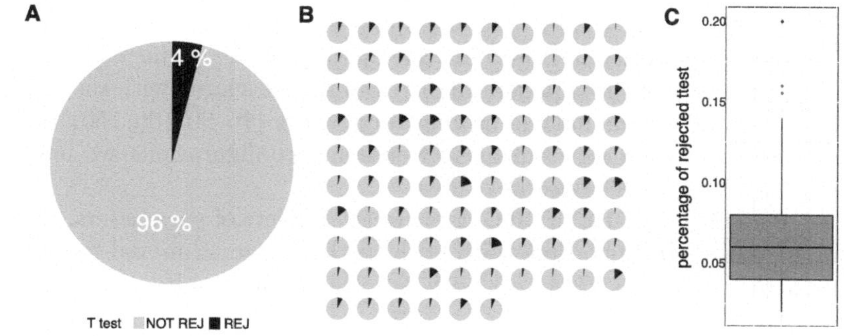


**Fig. 2.** A graphical representation of the t-test with significance level $\alpha = 0.05$. (A) The percentage of rejected tests for all tuples is very close to the significance level. (B) The graphical representation of the division of 50 groups of 200 of the 10.000 runs, and each pie shows for each tuple the percentage of groups rejecting the hypothesis test. The proportion of tests in each group rejecting the null hypothesis is again very close to the significance level as can be seen also in the boxplot of (C).

and *Recovery* transitions and compared results obtained using the classical SSA and the extended SSA forced to use the future event list.

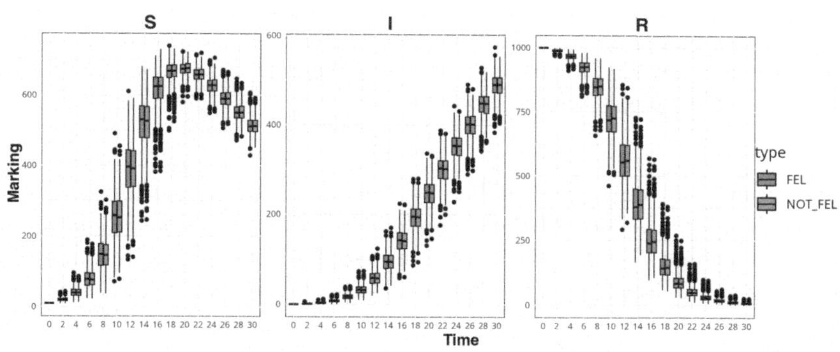


**Fig. 3.** Boxplots showing the variation of the marking (on the x-axis) through time (on the y-axis) of the three places S, I and R generated using the classical SSA (NOT_FEL) and the extended SSA (FEL). The values used for this simulation are ($s_0 = 1000$, $i_0 = 10$, $k = 0.0004$, $\theta = 0.04$). As can be seen, the data are comparable.

In Fig. 3 it is possible to see that the transient evolution of the system is comparable between the two algorithms. To further exploit a statistically significant comparison of the two evolutions, we performed an FDA test through the R package called fdatest (Functional Data Analysis) [22]. The FDA test takes into consideration two, eventually multidimensional, different groups of traces (either simulated or coming from experimental observations) and it tests for the null hypothesis that the underlying stochastic process is the same for the two of them. In order to do this, it exploits spline interpolations for each variable (place, in our framework) and tests for the equality of the coefficients.

We fixed the rate of the exponential distribution associated with the *Infection* transition at 1 and let the initial marking vary across $\{10, 50, 100, 150\}$ ($s_0$) and $\{1, 10, 25, 50\}$ ($i_0$), then for each of such starting configurations we made the *Recovery* transition take values in
$\{0.01, 0.1, 0.2, 0.5, 1, 2, 5, 10, 50, 100\}$, getting 144 triplets of parameters. For each of them, we ran 1000 simulations with both the FEL algorithm and the classical SSA, then performed an FDA test with $\alpha = 0.05$. The results of the tests are shown in Fig. 4.

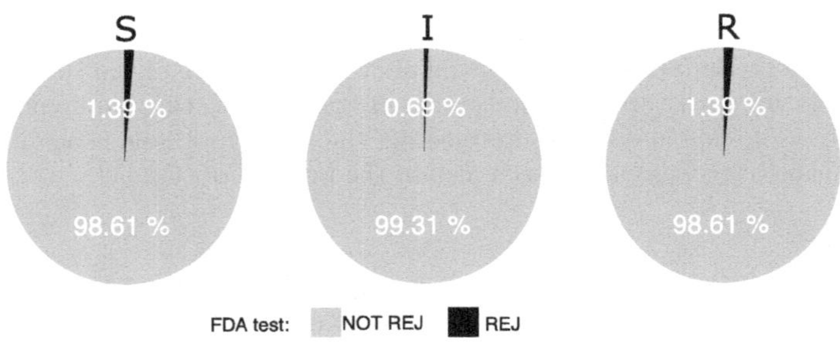


**Fig. 4.** Results of the FDA tests run for comparing the FEL algorithm traces and the SSA traces over 144 different triplets of parameters. The number of rejected tests is lower than the p-value, according to the hypothesis that the two algorithms are equivalent in the stochastic process produced.

Furthermore, an improvement has been made in the execution time of the algorithm. The first version of the extended SSA was several times slower than the classic reaching the 2 h execution compared to just a few seconds of the classic. Now the deletion of an event inside the future event list according to the modification of the enabling degree of the $T_{\neg exp}$ was speeded up making the execution comparable with the classical SSA. The analysis of the execution times is reported in Table 1, where the two algorithms have been executed 10 times on the nets of Fig. 1 only changing the value of the initial marking of S, $s_0$ and maintaining the same parameters, $i_0 = 1$, $k = 1$, $\theta = 0.25$, where $k$ and $\theta$ are respectively the rate of *Infection* and *Recovery*.

### 4.2 Parallel-Producer-Consumer Model

The *Parallel-Producer-Consumer* (PPC) as classically intended (i.e., Fig. 5A) represents two producers engaged in parallel manufacturing of two parts, intended for simultaneous consumption by a single consumer. Consumption commences only upon the availability of both parts, and the production cycle restarts once the previous parts have been consumed.

**Table 1.** Table showing the analysis of the execution times (expressed in seconds) using classical SSA (not FEL) or extended SSA (FEL). The table shows the mean and the standard deviation on ten times of ten different runs with the same parameters.

| $S_0$ | Mean FEL | $\sigma$ FEL | Mean Not FEL | $\sigma$ not FEL |
|---|---|---|---|---|
| 10 | 11.29 | 0.04 | 11.31 | 0.02 |
| 100 | 11.29 | 0.04 | 11.29 | 0.02 |
| 500 | 11.34 | 0.14 | 11.32 | 0.06 |
| 1000 | 11.30 | 0.05 | 11.30 | 0.03 |

On the contrary to the SIR model, the PPC one (Fig. 5) is a good example of a regenerative process, both when a single token is starting in $P_2$ [14], that when two of them are (at least once appropriate choices for $\mathcal{D}_1$, $\mathcal{D}_2$, $\mathcal{D}_3$ and enabling policy are made).

When more than one token is starting in $P_2$, as represented in Fig. 5B, it is possible to assign a *single-server semantic* for enabling rule. Such a choice lets us prove that not only the algorithm we implemented is in agreement with results derived from the theory of regenerative stochastic PNs, but also it is well adaptable to several enabling policies.

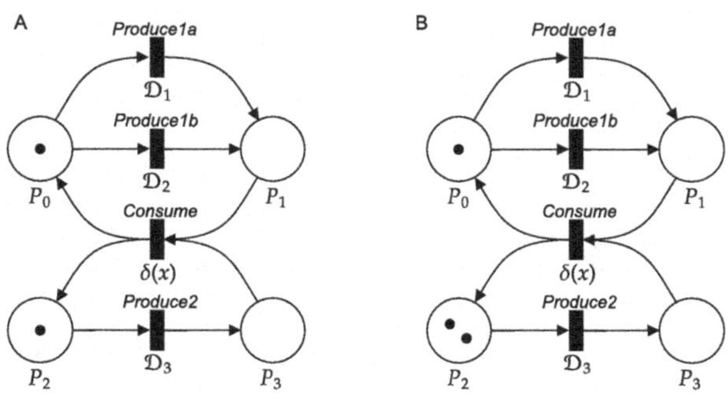


**Fig. 5.** (A) The PPC model with generally distributed times of production (distribution $\mathcal{D}_1$ and $\mathcal{D}_2$ for the upper producer and $\mathcal{D}_3$ for the bottom producer) and a deterministic holding time $x$ to be waited once both the goods have been completely produced before restarting of the process. [14]. (B) A modification of it, starting with two tokens in $P_2$ instead of one.

Therefore, we run two different experiments with the PPC model depicted in Fig. 5. The model 5A as studied in [14] has two regeneration points: the first one corresponding to the marking $(P_0, P_2)$ as firing of the *Consume* transition guarantees for all of the generally distributed holding times to be sampled, and the second one corresponding to the marking $(P_1, P_3)$, where the transitions *Produce1a, Produce1b* and *Produce2* are disabled. In particular, we computed through ORIS [20] the stationary distribution of the stochastic process associated with the PN with distributions of the first producer $\mathcal{D}_1 \sim Unif(0.5, 1.5)$ and $\mathcal{D}_2 \sim Unif(1, 2)$, distribution of the second producer $\mathcal{D}_3 \sim Unif(0, 2)$ and deterministic holding time before consumption $x = 3$ (see Fig. 6A).

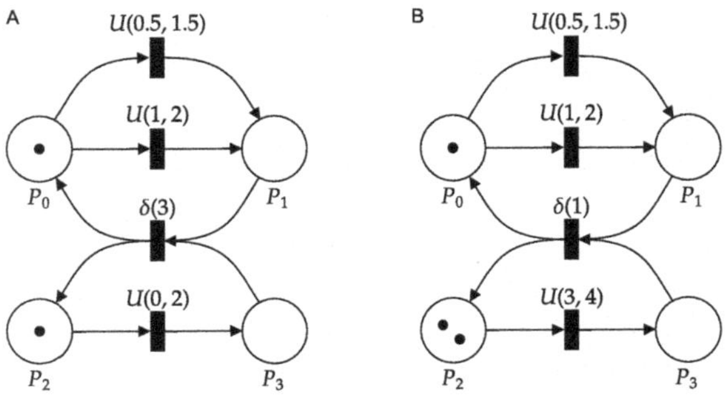


**Fig. 6.** Two different configurations of the Parallel-Producer-Consumer model.

We run 10.000 experiments using the FEL algorithm and computed the average frequency at which each state of the system is visited at each time step. As visible in Fig. 7, the simulations are in agreement with the theoretical results.

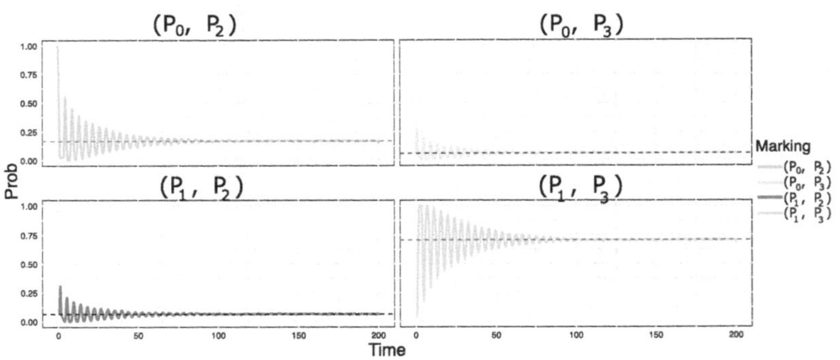


**Fig. 7.** The average among 10.000 simulated transient probability of each state compared with the theoretical stationary distribution (dashed) of the Parallel-Producer-Consumer in Fig. 6.

We further tested the PPC on a different setting (i.e. Fig. 5B) in order to make it relevant to change the enabling policy into a *single server semantic*. We started with 2 tokens in $P_2$ setting the distributions $\mathcal{D}_1 \sim Unif(0.5, 1.5)$, $\mathcal{D}_2 \sim Unif(1, 2)$, $\mathcal{D}_3 \sim Unif(3, 4)$ and $x = 1$ (Fig. 6B). Such a PN with *single server semantic* has still a Markov-regenerative process associated with it, as the firing of *Produce2* that determines the shift from $(P_1, 2P_2)$ to $(P_1, P_2, P_3)$ represents a regenerative event (i.e. all the non-exponential transitions are resampled in such

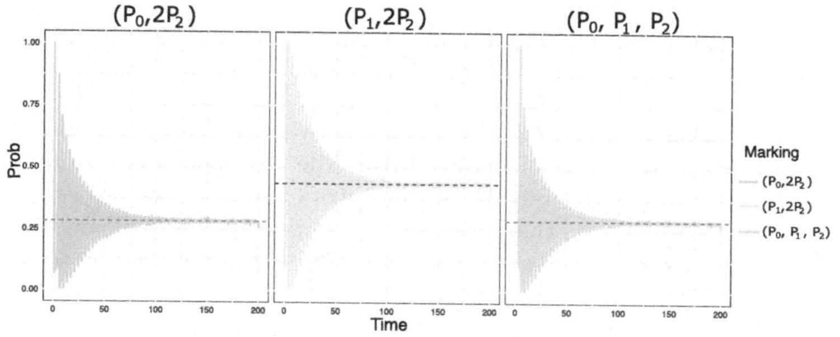


**Fig. 8.** The average among 10.000 simulated transient probability of each state compared with the theoretical stationary distribution (dashed) of the Parallel-Producer-Consumer in Fig. 6B run with *single server semantics*. The omitted states are reached from the system with probability 0.

an instant). Even in this case, it is possible to appreciate (Fig. 8) that the FEL algorithm correctly reaches the stable state of the system computed by ORIS.

## 5 Conclusion

In this paper, we have introduced a refined version of the GreatMod framework, enabling the modeling of behaviors that cannot be approximated successfully by exponential distributions. The ability to use distributions beyond the exponential is essential for accurately modeling systems, as not all behaviors can be fully captured by an exponential distribution alone. A series of example can be found, like the analytical methods for the performance evaluation of manufacturing systems where the failures of the systems are fully described by an exponential but the repairings are not using instead the phase-type distribution [19], or in epidemiology [16] to model part of the disease spread with other distributions. This enhancement necessitated the development of a novel graphical formalism within the Petri Net formalism, called GDESPN, accommodating events with general distributions. Additionally, we extended the well-known Stochastic Simulation Algorithm to handle such models efficiently, utilizing binary heaps for implementation. Our validation of this extended framework centered on two distinct models: the Susceptible-Infected-Recovered one and the Parallel-Producer-Consumer one. By applying the extended GreatMod to simulate the dynamics of these systems, we demonstrated the accuracy, correctness and efficacy of our approach. Specifically, through comparison with another state-of-the-art tool and examination of relevant theoretical literature, our results confirmed that GDESPN and GreatMod are adept at capturing and simulating the dynamics of processes with general characteristics.

Through this validation, we not only showcased the capability of GreatMod to handle non-Markovian dynamics but also underscored the robustness of the GDESPN formalism in representing a wide range of systems. This validation lays a solid foundation for future applications of the GreatMod framework in modeling and simulating complex systems characterized by non-exponential behavior.

**Funding Information.** This work was supported by grants from "Creation of a computational framework to model and study West Nile Disease" (Cod. ROL 67410, PI Marco Beccuti) funded by CRT foundation. This work is part of the project TrustAlert which has received funding from the "Fondazione Compagnia San Paolo and Fondazione CDP" under the "Artificial Intelligence" call (ID: CSP-AI 2; number: 2022.1700).

**Conflict of Interests.** All authors declare that they have no conflicts of interest.

## References

1. Marsan, M.A., Balbo, G., Conte, G., Donatelli, S., Franceschinis, G.: Modelling with Generalized Stochastic Petri Nets. Wiley, Hoboken (1995)
2. Basnarkov, L., Tomovski, I., Sandev, T., Kocarev, L.: Non-Markovian sir epidemic spreading model of COVID-19. Chaos, Solitons Fractals **160**, 112286 (2022)
3. Boguñá, M., Lafuerza, L.F., Toral, R., Serrano, M.A.: Simulating non-Markovian stochastic processes. Phys. Rev. E **90**, 042108 (2014). https://doi.org/10.1103/PhysRevE.90.042108
4. Cai, X.: Exact stochastic simulation of coupled chemical reactions with delays. J. Chem. Phys. **126**(12) (2007). https://doi.org/10.1063/1.2710253
5. Castagno, P., et al.: A computational framework for modeling and studying pertussis epidemiology and vaccination. BMC Bioinf. **21**, 1–32 (2020). https://doi.org/10.1186/s12859-020-03648-6
6. Choi, H., Kulkarni, V.G., Trivedi, K.S.: Markov regenerative stochastic petri nets. Perform. Eval. **20**(1), 337–357 (1994). https://doi.org/10.1016/0166-5316(94)90021-3, https://www.sciencedirect.com/science/article/pii/0166531694900213, performance '93
7. Ciardo, G., Miner, A.: Smart: simulation and Markovian analyzer for reliability and timing. In: Proceedings of IEEE International Computer Performance and Dependability Symposium, p. 60 (1996). https://doi.org/10.1109/IPDS.1996.540204
8. Clancy, D.: Sir epidemic models with general infectious period distribution. Stat. Probab. Lett. **85**, 1–5 (2014)
9. Clark, G., et al.: The mobius modeling tool. In: Proceedings 9th International Workshop on Petri Nets and Performance Models, pp. 241–250 (2001). https://doi.org/10.1109/PNPM.2001.953373
10. Dugan, J.B., Trivedi, K.S., Geist, R., Nicola, V.F.: Extended stochastic petri nets: applications and analysis. In: Proceedings of the Tenth International Symposium on Computer Performance Modelling, Measurement and Evaluation, pp. 507–519. Performance 1984, North-Holland Publishing Co., NLD (1984)
11. German, R.: Non-Markovian analysis. In: Lectures on Formal Methods and Performance Analysis, pp. 156–182. Springer, Cham (2001)
12. Gillespie, D.T.: Exact stochastic simulation of coupled chemical reactions. J. Phys. Chem. **81**(25), 2340–2361 (1977)
13. Grossmann, G., Bortolussi, L., Wolf, V.: Efficient simulation of non-Markovian dynamics on complex networks. PLoS ONE **15**(10), e0241394 (2020). https://doi.org/10.1371/journal.pone.0241394

14. Horváth, A., Paolieri, M., Vicario, E.: Equilibrium analysis of Markov regenerative processes. In: Jansen, N., Tribastone, M. (eds.) Quantitative Evaluation of Systems, pp. 172–187. Springer Nature Switzerland, Cham (2023)
15. Kurtz, T.: Solutions of ordinary differential equations as limits of pure jump Markov processes. J. Appl. Probab. **1**(7), 49–58 (1970)
16. Lauro, F.D., KhudaBukhsh, W.R., Kiss, I.Z., Kenah, E., Jensen, M., Rempala, G.A.: Dynamic survival analysis for non-Markovian epidemic models (2022). https://arxiv.org/abs/2202.09948
17. Masuda, N., Rocha, L.: A gillespie algorithm for non-Markovian stochastic processes. SIAM Rev. **60**(1), 95–115 (2018). https://doi.org/10.1137/16m1055876
18. Munro, J., Suwanda, H.: Implicit data structures for fast search and update. J. Comput. Syst. Sci. **21**(2), 236–250 (1980). https://doi.org/10.1016/0022-0000(80)90037-9, https://www.sciencedirect.com/science/article/pii/0022000080900379
19. Özdoğru, Ü., Altiok, T.: Analysis of two-valve fluid-flow systems with general repair times. In: Gershwin, S.B., Dallery, Y., Papadopoulos, C.T., Smith, J.M. (eds.) Analysis and Modeling of Manufacturing Systems. ISOR, vol. 60, pp. 255–288. Springer, Boston (2003). https://doi.org/10.1007/978-1-4615-1019-2_11
20. Paolieri, M., Biagi, M., Carnevali, L., Vicario, E.: The ORIS tool: quantitative evaluation of non-Markovian systems. IEEE Trans. Softw. Eng. **47**(6), 1211–1225 (2021). https://doi.org/10.1109/TSE.2019.2917202
21. Pelissier, A., Phan, M., Beerenwinkel, N., Martinez, M.R.: Practical and scalable simulations of non-Markovian stochastic processes (2023). https://arxiv.org/abs/2212.05059
22. Pini A., Vantini, S.: The interval testing procedure: inference for functional data controlling the family wise error rate on intervals (2013). https://www.mate.polimi.it/biblioteca/add/qmox/13-2013.pdf
23. Sahner, R.A., Trivedi, K.S.: Reliability modeling using SHARPE. IEEE Trans. Reliab. **36**(2), 186–193 (1987). https://doi.org/10.1109/TR.1987.5222336

# Identifying Damage-Related Features in scRNA-seq Data

Giovanni Marteletto[1], Bruno G. Galuzzi[2,4,5](✉), and Chiara Damiani[3,4]

[1] Immunobiology Department, Yale School of Medicine, New Haven, USA
[2] Institute of Bioimaging and Complex Biological Systems (IBSBC), Segrate, Italy
[3] Department of Biotechnology and Biosciences, University of Milano-Bicocca, Milan, Italy
chiara.damiani@unimib.it
[4] SYSBIO Centre of Systems Biology/ ISBE.IT, Milan, Italy
[5] National Biodiversity Future Center (NBFC), Palermo, Italy
brunogiovanni.galuzzi@cnr.it

**Abstract.** Quality control (QC) is fundamental in single-cell RNA sequencing (scRNA-seq) data analysis pipelines to ensure data reliability. A critical QC step involves identifying damaged cells using quality metrics like the percentage of mitochondrial genes or the total number of reads. However, automatically determining the threshold of these metrics for filtering damaged cells can be challenging. Moreover, using this metric alone may result in the removal of biologically meaningful cells. This study aims to find alternative biomarkers to improve the identification of damaged cells, focusing on gene lists other than mitochondrial genes. We hypothesized that genes localized within other organelles, particularly the nucleus, would exhibit similar enrichment patterns as mitochondrial genes. To test this hypothesis, we used a public scRNA-seq dataset where damaged cells were labelled via optical inspection. We considered as potential descriptors the percentage of genes from various lists, in particular lists of transcripts detected within the nucleus. We built a binary logistic regression model to differentiate damaged cells from good cells and evaluated its performance. Our results showed that the traditional criteria, such as mitochondrial genes, number of genes, and total counts, successfully identified damaged cells but tended to overestimate damage. Our findings suggest that although standard features are effective, their poor precision can be problematic. Incorporating other gene lists, particularly those related to nuclear transcripts, into classification models can improve the prediction of damaged cells. Further investigation is needed to understand the underlying mechanisms driving these relationships.

**Keywords:** scRNA-seq data analysis · Quality control · Damaged cells

## 1 Introduction

Quality control (QC) is an essential step in scRNA-seq data analysis pipelines to ensure the reliability and validity of the data [9]. During the QC step, low-quality samples need to be excluded from the downstream analysis, such as


M. Vettoretti et al. (Eds.): CIBB 2023, LNBI 14513, pp. 192–201, 2025.
https://doi.org/10.1007/978-3-031-90714-2_14

doublets, empty chambers or broken cells. The classical approach to filtering out low-quality samples requires arbitrarily setting filtering thresholds on specific quality metrics. For instance, to identify potentially broken or dead cells, the percentage of mitochondrial genes is widely used due to the preferential loss of free cytoplasmatic components in comparison to the RNAs confined within mitochondrial membranes in damaged cells. Thus, the mitochondrial RNA is enriched and its fraction increases. The selection of an appropriate threshold is typically based on a visual examination of the distribution of mitochondrial gene percentages across cells, with the aim of excluding cells in the upper tail of the distribution. However, the choice of the cutoff selection for some biological descriptors could be very dataset dependent or biased from the user's choice. For example, there are datasets in which the hypothetical threshold to divide damaged from good cells, using the percentage of mitochondrial genes, is around 5% and other ones for which is around 10%. Actually, there are no standardized procedures to determine an optimal threshold for the percentage of mitochondrial genes in order to distinguish damaged from suitable cells. The lack of a flexible yet reliable and fast automatic procedure for determining damaged cells could also be a major obstacle to the implementation of automated pre-processing pipelines. Moreover, there are several problems associated with using the percentage of mitochondrial genes as a quality control marker. Firstly, the percentage of mitochondrial genes can be influenced by biological factors, such as cell type, as some cell types naturally have higher levels of mitochondrial content, e.g.: myocytes [4], or mitochondrial deregulation in cancer cells [10]. Furthermore, the high percentage of mitochondrial genes could be a hallmark of metabolically deregulated populations. Therefore using the percentage of mitochondrial genes alone may result in the removal of biologically meaningful cells from the analysis, especially when studying the metabolism. In a recent study [5], the authors used several single-cell RNA-seq datasets to train and test a machine-learning model to distinguish low-quality (doublet, empty chamber, and damaged cell) from good cells. However, the pipeline requires raw sequencing data as input and large computational resources.

In this paper, we attempted to identify alternative biomarkers to improve the identification of damaged cells, with a particular focus on the percentage of other gene lists in addition to the traditional mitochondrial genes list [3]. Our hypothesis is that transcripts localized within other organelles should follow the same trend as mitochondrial genes by means of the same enrichment mechanism but, at the same time, should be robust to biological variations. We focused our analysis on genes whose transcripts are confined within the nucleus. In order to assess the hypothesis we took advantage of a dataset where damaged cells were labeled [7]. We investigated the possibility to use a classification model based on the combination of several biological descriptors, avoiding any prior threshold computation based on visual inspection of the data. The aforementioned approach of using known gene lists in addition to the standard features in a classification model could provide an effective and simple solution to the

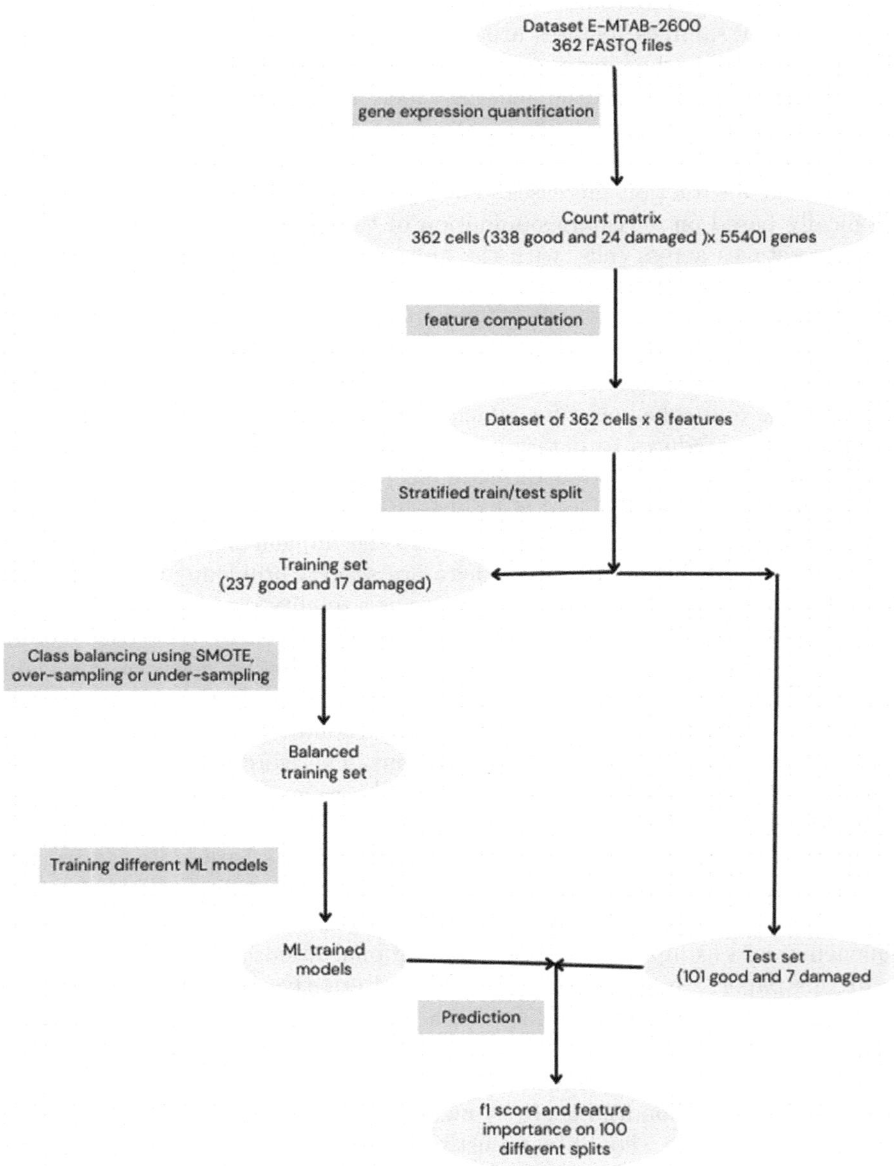


**Fig. 1.** Workflow for the pipeline used to build the classification model and feature importance analysis.

problem of damaged cells. We reported the workflow for the pipeline used to build the classification model and feature importance analysis in Fig. 1.

# 2 Data and Methods

## 2.1 Genes Lists

In addition to the list of 13 mitochondrial genes (MT), we considered four other candidate lists in this work. We started from results in literature obtained from two sequencing techniques able to distinguish locally enriched transcripts within the subcompartments of the cell, namely APEX and frac-seq [6]. We refer to the corresponding lists of candidate genes briefly as APEX and frac-seq. From these lists, we filtered out mitochondrial genes and kept the genes for which the log2 enrichment between the cytosolic and the nuclear fraction was above a threshold of 4.0 and 5.0, respectively. After this filtering, 267 genes over 5740 were retained in the APEX list, and 823 over 5363 genes in the frac-seq list. We also considered the list APEX/frac-seq formed by the intersection of these two lists, for a total of 46 genes. The fourth list consists of the 120 ribosomal protein genes (briefly ribo). In addition to the lists above, 50 lists composed of 500 random genes , not included in the other lists, were extracted to serve as control lists.

## 2.2 Dataset

We considered a scRNA-seq dataset [7], where the authors optically inspected the presence of low-quality cells within the capture chambers of the chips. The tagged cells serve as the ground truth for our analysis. This dataset is composed of 863 cells from Mouse Embryonic Stem Cells (mESCs). A total number of 338 were tagged as good, 14 as doublets, 24 as damaged, 5 as empty and 482 cells were not classified. From this dataset, we extracted the damaged and the good cells only, for a total of 362 cells. We labelled the damaged cells with 0 and the good cells with 1.

The single-cell RNA-seq data, provided as FASTQ files, were downloaded from European Bioinformatic Institute (EBI) using the access number E-MTAB-2600. We generated the count matrix using the RNAseq pipeline [11] provided via nf-core framework [2]. Cell metadata was associated with the correct sample via a custom R script. The final matrix includes 362 cells and 55401 genes. Genes expressed in fewer than three cells were filtered out and gene names were translated to human orthologs because we derived the candidate gene lists from human datasets.

From this dataset, we considered eight possible descriptors: the number of expressed genes, the total counts, the percentage of MT genes, the percentage of ribo genes, the percentage of APEX genes, the percentage of frac-seq genes, the percentage of APEX/fraq-seq genes, and the mean of the percentages of the 50 control lists. For any of these descriptors, we performed a preliminary differential expression analysis using the Welch's t-test, to check if the two groups (i.e. the damaged and good cells) have a statistically significant difference. Apart from the control lists ($p = 0.37$) and the APEX/fraq-seq genes list ($p = 0.17$), all the other descriptors result differential expressed ($p \leq 0.05$).

To add more complexity to the models, we created also 36 new descriptors, by applying a polynomial feature transformation of degree 2 to the original features, i.e. 8 squared terms and 28 interaction terms, with the aim to test if their use improves the performances.

### 2.3 The Train-Test Validation Procedure

We applied a stratified train-test split, in which we used 70% of the data to train the model, and the remaining 30% as a test set. This variant of the classical train test split preserves the same percentage of samples of each target class in both train and test set. We repeated the procedure 100 times, and we evaluated the performance in terms of mean and standard deviation of several metrics, among which precision, recall and F1-score. Precision indicates how many of the predicted positives (i.e. damaged) are actually positive. Recall shows how many real positives were correctly predicted. Finally, $F1$-score is the harmonic mean between precision and recall, providing a single metric that accounts for both false positives and false negatives.

### 2.4 Pre-processing of the Data

Before training, we standardized the features using information of training data only, by removing the mean and scaling to unit variance. Moreover, since training an ML model on heavily imbalanced dataset tend to be biased on the majority class, we employed three different strategies from the Python Imbalanced-learn library [8]. The first technique we considered, is the oversampling of the minority class, in which new instances for the minority class are created by picking at random with replacement. Therefore, the training set is augmented to obtain the same number of instances for both the classes. The second technique is the under-sampling of the majority class, in which a subset of instances from the minority class are selected by picking at random with replacement. In this case, the training set becomes smaller, to obtain the same number of instances for both the classes. The last technique is the Synthetic Minority Over-sampling Technique (SMOTE) [1]. In the SMOTE approach, new instances for the minority class are created in feature space, and as in the previous oversampling technique, the training set is augmented to obtain the same number of instances for both the classes. To obtain a new instance, the algorithm selects a minority class instance at random and finds its $k$ nearest minority class neighbors. The synthetic instance is then created by choosing one of the $k$ nearest neighbors at random and connecting them to form a segment in the feature space. The synthetic instance is generated as a convex combination of the two chosen instances.

### 2.5 The Classification Models

We considered the Logistic Regression (LR), which is one of the simplest parametric methods to model the relationship between the features (i.e. the gene lists descriptors) and the dependent variable (i.e., good or damaged information).

The probability that the dependent variable belongs to a particular category is established as follows:

$$p(y = 1|x) = \frac{1}{1 + e^{-(w_0 + w_1 x_1 + \ldots + w_n x_n)}} \quad (1)$$

where $n$ is the number of input variables, $x_i$ is the $i$-th feature, and $w_i$ is the i-th model parameter. After calculating the probability, a threshold of 0.5 is considered to associate the class 1 (good cell) or 0 (damaged cell). LR is relatively easy to implement, interpret, and very efficient to train.

# 3 Results

## 3.1 MT Gene List Outperforms the Other Lists but Tends to Overestimate Damaged Cells

We evaluated if the percentage of genes belonging to the proposed lists can be used to identify damaged cells by applying a simple arbitrary cut-off. We reported in Fig. 2a the distribution of the percentages of the MT list, where a threshold is applied to filter out the cells considered as "outliers" of the distribution. We applied the same approach to APEX, frac-seq, and APEX/frac-seq lists. For each feature, the cut-offs were selected by visually inspecting the upper tail of the list percentage's distribution in the cell population. The effectiveness of the above approach was evaluated by calculating the probability of achieving the same results by chance assuming a binomial distribution. Using the Venn diagrams, we depict the intersection level between the true "damaged" cell list (red set) and the lists of predicted damaged cells, obtained through the considered gene list enrichment. The outcomes reveal that the traditional criteria, namely, the MT genes, number of genes and total counts, (Fig. 2g) successfully identifies damaged cells better than expected by chance ($p < 0.01$) and show sensitivity to damaged cells ($recall = 0.91$). However, they tend to overestimate damage by selecting 54 cells that do not exhibit any apparent damage, resulting in poor precision ($precision = 0.22$). Conversely, the alternative lists using simple cutoffs fail to distinguish damaged cells from healthy cells more effectively than randomly selecting the same number of cells, as shown in Fig. 2h-j. Neither the frac-seq nor apex lists selected damaged cells, and the gene list shared by both techniques performs no better than random selection from the dataset ($p = 0.91$).

## 3.2 The Use of Multiple Gene Lists Improves the Classification of Damaged Cells

We used the LR model to study the possible impact of any feature on the classification task. To do this, we built several LR models from any possible combinations of features, for a total of 255 feature sets and equally many LR models. For each feature set, the performance of the model is computed 100 times, using 100 different train-test splits, to provide a more robust evaluation of model performance. Then, the resulting metrics were averaged across the runs. By employing

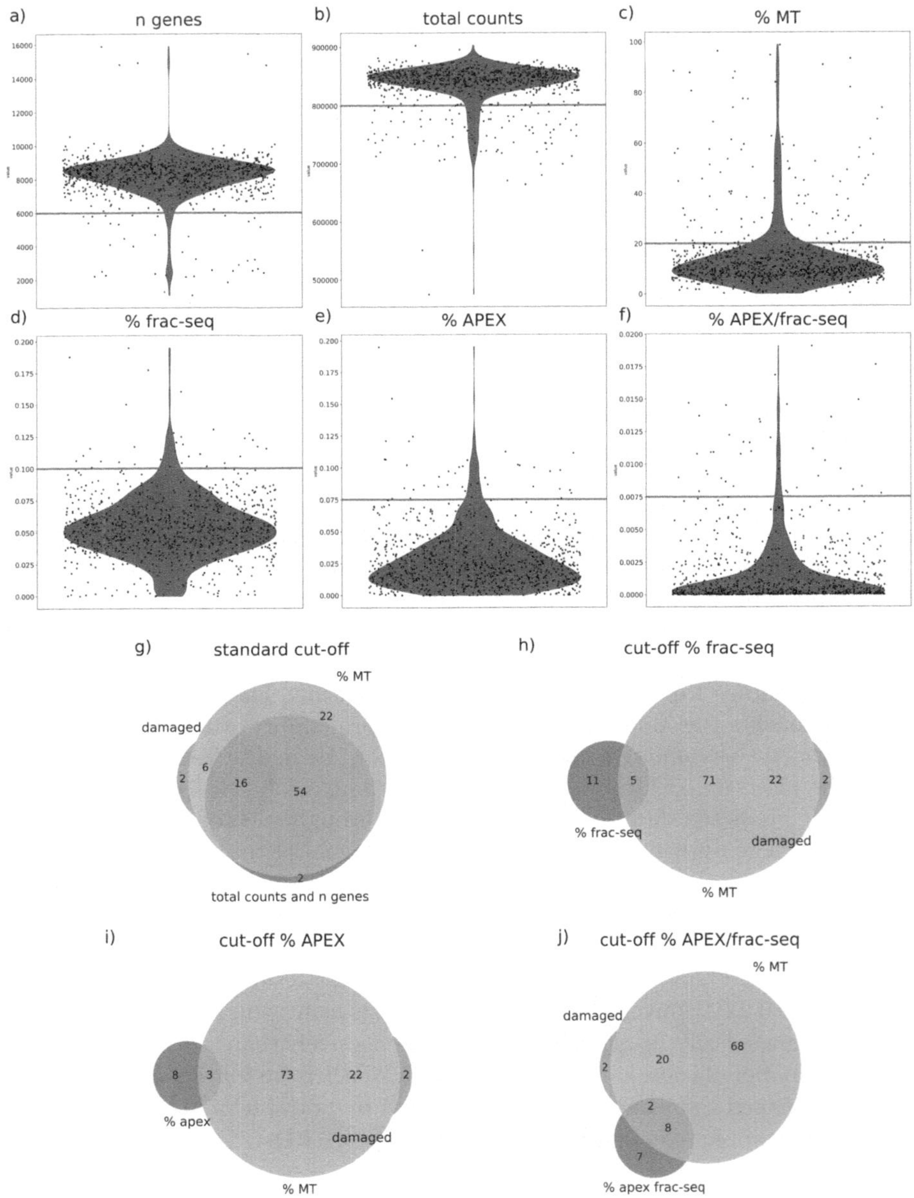


**Fig. 2.** a–f) Metrics distributions and corresponding thresholds. g–j) Venn plots representing the level of intersection between the true "damaged" cell list (red set) and the lists of predicted damaged cells. For visualization reasons, total counts (subplot g) and the number of genes were united as a single selection criterion (Color figure online).

this comprehensive approach, we ranked the feature combinations on the base of the $F1$ score.

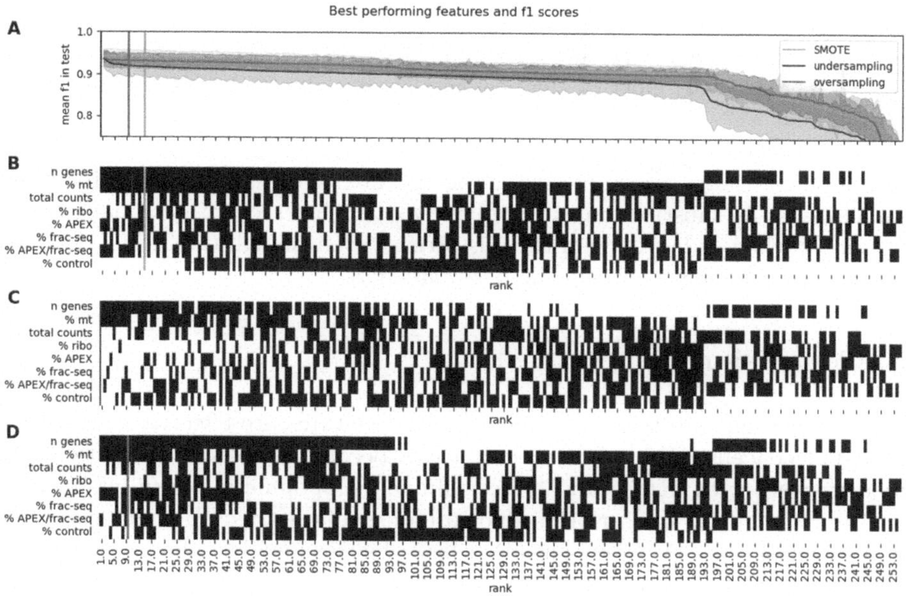


**Fig. 3.** (A) Ranks of the best-performing models and their respective mean F1 scores in the test set. The vertical line indicates the model with the standard features, i.e. total counts, number of genes and % MT. (B–D) Heatmaps showing the presence (black) or absence (white) of a specific feature in the classification model, using the three different balancing strategies, i.e. SMOTE (b), undersampling (c) and oversampling (d). (Color figure online)

In Fig. 3a, we report three line-plots showing the mean and standard deviation of the $F1$ score (axis y) across the 100 train-test splits of any of the 255 LR models (axis x), using the SMOTE (red), the oversampling (green) and the undersampling (blue) strategy, respectively. The models are ranked as a function of the corresponding mean $F1$ score. In Fig. 3b–d, we reported three heatmaps showing the presence (black) or absence (white) of a specific feature for any LR model using the SMOTE (b), undersampling (c) and oversampling (d) strategy. The models are sorted as a function of the corresponding $F1$ score reported in Fig. 3a.

The SMOTE technique is the best sampling strategy, closely followed by the oversampling approach. The undersampling strategy lags significantly behind the first two, showing a much weaker performance both in terms of mean and standard deviation. Taking together, this figure reveals that the number of genes and percentage of mitochondrial genes consistently appeared among the best-scoring models. Although the standard features, namely % MT, number of genes and total counts are present in most of the best models, the model having only these features is not in the first positions both for the SMOTE and the oversampling strategies (Fig. 3b and 3d). For example, the top oversampling and SMOTE models are significantly different from the standard model (t-test $p < 0.01$) and have both the following features: number of genes, % MT, % APEX/fraq-seq, %frac-seq. Moreover, most of the best models has the % APEX descriptors as feature.

All the models also suffer a great decline in the overall F1-score as soon as the mitochondrial genes is not among the available features. Another interesting behaviour is the substitution of % MT for $n_{genes}$ that seems to compensate for each other when at least one of them is present even though the % MT tends to appear most frequently among the best classifiers.

## 4 Conclusion

Our findings indicate that although standard features are effective in identifying damaged cells, their poor precision can become problematic, particularly when working with datasets where certain cell subpopulations exhibit upregulated mitochondrial transcription due to biological reasons. To compensate for this overestimation, we posit that other lists of genes may be valuable when predicting damaged cells. In this work, we hypothesised that nuclear transcripts, detected with techniques capable of distinguishing their subcellular localization, would exhibit a similar pattern to mitochondrial genes. However, the simple cutoff results indicate otherwise. Upon closer examination, these gene lists demonstrated their usefulness when incorporated into a classification model for predicting broken or dead cells. This work has some possible limitations. The main concern is the small size of the dataset, which clearly poses some issues about the generalization of the results. Even if the train/test split was repeated on the analysis 100 times, and using different balancing strategies, the testing set remains too small to be sure that the results are completely generalizable. To this aim, as further investigation, we will search for other datasets for training and testing a classification model based on the use of these gene lists, possibly considering also other classification models, such as Random Forest or Gradient Boosting.

**Acknowledgements.** We warmly thank Davide Maspero for his useful suggestions.

**Data and code availability.** The dataset used in this report is public available on European Bioinformatic Institute (EBI) using the access number E-MTAB-2600. We developed all our computational tests in Python. For the estimator LR, we used the implementation provided by Scikit-Learn. For the balance strategies, we used the implementation provided by the Imbalanced-learn library. The codes and data used in this work are publicly available at the GitHub repository https://github.com/CompBtBs/DeadCell_analysis.

## References


1. Chawla, N., Bowyer, K., Hall, L., Kegelmeyer, W.: SMOTE: synthetic minority over-sampling technique. J. Artif. Intell. Res. **16**, 321–357 (2002)
2. Ewels, P., et al.: The nf-core framework for community-curated bioinformatics pipelines. Nat. Biotechnol. **38**(3), 276–278 (2020)
3. Fazal, F., et al.: Atlas of subcellular RNA localization revealed by APEX-Seq. Cell **178**, 473–490 (2019)
4. Galow, A., et al.: Quality control in scRNA-Seq can discriminate pacemaker cells: the mtRNA bias. Cell. Mol. Life Sci. **78**, 6585–6592 (2021)
5. Ilicic, T., et al.: Classification of low quality cells from single-cell RNA-seq data. Genome Biol. **17**, 1–15 (2016)
6. Kaewsapsak, P., Shechner, D., Mallard, W., Rinn, J., Ting, A.: Live-cell mapping of organelle-associated RNAs via proximity biotinylation combined with protein-RNA crosslinking. Elife **6**, e29224 (2017)
7. Kolodziejczyk, A., et al.: Single cell RNA-sequencing of pluripotent states unlocks modular transcriptional variation. Cell Stem Cell **17**, 471–485 (2015)
8. Lemaître, G., Nogueira, F., Aridas, C.: Imbalanced-learn: a python toolbox to tackle the curse of imbalanced datasets in machine learning. J. Mach. Learn. Res. **18**, 1–5 (2017). http://jmlr.org/papers/v18/16-365.html
9. Luecken, M., Theis, F.: Current best practices in single-cell RNA-seq analysis: a tutorial. Mol. Syst. Biol. **15**, e8746 (2019)
10. Nguyen, N., Kim, S., Jo, Y.: Deregulated mitochondrial DNA in diseases. DNA Cell Biol. **39**, 1385–1400 (2020)
11. Patel, H., et al.: nf-core/rnaseq: nf-core/rnaseq v3.11.2 - Resurrected Radium Rhino. (Zenodo, 2023, 4). https://doi.org/10.5281/zenodo.7862091

# A Benchmark Study of Gene Fusion Prioritization Tools

Francesca Miccolis, Marta Lovino(✉), and Elisa Ficarra

Enzo Ferrari Engineering Department, University of Modena and Reggio Emilia, Via P. Vivarelli 10, 41125 Modena, Italy
{francesca.miccolis,marta.lovino,elisa.ficarra}@unimore.it

**Abstract.** A gene fusion is a chromosomal aberration from juxtaposing separate genes. Since some gene fusions are involved in tumorigenesis, proper gene fusion investigation and analysis are crucial in the literature. After DNA/RNA sample extraction, detecting gene fusions requires first gene fusion detection tools, which usually provide many false positives. Given the high experimental costs in wet lab validation of a single fusion, gene fusion prioritization tools were made available over the years to significantly narrow down candidate gene fusions for validation (e.g., Oncofuse, Pegasus, DEEPrior, ChimerDriver). Although a few reviews about gene fusion detection tools are available, a benchmark on prioritization tools is not available yet in the literature. The aim of this paper is twofold: 1. to provide a curated dataset for a fair gene fusion prioritization tool evaluation. 2. to develop a proper comparison based on time, resources, and tool confidence on selected gene fusions. Based on this benchmark, it can be stated that ChimerDriver is the most reliable tool for prioritizing oncogenic fusions.

**Keywords:** benchmark · gene fusions · kinases · oncogenes · prioritization

## 1 Introduction

Gene fusions play a significant role in oncogenesis and tumorigenesis, making the investigation of such mutations a pivotal area of study in the scientific literature [3]. The analysis of gene fusions is a complex process involving several steps. Firstly, the DNA or RNA sample is extracted. Then, the sequencing procedure produces text files containing the sample information reported line by line (called reads). By exploiting suitable gene fusion detection tools (e.g., StarFusion [7], Arriba [17]), it is possible to obtain the list of potential gene fusions in the samples [14]. However, such tools are prone to numerous false positives, leading to increased complexity in selecting gene fusions for wet lab validation (which is notoriously costly).

Overall, the combination of genomic complexity, detection method sensitivity, technical factors, and tumor heterogeneity collectively contribute to the higher


M. Vettoretti et al. (Eds.): CIBB 2023, LNBI 14513, pp. 202–211, 2025.
https://doi.org/10.1007/978-3-031-90714-2_15

occurrence of false positives in gene fusion detection. Therefore, it is crucial to carefully validate and confirm fusion events in research and clinical settings. For this purpose, machine and deep learning methods proved to be effective in processing many biological data [5,15]. In this context, gene fusion prioritization tools (e.g., Oncofuse, Pegasus, DEEPrior, ChimerDriver) are meant to shorten the gene fusion list outputted by gene fusion detection tools, providing which ones are more likely to be involved in oncogenic mechanisms or significantly impact disease development [1,11,12,16] exploiting statistical, dimensionality reduction, and deep learning techniques [4,9].

Despite the considerable number of available prioritization tools, there is currently no curated state-of-the-art benchmark that highlights and compares each tool's peculiarities and predictive performance. Indeed, a few reviews in the literature address the gene fusion detection problem, but they do not focus specifically on prioritization tools [10,14].

This work aims to provide a systematic benchmark of prioritization tools offering a dual contribution: firstly, it provides a curated dataset for the evaluation, and secondly, it describes the computational requirements, execution time, and tool confidence as evaluation criteria for such tools.

## 2 Data and Methods

The investigation of gene fusions is crucial in scientific literature due to their significant involvement in oncogenesis and tumorigenesis. Given the considerable number of false positives usually generated from the detection step, it becomes essential to implement effective filtering strategies during the gene fusion prioritization step, with the aim to reduce the number of false positives. This benchmark focuses on online available prioritization tools, which are all open source and available on GitHub: ChimerDriver [12], DEEPrior [11], Oncofuse [16], Pegasus [1].

### 2.1 Data

A proper dataset has been built to compare gene fusion prioritization tools fairly. Indeed, it aims to test tools' effectiveness on known oncogenic and not oncogenic gene fusions. The dataset comprises 15500 gene fusions (4500 and 11000, respectively, oncogenic and not oncogenic). Oncogenic fusions are selected from Fusiongdb [8], an extensive database collecting fusions from multiple sources. In contrast, not oncogenic fusions are selected from Babicenau *et al.* paper [2], which reports the most complete list of gene fusion occurring in non-tumor samples.

The dataset is made up of five subsets, based on the relevance and consequences that gene's role (e.g., kinase, cancer-related) has in the gene fusion prioritization problem:

(A) **500 oncogenic gene fusions involving kinases**. Kinases are particularly involved in signal transduction pathways, transmitting signals within cells and coordinating responses such as growth and differentiation. They also contribute to cell cycle regulation, ensuring proper cell division. Hence, abnormal kinase activity can lead to diseases, including cancer, making them important targets for drug development. Therefore, fusions involving gene kinases are particularly scrutinized in cancer research and missing their detection is critical [13]. The number of gene fusions in the dataset is proportional to humans' overall number of kinase genes.

(B) **1000 oncogenic gene fusions involving cancer genes**. Another critical gene group to carefully consider in prioritizing gene fusions is cancer-related. Therefore, we included 1000 oncogenic fusions in which one of the two genes is cancer-related, exploiting the COSMIC [6][1] database annotations (the number of cancer-related genes is higher with respect to the kinases ones according to the proportion to humans' overall number of cancer-related genes).

(C) **3000 other oncogenic gene fusions**. We selected three thousand fusion genes to evaluate the ability to recognize oncogenic fusions in which genes are not already known to be associated with cancer. This point is still important in cancer research.

(D) **1000 not oncogenic gene fusions involving cancer genes**. Fusions involving cancer-related genes do not necessarily imply that the fusion is oncogenic. Indeed, non-oncogenic fusions containing cancer-related genes can also be found in healthy samples. For this reason, we selected 1000 fusions that contain cancer-related genes even though the fusion is non-oncogenic.

(E) **10000 other not oncogenic gene fusions**. Evaluating non-oncogenic fusions that do not contain kinases or cancer-related genes is crucial too. The number reflects the known ratio of oncogenic to non-oncogenic fusions.

### 2.2 Methods

An overview of the paper contributions is represented in Fig. 1. Nextflow has been used as a framework[2] to ensure the isolated execution of each tool in its specific container. The prioritization tools have been analyzed to evaluate the main characteristics, prediction performances, time, and memory requirements.

The main features evaluated in this benchmark are input type to the prioritization tool (RNA or DNA data or both), accepted genome versions (GRCh36/37/38), ease of use, year of the last updates to consider tool maintenance and any other noteworthy feature. When considering ease of use, several factors are taken into account, such as the level and detail of available documentation, the ease of creating an isolated environment to run the tool, the number of modifications and steps required for execution, and the number of gene fusion

[1] *Catalogue of Somatic Mutation in Cancer.*

[2] The only prerequisites are Bash 3.2 (or later) and Java 11 (or later, up to 20).

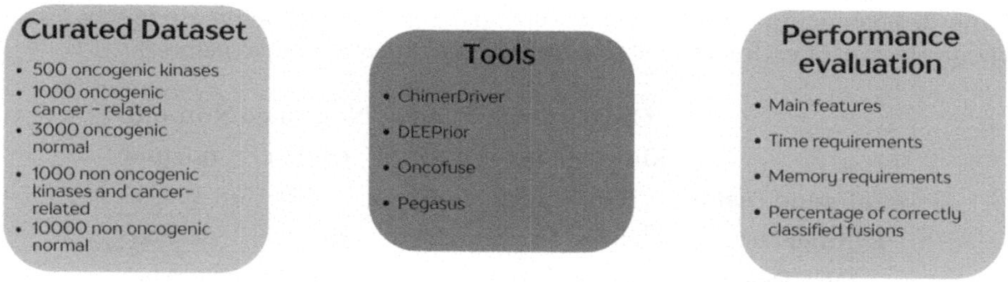


**Fig. 1.** Overview of the benchmark analysis: green rectangles contain the work contributions; blue rectangle contains evaluated prioritization tools (Color figure online)

detection tools whose inputs are accepted by each prioritization tool. These factors play a role in assessing how user-friendly and accessible a prioritization tool is for researchers and practitioners. Based on all this information, each tool has been defined as poor, fair, good, or excellent.

Each prioritization tool takes each subset of gene fusions as input and provides a score for each fusion representing its oncogenic probability. For accuracy evaluation, the chosen metric was the percentage of high confidence correctly classified fusions out of the total within each subset. For oncogenic fusions, fusions with a score above the threshold of 0.8 were considered correctly classified, while for non-oncogenic fusions, those with a score below 0.2 were considered correctly classified. 0.8 and 0.2 thresholds were chosen according to the literature [11,16].

**Table 1. Main features of prioritization tools.** *RNA* and *DNA* columns stands for the native input accepted from the tool. The *compatible genome version* column contains the list of the accepted genome versions. The *ease of use* column contains a parameter of user-friendliness of the tool based on several considerations explained in Sect. 2. The *last update* column contains the year of the last release for each tool. The *other* column contains other relevant feature.

| | RNA | DNA | Compatible genome version | Ease of use | Last update | Other |
|---|---|---|---|---|---|---|
| **ChimerDriver** | Yes | Yes | GRCh37, GRCh38 | Good | 2023 | Retraining mode |
| **DEEPrior** | Yes | Yes | GRCh37, GRCh38 | Excellent | 2022 | Retraining mode |
| **Oncofuse** | Yes | Yes | GRCh36, GRCh37, GRCh38 | Excellent | 2016 | None |
| **Pegasus** | Yes | No | GRCh37 | Poor | 2016 | None |

## 3 Results

The primary objective of this paper consists of two key aspects. Firstly, it aims to offer a curated dataset that ensures a fair evaluation of gene fusion prioritization

**Table 2. Execution time.** Columns correspond to each of the input subsets. Rows contain the size of each input subsets and the prioritization tools' names.

| | **Onco kinases** | **Onco cancer** | **Onco normal** | **Non onco cancer** | **Non onco normal** |
|---|---|---|---|---|---|
| **# of gene fusions** | 500 | 1000 | 3000 | 1000 | 10000 |
| **ChimerDriver** | 26m15s | 41m12s | 2h3m | 46m25s | 9h25m |
| **DEEPrior** | 38m48s | 1h19m | 3h38m | 46m57s | 6h7m |
| **Oncofuse** | **11.7s** | **12.1s** | **24.5s** | **11.9s** | **17.3s** |
| **Pegasus** | 31m11s | 1h3m | 1h53m | 45m12s | 7h29m |

**Table 3. Computational resources.** Columns correspond to each of the input subsets. Rows contain the size of each input subset and the prioritization tools' names.

| | **Onco kinases** | **Onco cancer** | **Onco normal** | **Non onco cancer** | **Non onco normal** |
|---|---|---|---|---|---|
| **# of gene fusions** | 500 | 1000 | 3000 | 1000 | 10000 |
| **ChimerDriver** | 3.1 GB | **3.2 GB** | **2.4 GB** | 3.2 GB | 5.0 GB |
| **DEEPrior** | **2.9 GB** | 3.5 GB | 4.2 GB | **2.9 GB** | 5.0 GB |
| **Oncofuse** | 5.1 GB | 5.1 GB | 3.8 GB | 5.1 GB | 5.2 GB |
| **Pegasus** | 4.1 GB | 4.5 GB | 3.1 GB | 4.1 GB | **4.6 GB** |

**Table 4. Accuracy.** Columns correspond to each of the input subsets. Rows contain the size of each input subset and the prioritization tools' names.

| | **Onco kinases** | **Onco cancer** | **Onco normal** | **Non onco cancer** | **Non onco normal** |
|---|---|---|---|---|---|
| **# of gene fusions** | 500 | 1000 | 3000 | 1000 | 10000 |
| **ChimerDriver** | **91.9 %** | **83.5 %** | **59.8 %** | 30.0 % | **66.8 %** |
| **DEEPrior** | 35.5 % | 46.0 % | 30.6 % | 19.4 % | 23.7 % |
| **Oncofuse** | 29.1 % | 23.5 % | 9.4 % | 45.5 % | 47.2 % |
| **Pegasus** | 5.4 % | 0.6 % | 0.4 % | **70.1 %** | 55.7 % |

tools. This dataset will serve as a reliable foundation for assessing the performance and efficacy of various tools in accurately identifying and prioritizing gene fusions. Secondly, the paper provides an evaluation framework beyond mere accuracy assessment, enabling a deeper understanding of the practical implications and trade-offs associated with different gene fusion subsets.

An overview of the *main features* of the analyzed prioritization tools is reported in Table 1. Several aspects are taken into consideration, including the type of input the tools can accept, compatible genome versions, the latest tool update (to consider tool maintenance), ease of use, and any other noteworthy features. All the included information is obtained from each tool documentation, except for the ease of use, defined in Sect. 2 and determined based on the following considerations. ChimerDriver and DEEPrior provide a file with a list of requirements for setting up the execution environment. For ChimerDriver, the instructions for obtaining predictions involve building the model and applying it to the input data. The only drawback is that it is incompatible with any fusion detection tool outputs. However, the reference input is straightforward to produce. ChimerDriver also supports the **cross-validation** with the k-fold method[3] (k = 10). DeepPrior is compatible with the output of a few fusion detection tools such as DeFuse, STAR-Fusion, ChimPIPE, EricScript, FusionCatcher, InFusion, JAFFA, SOAPfuse, and TopHat. DEEPrior, as Oncofuse, allows the option to construct custom input by providing only the chromosomes and breakpoints of the two genes involved in the fusion; also, the execution can be initiated with a single command in the command prompt. Oncofuse only requires Java (1.7.0 or higher) to be executed and is also compatible with the output of TopHat, RNASTAR, and FusionCatcher. Pegasus is the most complex tool to be used since it requires manual installation of requirements and modification of several options and files, making the setup process and error comprehension more challenging. Pegasus supports the output of Chimerascan, DeFuse, and Bellerophontes. As an additional feature, DEEPrior and ChimerDriver also support the **retraining mode**, producing the weights and the architecture of the newly trained model that could be used instead of the default deep learning model. In this way, the prediction is optimized for the specific biological case given as input. Also, in Oncofuse there is a supplemental parameter that allows the user to specify the tissue among EPI (epithelial), HEM (hematopoietic), MES (mesenchymal), or AVG (for all other cases).

Regarding the *execution time*, it can be stated that Oncofuse is the fastest tool for each of the 5 different subsets, regardless of the dataset size, taking a maximum of 25 s to perform the execution. The longest run overall was performed by ChimerDriver on the largest subset, taking 9 h and 25 min (compared to 6 h and 7 min for DEEPrior and 7 h and 29 min for Pegasus). Regarding the remaining four subsets, DEEPrior consistently appears slightly slower than ChimerDriver and Pegasus on all the oncogenic subsets, requiring from 7 min to 1 h and a half of additional time running on the same subsets. In the subset with non-oncogenic gene fusions involving cancer genes, the three tools show a very similar execution time, around 46 min. Results are shown in Table 2.

Regarding *computational resources*, Oncofuse has the highest demand, averaging around 4.47 GB of RAM. On the contrary, ChimerDriver has the lowest demand, averaging about 3.54 GB of RAM. DEEPrior requires around 4.28 GB of RAM, while Pegasus typically needs slightly less, around 4.28 GB. Results are shown in Table 3.

[3] K-fold cross-validation is a technique that involves dividing a dataset into k subsets, using k-1 subsets for training and the remaining subset for validation in each iteration, to evaluate and improve the performance of a machine learning model.

As mentioned in Sect. 2, one of the metrics used to evaluate performances is the *percentage of high confidence* correctly classified fusions within each subset. For oncogenic fusions, those with a score higher than the 0.8 threshold were deemed correct, whereas, for non-oncogenic fusions, those with a score lower than 0.2 were considered correct. Results are shown in Table 4.

## 4 Discussion

About the execution time, Oncofuse outstands its competitors by running in a bunch of seconds for each subset. However, it seems sensitive to subset types, as execution times increase in runs involving subsets of genes not annotated as cancer-related or kinases. Among the remaining trio of tools, the time taken for execution is notably contingent on the size of the dataset, with an average time for each fusion of $2.86 \pm 0.35$ s for ChimerDriver, $3 \pm 0.54$ s for Pegasus and $3.76 \pm 1$ s for DEEPrior. As a result, DEEprior appears to be slower than both ChimerDriver and Pegasus when considering the estimated runtime for fusion as well. ChimerDriver, which exhibits the lowest variance, appears to be the tool where execution time is more reliant on input dimensionality. DEEPrior, such as Oncofuse, seems to be somehow sensitive to the subset type. Indeed, the average execution time for fusion for DEEPrior is 4.5 s among subsets with oncogenic fusions, as opposed to 2.5 s (almost half the time) in the two subsets containing non-oncogenic fusions.

For what concerns the computational resources required from each tool, ChimerDriver is generally the most convenient tool with only 3.4 GB on average. However, the resource demands of other tools are not substantially higher, making them still viable options despite slight variations.

As mentioned in Sect. 3, the Table 4 shows the accuracy of each tool for all the subsets. ChimerDriver has good performance with all oncogenic subsets, especially the ones with kinases and cancer-related genes, with respectively 91.9% and 83.5% accuracy. The performance is relatively lower, but still acceptable, for subsets of fusion events (both oncogenic with 59.8% and non-oncogenic with 66.8%) involving genes that do not have kinase function or cancer-related associations. The only subset where ChimerDriver does not provide adequate accuracy is the one with non-oncogenic fusions involving genes annotated as kinases or cancer-related, with a drop of the accuracy to 30.0%. Considering ChimerDriver's focus on the functional roles of genes, it is anticipated that its accuracy diminishes for non-oncogenic fusions involving kinases or cancer-related genes. This emphasis on functionally significant genes, irrespective of their oncogenic potential, likely accounts for the observed decrease in accuracy within this subset.

The results of DEEPrior are generally poor across all analyzed subsets, around 31%. It can be noted that the accuracy value is slightly higher on subsets of oncogenic fusions with an average of 37.4% and a peak in the classification of fusions involving cancer-related genes of 46%. The recognition of non-oncogenic fusions is even poorer, especially for fusions involving cancer-related genes and kinases, where the accuracy is 19.4%. For fusions that do not involve these genes, the accuracy is around 23.7%.

The result of Oncofuse is similar to that of DEEPrior in terms of overall performance, which is 30.94%. However, an opposite situation is observed when considering the type of inputs. Oncofuse performs better in recognizing non-oncogenic fusions, with accuracies of 45.5% in the subset where the fused genes are kinases or cancer-related, and 47.2% in the subset with non-functional genes. Conversely, the performance in subsets of oncogenic fusions is significantly lower, with 29.1% for the cancer-related subset, 23.5% for the subset with kinases, and even 9.4% in the subset with non-functional genes.

Pegasus seems to achieve decent performances discriminating non-oncogenic gene fusions with cancer-related or functional genes (such as kinases), where all the other tools failed. Indeed, for this subset, its performance of 70.1% is the best among every tool by far (45.5% by Oncofuse, 30% with ChimerDriver, and 19.4% by DEEPrior). On the other hand, Pegasus is the tool that exhibits the lowest performance due to its specific purpose. Its performance on oncogenic subsets is negligible, reaching a maximum of 5.4% in the subset with oncogenic fusions involving kinases, and less than 1% in the subset of oncogenic cancer-related fusions.

## 5 Conclusion

In conclusion, the runs of prioritization tools on different subsets lead to different deductions. Regarding the main characteristics of the tools, the most recommended ones are ChimerDriver and DEEPrior, which have the most recent maintenance update (respectively in 2023 and 2022), compatibility with multiple versions of the reference genome, and additional features compared with the other tools.

Concerning the time required for the execution, Oncofuse has an extreme prediction speed, while all the other tools are dependent on the input dimensionality.

From the perspective of the computational resources required for execution, all prioritization tools behave similarly having a low demand that is always less than 6 GB.

Since the main objective of these tools is to filter fusions and effectively reduce false positives correctly, the prediction performance is undoubtedly the most relevant parameter for this analysis.

ChimerDriver appears to be the tool with the highest prediction performances overall, especially regarding fusions with annotated genes involved, regardless of the gene types (91.9% and 83.5% respectively for kinases and cancer-related genes). The prediction accuracy is still acceptable when the fused genes are not kinases or cancer-related but fail to recognize non-oncogenic fusions involving these genes. Indeed, these performances could be explained by ChimerDriver's feature set since it relies on gene functionalities (e.g. kinases, tumor suppressors, druggable genes) to provide the final oncogenic score.

Regarding DEEPrior and Oncofuse, similar considerations can be made. The prediction performances are generally poor across all subsets: there is no run where these two tools achieved performance above 50%. DEEPrior has the highest results on the subset with oncogenic fusions made of cancer-related genes, recognizing 46% of the listed fusions.

Pegasus seems to be the most reliable tool for non-oncogenic fusions. It exhibits the highest performance (70.1%) on the non-oncogenic subset with kinases and cancer-related genes. However, Pegasus provides very low scores on oncogenic fusions regardless of the subset; therefore, it is not the most reliable performing tool overall.

As mentioned earlier, the primary objective of prioritization tools should be to filter out false positives as much as possible, highlighting and preserving oncogenic fusions. In this regard, ChimerDriver is undoubtedly the most reliable tool. In addition to having the best accuracy results, it is easy to install, user-friendly, and regularly updated.

**Funding Information.** This study was funded by the European Union's Horizon 2020 research and innovation program DECIDER under Grant Agreement 965193.

**Data Availibility Statement.** The oncogenic fusions dataset is available at the following URL: https://compbio.uth.edu/FusionGDB2/combined_tables/combinedFGDB2genes_ORF_analyzed_gencode_h19v19_real_Inframe_only.txt
The non oncogenic fusion dataset is available at the following URL: https://academic.oup.com/nar/article/44/6/2859/2499453#supplementary-data.
The curated dataset is provided at the following URL https://drive.google.com/drive/folders/1kBpYJTvIjML9QWvfMq4YE292fipn1Tbr?usp=sharing

**Conflict of Interests.** The authors declare there are no conflicts of interest.

## References

1. Abate, F., et al.: Pegasus: a comprehensive annotation and prediction tool for detection of driver gene fusions in cancer. BMC Syst. Biol. **8**, 97 (2014). https://doi.org/10.1186/s12918-014-0097-z
2. Babiceanu, M., et al.: Recurrent chimeric fusion RNAS in non-cancer tissues and cells. Nucleic Acids Res. **44**, 2859–2872 (2016). https://doi.org/10.1093/nar/gkw032
3. Bruno, R., Fontanini, G.: Next generation sequencing for gene fusion analysis in lung cancer: a literature review. Diagn. MDPI **10** (2020). https://doi.org/10.3390/diagnostics10080521
4. Cirrincione, G., Randazzo, V., Pasero, E.: The growing curvilinear component analysis (GCCA) neural network. Neural Netw. **103**, 108–117 (2018). https://doi.org/10.1016/j.neunet.2018.03.017
5. Cirrincione, G., Randazzo, V., Pasero, E.: A neural based comparative analysis for feature extraction from ECG signals. In: Esposito, A., Faundez-Zanuy, M., Morabito, F.C., Pasero, E. (eds.) Neural Approaches to Dynamics of Signal Exchanges. SIST, vol. 151, pp. 247–256. Springer, Singapore (2020). https://doi.org/10.1007/978-981-13-8950-4_23

6. Forbes, S., et al.: Cosmic: mining complete cancer genomes in the catalogue of somatic mutations in cancer. Nucleic Acids Res. **39**, 945–950 (2010). https://doi.org/10.1093/nar/gkq929
7. Haas, B.J., Dobin, A., Li, B., Stransky, N., Pochet, N., Regev, A.: Accuracy assessment of fusion transcript detection via read-mapping and de novo fusion transcript assembly-based methods. Genome Biol. **20** (2019). https://doi.org/10.1186/s13059-019-1842-9
8. Kim, P., Zhou, X.: FusionGDB: fusion gene annotation database. Nucleic Acids Res. **47**(D1), D994–D1004 (2009). https://doi.org/10.1093/nar/gky1067
9. Kumar, R.R., Randazzo, V., Cirrincione, G., Cirrincione, M., Pasero, E.: Analysis of stator faults in induction machines using growing curvilinear component analysis. In: 2017 20th International Conference on Electrical Machines and Systems (ICEMS), pp. 1–6. IEEE (2017). https://doi.org/10.1109/ICEMS.2017.8056240
10. Latysheva, N.S., Babu, M.M.: Discovering and understanding oncogenic gene fusions through data intensive computational approaches. Nucleid Acids Res. **44**, 4487–4503 (2016). https://doi.org/10.1093/nar/gkw282
11. Lovino, M., Ciaburri, M.S., Urgese, G., Di Cataldo, S., Ficarra, E.: DEEPrior: a deep learning tool for the prioritization of gene fusions. Bioinformatics **36**, 3248–3250 (2020). https://doi.org/10.1093/bioinformatics/btaa069
12. Lovino, M., Montemurro, M., Barrese, V.S., Ficarra, E.: Identifying the oncogenic potential of gene fusions exploiting miRNAs. J. Biomed. Inform. (2022). https://doi.org/10.1016/j.jbi.2022.104057
13. Matter, M.S., Chijioke, O., Savic, S., Bubendorf, L.: Narrative review of molecular pathways of kinase fusions and diagnostic approaches for their detection in non-small cell lung carcinomas. Transl. Lung Cancer Res. **9**(6) (2020). https://doi.org/10.21037/tlcr-20-676. https://tlcr.amegroups.com/article/view/45595
14. Mertens, F., Johansson, B., Fioretos, T., Mitelman, F.: The emerging complexity of gene fusions in cancer. Nat. Rev. Cancer **15**, 371–81 (2015). https://doi.org/10.1038/nrc3947
15. Paviglianiti, A., Randazzo, V., Pasero, E., Vallan, A.: Noninvasive arterial blood pressure estimation using ABPNet and vital-ECG. In: 2020 IEEE International Instrumentation and Measurement Technology Conference (I2MTC), pp. 1–5. IEEE (2020). https://doi.org/10.1109/I2MTC43012.2020.9129361
16. Shugay, M., de Mendíbil, O., Vizmanos, J., Novo, F.: Oncofuse: a computational framework for the prediction of the oncogenic potential of gene fusions. Bioinformatics **29**, 2539–2546 (2013). https://doi.org/10.1093/bioinformatics/btt445
17. Uhrig, S., et al.: Accurate and efficient detection of gene fusions from RNA sequencing data. Genome Res. (2021). https://doi.org/10.1101/gr.257246.119

# Improving the Reliability of Tree-Based Feature Importance via Consensus Signals

Bastian Pfeifer(✉) and Michael G. Schimek

Institute for Medical Informatics, Statistics and Documentation, Medical University of Graz, Graz, Austria
bastian.pfeifer@medunigraz.at

**Abstract.** Feature (variable) selection methods are used to detect the most important features (variables) within high-dimensional data. Tree-based models, such as Random Forests, are often exploited for this purpose, as they provide a built-in mechanism to quantify feature importance. However, the stochastic sampling strategies used in these models can lead to unstable feature importance rankings, particularly when the number of trees is low. In our study, we investigate the extent to which these unstable feature rankings can be consolidated through rank aggregation and consensus signal techniques. We propose to compute consensus values from multiple feature selection runs, where each run generates a ranked list of features. We have evaluated our approach while varying a spectrum of hyperparameters such as the number of trees and the number of features available for splitting a node. Our results suggest that consensus ranks provide a more accurate and robust selection of features compared to single-run feature selection procedures. The proposed approach is especially relevant for biomarker discovery, as it can improve the accuracy and reliability of feature selection, leading to the identification of the most informative and relevant biomarkers associated with a particular disease or condition. By consolidating the results from multiple feature selection runs, our approach may help to overcome problems associated with noisy or complex data because it can provide more robust and accurate estimates of the feature importance rankings.

**Keywords:** Random Forest · Feature Selection · Consensus Signals · Rank Stability · Biomarker Discovery

## 1 Introduction

Feature selection is an important preprocessing step in many machine learning applications and has long been a fundamental challenge. Classic goals of feature selection include improving predictive performance, providing more effective predictors, and better understanding the underlying data [1]. The latter is of increasing importance, especially for applications in the medical field, due to the increasing need for interpretability of complex models. In the biomedical field, feature selection is commonly used for data-driven biomarker discovery. Some


M. Vettoretti et al. (Eds.): CIBB 2023, LNBI 14513, pp. 212–225, 2025.
https://doi.org/10.1007/978-3-031-90714-2_16

of the most common feature selection methods are based on the random forest (RF) [2] classifier because it provides an interpretable mechanism for computing feature importance. RF is a supervised classifier that consists of multiple decision trees, each of which is created based on bootstrap samples taken from the input feature space. At each tree node, a feature is selected in order to split the samples into their associated classes. Depending on how well a feature separates the samples, the more importance is assigned to it. Typically, the Gini index is used as a selection criteria for the best candidate features.

Very important aspects of trustworthy machine learning are robustness and explainability [3,4]. A robust feature selector should (1) report on a constant set of relevant features when executed on exactly the same data set multiple times (stability), and (2) select the most relevant features for the modelling process of interest. However, RFs suffer from unstable feature importance ranks [5]. Furthermore, the bootstrap mechanism for building the RF may not select highly correlated features with equal frequency. Consequently, highly correlated features may not have the same Gini scores. To address this shortcoming, several optimised feature selection algorithms have been proposed [6–8]. These methods use a RF as a wrapper, but relate the importance of a feature to its permuted versions.

In this work, we analysed the robustness of three feature selection methods which are widely used for data-driven biomarker discovery, namely the BORUTA algorithm [8], the VITA algorithm [7], the RRF algorithm [9], and an improved Gini impurity measure [10]. We validated the extent to which consensus scores from multiple feature selection runs can be utilised to improve the result. In the context of biomarker discovery, the goal is to identify specific biomarkers that are associated with a particular disease or condition. These biomarkers can be used to develop diagnostic tests, predict disease outcomes, and guide treatment decisions. To identify these biomarkers, various feature selection algorithms are employed to identify the most relevant features. However, these algorithms can be prone to instability, meaning that they may produce different rankings of feature importance when applied to the exact same dataset multiple times.

To address this instability, we propose to run these feature selection algorithms multiple times and to consolidate the observed rank variations of importance scores through rank aggregation and signal reconstruction techniques. In doing so, we can obtain a more robust and reliable estimate of the true feature importance rankings, which to this end could improve the accuracy and reliability of biomarker discovery.

The paper is organized as follows: Sect. 2 outlines the overall motivation of the conducted research and discusses related work. Section 3.1 presents some widely used RF-based feature selection algorithms and Sect. 3.2 suggests techniques to consolidate the fluctuations of the resulting feature importance ranks. Section 4 describes the strategy and the data sets used for evaluation. The results are discussed in Sect. 5, and Sect. 6 concludes by outlining limitations and potential directions for future work.

## 2 Motivation of Research and Related Work

Stabilising feature selection procedures have been mainly studied in terms of ensemble learning. Ensemble learning aims at improving the predictive accuracy by combining multiple machine learning models. Currently, this idea is influencing more and more the development of robust feature selectors [11,12]. Since RF models are built upon stochastic procedures (bootstrap samples for tree building and bootstrap feature candidates for node splitting), rank fluctuations of selected features are most likely. Some research has been done to study this problem [5], but no general solution has been found. To the best of our knowledge, this work is the first to provide a general solution to this problem using consensus scores.

The accuracy of RF-based feature selection algorithms were evaluated in a few studies. Degenhardt et al. (2019) [13] compared five RF-based feature selectors, namely the BORUTA algorithm [8], the VITA algorithm [7], recurrent relative variable importance (a permutation approach and its parametric variant) [6], as well as recursive feature elimination (RFE) [14]. They found that both VITA and BORUTA are most suited for high-dimensional data sets, whereas VITA is more robust compared to BORUTA when applied to real-world data sets. An additional comparative study was conducted by Kursa et al. (2014) [15]. The authors report about BORUTA as the method of choice when dealing with high-dimensional data, where the number of predictors is much larger than the number of observations.

In our study, the particular focus was on BORUTA and VITA. We evaluated whether and to which extent the stochastic fluctuation of the obtained feature ranks can be stabilised via consensus scores.

## 3 Materials and Methods

### 3.1 Feature Importance

**Random Forest-Based Feature Importance by Breiman (2001)** [2]**.** RFs make use of the Gini index (also called Gini impurity) to select the best feature out of a set of candidate features for splitting a tree node. The better a feature separates the samples into their labeled classes, the more importance is assigned to it. Here, we recall and adopt the definition of the Gini index as outlined in [9]. The Gini index at node $v$ is

$$Gini(v) = \sum_{k=1}^{K} \hat{p}_k^v (1 - \hat{p}_k^v), \tag{1}$$

where $\hat{p}_k^v$ is the proportion of samples belonging to class $k$ at node $v$ and $K$ is the number of classes. The Gini information gain obtained by a feature $X_i$ for splitting node $v$ is the difference between the Gini impurity at node $v$ and the weighted average of impurities at each child node of $v$. The Gini information gain is defined as

$$Gain(X_i, v) = Gini(v) - w_L Gini(v^L, X_i) - w_R Gini(v^R, X_i), \quad (2)$$

where $v^L$ and $v^R$ are the left and the right child nodes of $v$, respectively, and $w_L$ and $w_R$ are the proportions of $k$-class instances assigned to the left and to the right child nodes. At each node $v$, a random set of features out of total $P$ features is evaluated, and the feature with the maximum $Gain(X_i, v)$ is used for splitting the node. The importance score for variable $X_i$ can be calculated as

$$Imp_i = \frac{1}{ntree} \sum_{v \in S_{X_i}} Gain(X_i, v), \quad (3)$$

where $S_{X_i}$ is the set of nodes split by $X_i$ in the RF comprising $ntree$ trees. The RF importance scores are commonly used to select the most important features. However, importance scores purely based on the Gini index are known to be biased towards high cardinal features [10,16]. These features have many split points and thus appear more often in the tree leading to a "multiple testing" problem. To address this shortcoming, permutation based approaches were proposed. In this context, the importance score of a given feature is compared with the importance score of its randomised version. This is done by permuting the feature values within the out-of-bag (OOB) samples to measure the Mean Decrease in Gini (MDG) or the Mean Decrease Accuracy (MDA). The MDA is applicable to any model, not just to RFs. The following steps are taken to calculate permutation based importance scores:

- A RF model is trained and for the first decision tree the prediction accuracy is calculated using the OOB samples.
- Within the OOB samples, any correlation between the feature of interest $X_i$ and the response vector is diminished by permuting the values of $X_i$. Then the accuracy of the prediction is calculated again.
- The difference between the two accuracy values is the permutation importance of $X_i$.
- The aforementioned steps are repeated for each decision tree within the RF.
- The average of all importance values is the final permutation importance of a given feature $X_i$.
- Repeat the procedure for all features of interest.

Permutation importance is more reliable, albeit more computationally expensive compared to MDG.

**Regularized Tree-Based Feature Importance by Deng and Runger (2013)** [9]. The key idea of the RRF approach is to penalise the selection of a new feature for splitting a node when its information gain is similar to the features used in previous splits. The information gain of $X_i$ for splitting a tree node $v$ is defined as

$$Gain_R(X_i, v) = \begin{cases} \lambda \cdot Gain(X_i, v) & i \notin F \\ Gain(X_i, v) & i \in F \end{cases}$$

where $F$ is the set of indices of features used for splitting in previous nodes and is an empty set at the root node in the first tree. Here $\lambda \in (0, 1]$ is called the penalty coefficient. When $i \notin F$, the coefficient penalises the *ith* feature for splitting node $v$. A smaller $\lambda$ leads to a larger penalty. RRF uses $Gain_R(X_i, v)$ at each node, and adds the index of a new feature to $F$ if the feature adds enough predictive information to the selected features.

**Hold-Out Feature Importance by Janitza et al. (2018)** [7]**.** In this crossvalidation-based approach, permutation importance is used to evaluate the relevance of a feature. Instead of OOB, the authors apply a hold-out sample set, on which feature importance is verified. It is calculated as

$$Imp_i^{CV(l)} = \frac{1}{ntree} \sum_{t=1}^{ntree} \frac{1}{|S_l|} \sum_{j \in S_l} \{I(y_j \neq \hat{y}_{jt}^*) - I(y_j \neq \hat{y}_{jt})\}, \tag{4}$$

where $I(.)$ is the indicator function and $\hat{y}_{jt}$ and $\hat{y}_{jt}^*$ are denoting the prediction by the $t$-th tree before and after permuting the values of $X_i$. $S_l$ reflects the $l$-th hold-out sample set eliminated from the tree-building process. The *cross-validated* variable importance for feature $X_i$ is defined by

$$Imp_i^{CV} = \frac{1}{k} \sum_{l=1}^{k} Imp_i^{CV(l)}. \tag{5}$$

The proposed *hold-out* variable importance for feature $X_i$ is then defined by

$$Imp_i^{HO} = \frac{1}{2} \sum_{l=1}^{2} Imp_i^{CV(l)}. \tag{6}$$

To finally compute $p$-values from the importance scores the authors argue that all non-positive importance values are assumed to correspond to non-associated variables. The $p$-values are used to construct the distribution of importance under the null hypothesis of no association with the response. Since only the non-positive values of this distribution can be observed, the positive values are created by mirroring the negative distribution. The described approach is implemented in the R-package VITA [17] as well as in the R-package *ranger* [18]. In this study, we use the implementation realized in the *ranger* package because of its computational efficiency. Nevertheless, we point to the VITA algorithm throughout the manuscript when discussing the approaches introduced by Janitza et al. (2018).

**Comparing a Feature to its Random Probe: An Feature Importance Measure by Kursa and Rudnicki (2010)** [8]**.** The heuristic introduced by the authors quantifies the importance of a feature by the loss of accuracy of classification caused by randomly permuted probes (*shadow features*) of the original feature set. The underlying algorithm is implemented in the R-package BORUTA [8] and works as follows:

- Duplication of the data pool by adding copies of *all* features (so called *shadow features*).
- Permutation of the added *shadow features* to remove their correlations with the response vector.
- A random forest classifier is trained on the extended data pool (*original* + *shadow features*) and the $z$-scores of the Mean Decrease Accuracy (MDA) measures are computed.
- Calculation of the maximum $z$-score among *shadow features* (MZSA). A "hit" is assign to every *original* feature that scores better than MZSA.
- For each feature with undetermined importance a two-sided test of equality with the MZSA is executed.
- The *shadow features* which have importance significantly lower than MZSA are deemed as *unimportant* and are permanently removed from the data pool.
- The *shadow features* which have importance significantly higher than MZSA are deemed as *important*.
- All *shadow features* are removed.
- The aforementioned procedure is repeated until importance values are assigned to all features, or the algorithm stops in case it has reached the previously defined limit of random forest runs.

### 3.2 Consensus Calculation

**Borda's Method by Jean-Charles de Borda (1781)** [19]. Since its introduction various modifications and extensions were proposed, all labeled as Borda scores. Due to their simplicity, they are widely used in practice and are perfectly suited as a baseline for comparison with more sophisticated approaches. What Borda's methods have in common is the assignment of scores corresponding to the positions in which the objects (features) appear within each ranking. Then each object is sorted by its total score or by a score obtained from an aggregation (sorting) function. We used the following sorting functions: arithmetic mean (equivalent to L1-norm), L2-norm, median, and geometric mean. We applied all of them because of their simple deterministic structure and their linear computer time implementation. We employed Borda's methods as implemented in the R-package TopKLists [20].

**Robust Rank Aggregation by Kolde et al. (2012)** [21]. Various methods for rank aggregation based on probabilistic models have been developed. Probably the most widely used is a method called Robust Rank Aggregation (RRA), which is frequently applied in the field of bioinformatics for gene selection. An implementation of this method is realised in the R-package RobustRankAggreg. RRA follows a simple idea: Given an ordered rank vector $r$, so that $r_{(1)} \leq \ldots \leq r_{(n)}$, RRA verifies how probable it is to obtain $\hat{r}_{(k)} \leq r_{(k)}$. The probability that $\hat{r}_{(k)}$ is smaller or equal to $x$ is formulated via a binomial distribution

$$\beta_{k,n}(x) := \sum_{l=k}^{n} \binom{n}{l} x^l (1-x)^{n-l}. \quad (7)$$

A score for the rank vector $r$ is calculated as

$$p(r) = \min_{k=1,\dots,n} \beta_{k,n}(r). \tag{8}$$

The aggregated consensus rank is the vector $r$ ordered by the $p$ scores.

**Signal Reconstruction by Schimek et al. (2024)** [22]. The consensus rank in [22] is derived using a signal-plus-noise model for signal reconstruction. Signal reconstruction computes the true underlying ranking from a set of noisy rankings. By contrast, rank aggregation techniques typically do not take into account the specific noise characteristics of the ranking data and thus may be more prone to errors or inconsistencies. Moreover, the underlying methodology also provides measures of (un)certainty for the estimated consensus signals. Such measures are useful to quantify the reliability of results obtained from feature selection runs. The described methodology is implemented within the R-package TopKSignal (https://cran.r-project.org/web/packages/TopKSignal).

## 4 Evaluation Strategy

To evaluate the performance of our approach, we based our study on the Madelon datasets from the NIPS feature selection competition as provided by the UCI Machine Learning Repository [23]. The Madelon dataset is a synthetic dataset that consists of data points grouped in 32 clusters located on the vertices of a five-dimensional hypercube. Each data point is randomly labeled with either +1 or -1. The dataset contains 20 relevant and 480 irrelevant features, as well as ground-truth information, providing an ideal benchmark for evaluating the performance of feature selection algorithms. To assess the performance of the algorithms, we performed a step-wise reduction of the relevant signals by subsequentially down-sampling the data. We conducted an initial investigation using default parameter settings for all classifier-specific parameters.

The feature selection algorithms described in Sect. 3.1 were executed 20 times on exactly the same data set. The stability of the outcomes and the extent to which the consensus calculation (for the methods see Sect. 3.2) improves the outcomes, is reported.

To evaluate the stability of the feature selectors, coverage was used, defined as $R_s/R_t$, where $R_s$ is the number of relevant features which are successfully selected and $R_t$ is the total number of relevant features present in the data. For instance, when $R_t = 10$ relevant features are present within the data, and the evaluated feature selector ranks eight of these features within the top-10 ($R_s = 8$), the coverage is 8/10.

In a second experiment, we investigated the stability of BORUTA and VITA based on gene expression data of human breast cancer patient samples from The Cancer Genome Atlas (TCGA). The retrieved dataset comprised 981 patients and 8469 genes. The binary prediction task was to classify the samples into a group of patients with the luminal A subtype (499 samples) and patients with other breast cancer subtypes (482 samples). We randomly sampled 200 patients and 100 genes and split the resulting dataset into two equally-sized parts, so that each dataset reflected 100 randomly selected patients. We ran the feature selection methods on both datasets independently and calculated the Kendall's $\tau$ correlation of the independently computed rank lists.

## 5 Results and Discussion

The analysis of the Madelon data set indicates that VITA as well as BORUTA can be improved by consolidated consensus scores (see Fig. 1 and Table 1). When 50% of the data are randomly sampled, we can see the following: although most of the VITA runs have a coverage less than 1, the consensus ranks from the aggregation techniques are able to verify the complete set of relevant features (Fig. 1). Also for BORUTA it is beneficial to consolidate via consensus because some of the 20 runs miss relevant features, even when the full signal is available. As expected, when only 25% of the data are randomly sampled, the overall performance and robustness decreases.

We repeated the analysis, varying the number of trees within the forest. A detailed summary of this additional evaluation is provided in Table 1. For both BORUTA and VITA the consolidated consensus ranks show a much higher coverage compared to the median coverage of the 20 feature selection runs. This observation indicates that the consensus calculation may have a higher impact on robustness than an increased number of trees. Even for a number of trees as high as 1000, there is a notable difference in performance between the worst (min coverage) and the best (max coverage) run. Furthermore, we identify BORUTA as the most accurate algorithm, especially when the number of trees is low and the signal is weak. However, it is computationally more demanding than VITA (see Table 2). Finally, RRF is more stable compared to the other algorithms, but fails to select the most important features (Fig. 1). Compared to BORUTA and VITA, RRF does not benefit much from the consensus calculations.

To determine whether the outcomes are consistent, we repeated the above described experiments 20 times, leading to $20 \times 20$ feature selection runs. This yielded 400 coverage values for the non-consolidated BORUTA and VITA runs and 20 coverage values for the consolidated consensus ranks. The results are presented in Fig. 2. Evidently, the improvement is substantial across all feature selection methods and consensus computation techniques.

In an additional experiment we varied the number of features which are available for splitting a node. We subsequently increased the number up to the default value, which is the squared root of the total number of features; in our case $\sqrt{500} \approx 22$. In general, a low number of features available for splitting induces additional randomness into the feature selection process. As can be seen from Fig. 3, the available number of features and the overall performance of the studied feature selection methods correlates. In almost all cases the consolidated consensus feature ranks perform better compared to those generated by any single feature selection run. Moreover, the consensus signals of VITA compete with those obtained via BORUTA. These are intriguing results, since VITA is computationally much more effective (see Table 2). Furthermore, we could observe that

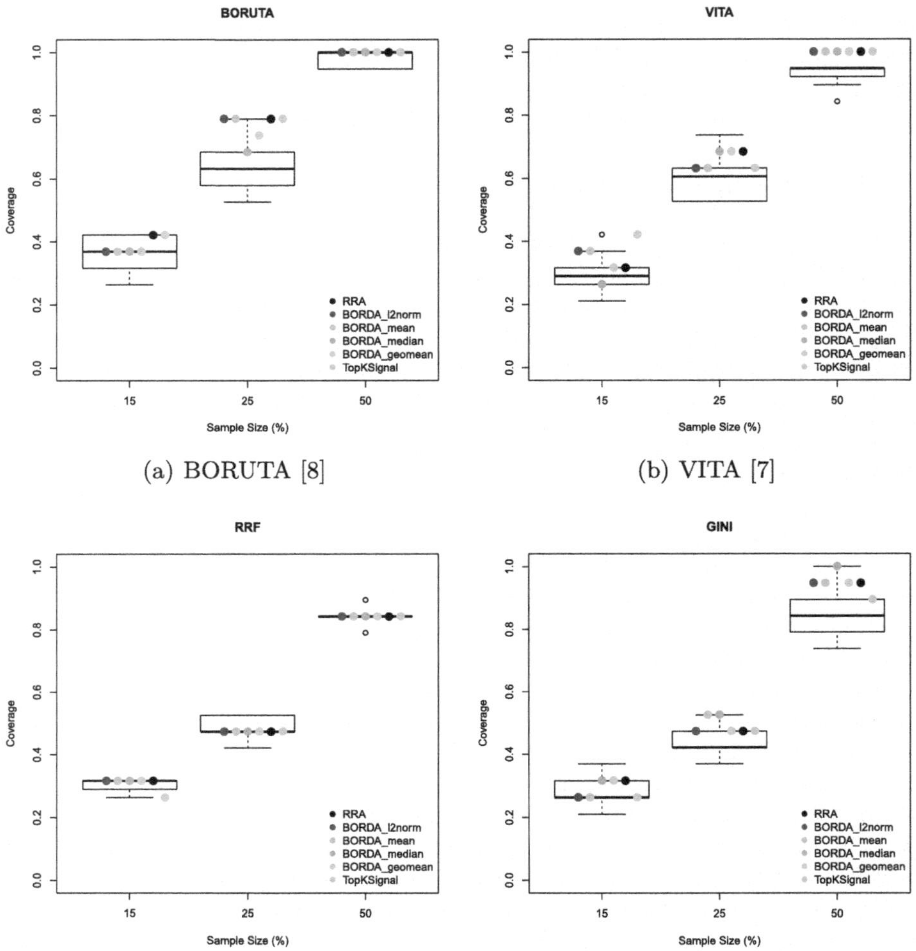


(a) BORUTA [8]

(b) VITA [7]

(c) RRF [9]

(d) GINI impurity [10]

**Fig. 1.** Madelon data set. The coverage for each of the 20 feature selection runs and the corresponding consensus ranks. The values from the feature selectors are all based on the same randomly sampled subset of the original data (15%, 25%, and 50%). The default for the number of trees is 500 for all methods.

the signal reconstruction approach might have difficulties in situations where the feature ranks are very noisy.

Our second experiment based on gene expression data of cancer patients indicates that consensus signals from multiple feature selection runs do stabilise the obtained feature ranks also in real-world applications (see Fig. 4(a)). The consolidated feature ranks are more robust compared to the ranks computed from single BORUTA runs. Furthermore, the signal estimates obtained from the TopKSignal package allow us to verify nine genes, namely C4A, CFB, STC2,

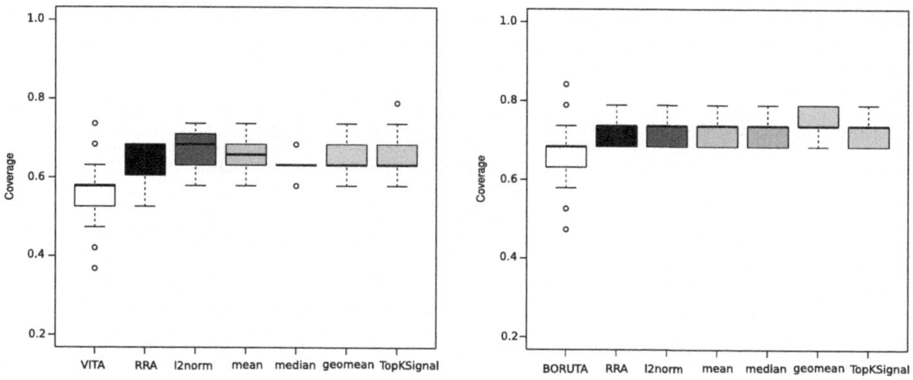


(a) VITA [7]

(b) BORUTA [8]

**Fig. 2.** Madelon data set. The coverage for 20 × 20 feature selection runs. For VITA and BORUTA there are 400 data points in total. The coverage values of the consensus ranks consist of 20 data points and are based on the consolidated ranks after each of the 20 feature selection iterations. The values from the feature selectors are all based on the same randomly sampled 25% subset of the original data. The number of trees was 500.

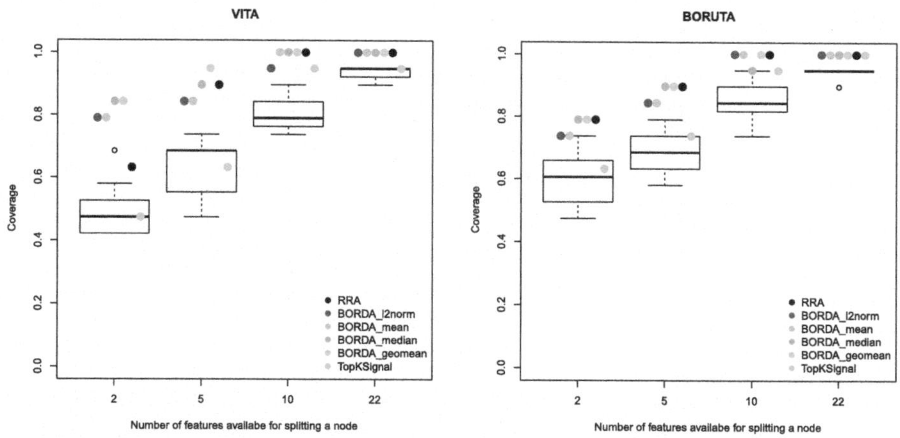


(a) VITA [7]

(b) BORUTA [8]

**Fig. 3.** Madelon data set. Varying the number of features available for splitting a node. The coverage is shown for each of the 20 feature selection runs and the corresponding consensus ranks. Values from the feature selectors are all based on the same randomly sampled 50% subset of the original data. The number of trees was 500.

PGR, CLIC6, AGR2, KRT17, CST5, and LRP2 (see Fig. 4(b)). Those genes have top signal levels and at the same time low signal variability compared to the rest of the genes. Overall, the lower the ranking of a specific gene, the more

**Table 1.** Madelon data set. Coverage of the consensus ranks with varying numbers of trees. The results are based on a 25% sampled subset of the original data and 20 feature selection runs.

| Method | trees | min/med/max | l2norm | mean | geomean | median | RRA | TopKSignal |
|---|---|---|---|---|---|---|---|---|
| BORUTA | *100* | 0.42/0.50/0.63 | 0.68 | 0.68 | 0.68 | 0.74 | 0.68 | 0.68 |
| | *500* | 0.53/0.68/0.79 | 0.74 | 0.74 | 0.74 | 0.74 | 0.74 | 0.74 |
| | *1000* | 0.58/0.68/0.74 | 0.68 | 0.68 | 0.74 | 0.68 | 0.74 | 0.68 |
| VITA | *100* | 0.26/0.42/0.58 | 0.63 | 0.68 | 0.68 | 0.63 | 0.68 | 0.21 |
| | *500* | 0.42/0.55/0.68 | 0.63 | 0.63 | 0.68 | 0.58 | 0.68 | 0.58 |
| | *1000* | 0.53/0.58/0.74 | 0.63 | 0.63 | 0.63 | 0.63 | 0.68 | 0.63 |
| RRF | *100* | 0.37/0.47/0.58 | 0.53 | 0.53 | 0.47 | 0.47 | 0.53 | 0.47 |
| | *500* | 0.42/0.47/0.58 | 0.47 | 0.47 | 0.47 | 0.47 | 0.42 | 0.47 |
| | *1000* | 0.42/0.47/0.53 | 0.47 | 0.47 | 0.47 | 0.47 | 0.47 | 0.47 |
| GINI | *100* | 0.26/0.37/0.47 | 0.58 | 0.58 | 0.58 | 0.53 | 0.58 | 0.37 |
| | *500* | 0.42/0.47/0.53 | 0.53 | 0.53 | 0.53 | 0.53 | 0.53 | 0.47 |
| | *1000* | 0.32/0.47/0.58 | 0.53 | 0.58 | 0.53 | 0.53 | 0.58 | 0.53 |

**Table 2.** Madelon data set. Computational speed (elapsed time in seconds) while varying the numbers of trees. The results are based on a 25% sampled subset of the original data and 20 feature selection runs.

| Method | trees | Borda | RRA | TopKSignal |
|---|---|---|---|---|
| BORUTA | *100* | 129.304 | 132.401 | 188.849 |
| | *500* | 656.369 | 645.298 | 700.946 |
| | *1000* | 1320.057 | 1382.462 | 1389.751 |
| VITA | *100* | 5.057 | 4.800 | 57.823 |
| | *500* | 19.622 | 19.474 | 71.314 |
| | *1000* | 38.393 | 37.947 | 87.498 |

uncertainty is connected to it, as can be easily seen in Fig. 4(b). In addition, we can see that genes PGR and CLIC6 have almost the same signal level, so that a distinct rank order can not be derived. The same is true for the genes KRT17, CST5, and LRP2. In contrast, the top-3 ranked genes are confident in their ranks (gene C4A, CFB, and STC2). Based on these gene-specific consensus signals and the corresponding errors, we are currently developing strategies for the *automated* selection of relevant and stable features such as genes.

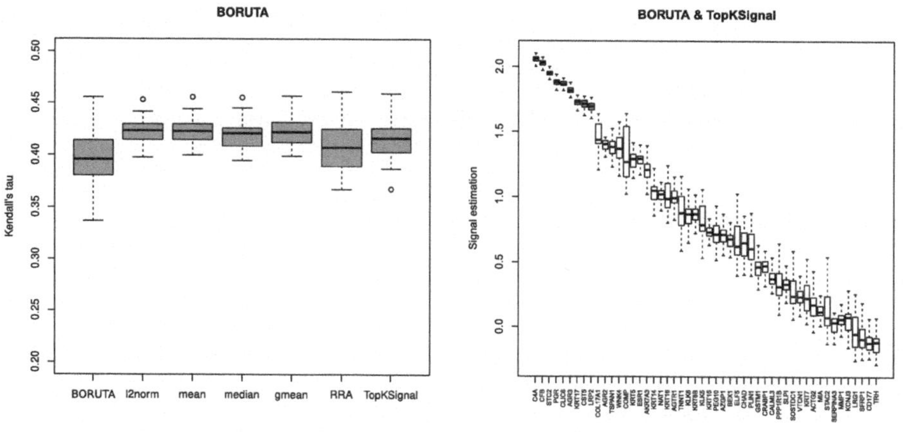


(a) BORUTA [8]

(b) TopKSignal [22]

**Fig. 4.** TCGA breast cancer data set. (a) The Kendall's $\tau$ correlation is shown for the feature ranks computed on the two splitted datasets. The very left boxplot refers to a randomly selected single run from BORUTA. The remaining boxplots show the correlations between the consolidated feature ranks. BORUTA was executed 20 times using 500 trees, which we then repeated 100 times. (b) The signal estimates and their standard errors computed by TopKSignal. The top 50 genes are shown.

## 6 Conclusion

In this paper we have investigated to which extent random forest-based feature importance ranks can be stabilised using rank aggregation and signal reconstruction techniques. We have evaluated the performance of consensus ranks of relevant features by means of multiple feature selection runs. Our results suggest that consensus ranks provide a more accurate and robust selection of features compared to single-run feature selection procedures. The consolidation of multiple feature selection runs reduces the impact of data errors and inconsistencies. Thus, we can obtain more comprehensive and representative feature sets. The herein presented post-hoc approach has the potential to improve the discovery of more reliable and informative biomarkers, in consequence enhancing diagnosis, disease treatment, and patient prognosis.

The presented work also opens avenues for further exploration and refinement. While the manuscript offers a thorough analysis of the approaches, emphasizing variations in the number of trees and the number of features used for node splitting, future research should consider extending the analysis to incorporate variations in additional parameters, which would provide further valuable insights. Moreover, while this study provides a robust framework for feature importance ranks, strategies are required for the automated selection of a well-defined set of relevant and stable features. Limiting the candidate set enhances interpretability, making it easier to understand underlying biological mechanisms, and supports scalability for analysing large omics datasets. These qualities are essential for translating biomarker candidates into reliable clinical applications.

**Conflict of Interests.** No conflict of interest declared.

## References

1. Guyon, I., Elisseeff, A.: An introduction to variable and feature selection. J. Mach. Learn. Res. **3**(3), 1157–1182 (2003)
2. Breiman, L.: Random forests. Mach. Learn. **45**(1), 5–32 (2001)
3. Nogueira, S., Sechidis, K., Brown, G.: On the stability of feature selection algorithms. J. Mach. Learn. Res. **18**(1), 6345–6398 (2017)
4. Barbieri, M.C., Grisci, B.I., Dorn, M.: Analysis and comparison of feature selection methods towards performance and stability. Expert Syst. Appl. 123667 (2024)
5. Wang, H., Yang, F., Luo, Z.: An experimental study of the intrinsic stability of random forest variable importance measures. BMC Bioinform. **17**(1), 1–18 (2016)
6. Altmann, A., Toloşi, L., Sander, O., Lengauer, T.: Permutation importance: a corrected feature importance measure. Bioinformatics **26**(10), 1340–1347 (2010)
7. Janitza, S., Celik, E., Boulesteix, A.L.: A computationally fast variable importance test for random forests for high-dimensional data. Adv. Data Anal. Classif. **12**(4), 885–915 (2018)
8. Kursa, M.B., Rudnicki, W.R.: Feature selection with the Boruta package. J. Stat. Softw. **36**(11), 1–13 (2010)
9. Deng, H., Runger, G.: Gene selection with guided regularized random forest. Pattern Recogn. **46**(12), 3483–3489 (2013)
10. Nembrini, S., König, I.R., Wright, M.N.: The revival of the gini importance? Bioinformatics **34**(21), 3711–3718 (2018)
11. Neumann, U., Genze, N., Heider, D.: EFS: an ensemble feature selection tool implemented as R-package and web-application. BioData Min. **10**, 1–9 (2017)
12. Bolón-Canedo, V., Alonso-Betanzos, A.: Ensembles for feature selection: a review and future trends. Inf. Fusion **52**, 1–12 (2019)
13. Degenhardt, F., Seifert, S., Szymczak, S.: Evaluation of variable selection methods for random forests and omics data sets. Brief. Bioinform. **20**(2), 492–503 (2019)
14. Díaz-Uriarte, R., De Andres, S.A.: Gene selection and classification of microarray data using random forest. BMC Bioinform. **7**(1), 1–13 (2006)
15. Kursa, M.B.: Robustness of random forest-based gene selection methods. BMC Bioinform. **15**(1), 1–8 (2014)
16. Farris, F.A.: The gini index and measures of inequality. Am. Math. Mon. **117**(10), 851–864 (2010). https://doi.org/10.4169/000298910X523344
17. Celik, E.: vita: Variable Importance Testing Approaches (2015). https://CRAN.R-project.org/package=vita, R package version 1.0.0
18. Wright, M.N., Ziegler, A.: Ranger: a fast implementation of random forests for high dimensional data in C++ and R. J. Stat. Softw. **77**(1), 1–17 (2017). https://doi.org/10.18637/jss.v077.i01
19. Borda, J.D.: Mémoire sur les élections au scrutin. Histoire de l'Academie Royale des Sciences (Paris) (1781)

20. Schimek, M.G., Budinská, E., Kugler, K.G., Švendová, V., Ding, J., Lin, S.: TopKLists: a comprehensive R package for statistical inference, stochastic aggregation, and visualization of multiple omics ranked lists. Stat. Appl. Genetics Mol. Biol. **14**(3), 311–316 (2015)
21. Kolde, R., Laur, S., Adler, P., Vilo, J.: Robust rank aggregation for gene list integration and meta-analysis. Bioinformatics **28**(4), 573–580 (2012)
22. Schimek, M.G., Vitale, L., Pfeifer, B., La Rocca, M.: Effective signal reconstruction from multiple ranked lists via convex optimization. Data Min. Knowl. Disc. **38**, 1125–1169 (2024)
23. Dua, D., Graff, C.: UCI machine learning repository (2017). http://archive.ics.uci.edu/ml

# Interpretable Machine Learning for Automated Cellular Population Analysis in Flow Cytometry

Riccardo Benedetti[1], Muhammad Suffian[2,3](✉), Alessandro Bogliolo[2], Barbara Canonico[1], Stefano Papa[1], Claudio Ortolani[1], and Sara Montagna[2]

[1] Department of Biomolecular Sciences, University of Urbino Carlo Bo, Urbino, Italy
r.benedetti3@campus.uniurb.it, {barbara.canonico,stefano.papa}@uniurb.it
[2] Department of Pure and Applied Sciences, University of Urbino Carlo Bo, Urbino, Italy
{alessandro.bogliolo,sara.montagna}@uniurb.it
[3] DIIES, Mediterranean University of Reggio Calabria, Reggio Calabria, Italy
m.suffian@campus.uniurb.it

**Abstract.** Given the significant improvement in flow cytometry technologies, a massive dataset has been acquired, measuring a larger number of cellular markers. Managing and analysing these data by hand, based on the sole expertise of human operators, is no longer feasible. Recent literature suggests exploiting machine learning algorithms to automate data analysis for extracting and counting sub-populations of cells and supporting diagnosis. In this paper, we applied Support Vector Machine, XGBoost, Decision Tree, Logistic Regression, and Multi-Layer Perceptron to identify three specific cellular types: Lymphocyte T, B, and T cytotoxic. Performances are promising across all models and experiments, with a balanced accuracy above 0.85. Moreover, when looking at the Recall and the F1 score, the Decision Tree is the unique model with values below 0.8 in the classification B Lymphocyte. Moreover, to improve the interpretability of the trained models, we computed the SHAP-based explanations for the XGBoost, Decision Tree and Multi-Layer Perceptron, obtaining a set of extracted features that domain experts recognised as significant for the three classification tasks, thus emphasising the viability of this approach in automating the gating process in flow citometry.

**Keywords:** Flow Cytometry · Supervised Machine Learning · Explainable Artificial Intelligence · Shapley Additive Explanations

## 1 Introduction

Flow cytometry (FC) is a technique that quantifies surface proteins or antigens for a single cell within a large population, providing measurements for hundreds


M. Vettoretti et al. (Eds.): CIBB 2023, LNBI 14513, pp. 226–236, 2025.
https://doi.org/10.1007/978-3-031-90714-2_17

of thousands of cells per second. It characterizes cell types and states measuring expression levels of pre-defined sets of surface and intracellular proteins in individual cells, using antibodies tagged with fluorochromes [1]. Gating is one of the most basic concepts of FC. It concerns the sequential identification and refinement of a cellular population of interest exploiting a graphical procedure that visualizes cells distributed in space according to the fluorescence spectrum of a panel of molecules [2]. Modern FC systems allow the detection of a great number of proteins (more than 20), enabling the identification of different cellular types. The high number of dimensions of obtained datasets has forced researchers to depart from manual gating strategies based on two-dimensional plots since it is very time-consuming and subjective [3]. Sub-setting single-cell data into populations and encircling data points in hand-drawn polygons in two-dimensional space has proved to be laborious, subjective, and difficult to standardize. The management of these high dimensional data opens the way to complex computer algorithms and machine learning (ML) applications to automate analysis and improve the investigator's ability to extract meaning from them [4]. Our literature review brings to different methods and approaches for the automation of FC data analysis using both supervised and unsupervised ML algorithms, to automate the gating process or to identify a diagnosis.

Computational pipelines, that are optimised for cytometry data – including FLOCK, flowSOM, flowMeans, flowMerge, SWIFT, PhenoGraph, and others – have been developed in biomedical research and use a combination of existing unsupervised ML methods and customised algorithms to optimize the analysis [5]. In [6], unsupervised methods for the diagnosis of bone marrow failure syndromes based on K-means clustering are proposed. The work in [7] uses both supervised and unsupervised approaches for the classification of acute leukemias. Unsupervised learning using the Gaussian mixture model and Fisher kernel methods was applied to FC list-mode data, followed by supervised support vector machine classification. A framework called *CytoPy* is developed in [4], which allows to work with FC data and that contains different kinds of ML algorithms extrapolated from *scikit-learn* Python library. Although the literature discusses a list of frameworks and application cases on the usage of ML in FC data analysis, our experiments aim at evaluating and comparing different models of supervised learning specifically for automatic gating. Moreover, our comparison not only considers the performance, but also evaluates the trustability, repeatability, and robustness of each algorithm.

This paper builds upon our prior work [8], extending the methodology to incorporate various supervised ML algorithms for three novel gating tasks: identifying total lymphocytes of *(i)* type T, *(ii)* type B, and *(iii)* type T cytotoxic. These tasks are managed as binary classification to classify these three cell types from the rest of the cell populations. Moreover, we employed Shapley Additive exPlanations (SHAP) [9] to enhance the post-hoc explainability of the ML models. Using SHAP, we understand the predictions of the model and quantify the contribution of each feature, thus transforming complex ML algorithms into interpretable tools. This advancement ensures that the results are not only accurate, but also meaningful to clinicians, bridging the gap between computational analysis and practical application in clinical decision-making.

This study experiments with a set of supervised ML algorithms. The selected models include Support Vector Machines (SVM), Logistic Regression (LR), Decision Trees (DT), XGBoost – an ensemble learning algorithm – and Multilayer Perceptron (MLP), within the framework of artificial neural networks. Results and performances are compared to understand the robustness of the algorithms dealing with FC data. Each algorithm provides promising results for each experiment, with balanced accuracy values over 0.9. However, the XGBoost and DT models revealed to be the most robust algorithm over different training sets. Moreover, the explainability analysis demonstrated that the features contributing most significantly to model predictions were closely aligned with those deemed most relevant during the manual gating process of cells.

The approach presented in this paper addresses the limitations of traditional FC workflows, that rely heavily on user intervention, by enabling accurate classifications across different gating tasks. Consequently, this research advances the state-of-the-art by automating the extraction of cellular populations, optimising high-dimensional data. Moreover, by moving beyond black-box models and emphasising explanation, this study establishes a significant methodological and practical contribution in FC data analysis.

## 2 Data and Methods

### 2.1 Data Pre-processing

The data used in this work have been acquired in the diagnostic activity of the Center of Cytometry at Urbino. After the acquisition, data was anonymised to make it impossible to identify the sources. Analyses were taken on the parameters of forward scatter (FSC), side scatter (SSC) and antigens expression CD3, CD4, CD8, CD19 and CD45. Then, data has been exported in *.csv* format, where every row represents a cell and every column one of the parameters cited above. We obtained 15 files from 15 experiments and integrated them into a unique dataset. Dataset labelling has been performed manually by domain experts, adopting the state-of-the-art methodology for gating tasks [10]. To reduce the impact of the operator-dependent bias, we involved different specialists who gated the cellular population according to the following rules.

Figure 1 illustrates the data preprocessing and ML model creation pipeline. FC data files are used for preprocessing. The preprocessed FC data are then subjected to hierarchical gating. Cell annotation is performed after the gating process: TL, BL, and CTL cells are separated from the rest of the cell populations. This supervised ML pipeline is used to train and validate the models. Trained ML models are tested over the test set to evaluate the performance of all models. These models are then subjected to interpretations using the SHAP explainer [9].

In particular, we are interested in automating the extraction of three cellular populations, and accordingly, we performed three experiments (each as a binary classification):

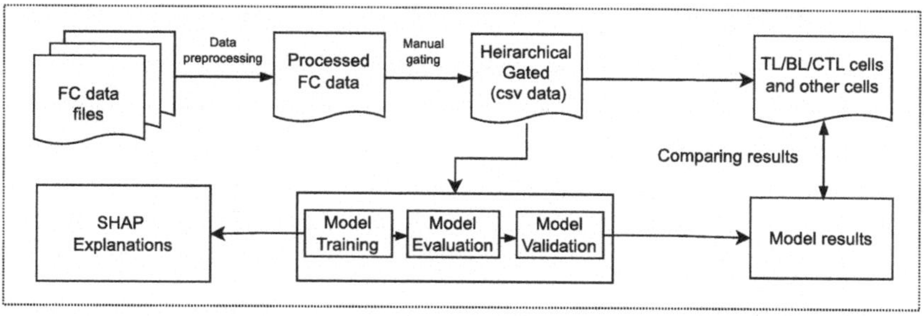


**Fig. 1.** Data processing and interpretable machine learning pipeline.

1. Classification of total T Lymphocytes (TL), events with low SSC, CD45+, CD3+ and CD19-. Lymphocytes T have been classified with class 1 (gated) and the rest as (ungated) 0.
2. Classification of total B Lymphocytes (BL), events with low SSC, CD45+, CD3- and CD19+. Lymphocyte B has been classified as class 1 (gated) and the rest as (ungated) 0.
3. Classification of Cytotoxic T Lymphocytes (CTL), events with low SSC, CD45+, CD3+, CD8+ and CD4-. Lymphocytes T cytotoxic have been classified with class 1 (gated) and the rest as (ungated) 0.

The resulting dataset was imbalanced for each of the three classification tasks, as reported in Table 1. However, we avoided data-balancing techniques to maintain the integrity of the original data distribution. Instead, the nested stratified cross-validation strategy allowed us to mitigate the impact of class imbalance without altering the underlying data structure.

**Table 1.** Amount of total and gated instances for each experiment. Differences in terms of the total number of rows depending on the pipeline of gating: some rows are not considered after the drop of missing values, codified as "not a number" (NaN).

| Experiment | Total Rows | Gated Rows (%) |
|---|---|---|
| T Lymphocytes (TL) | 395234 | 100043 (25.3%) |
| B Lymphocytes (BL) | 389682 | 35516 (9.1%) |
| Cytotoxic T Lymphocytes (CTL) | 390026 | 34593 (8.8%) |

## 2.2 Data Visualisation and Machine Learning Application

In order to evaluate the effect of all the extracted features, we performed a dimensionality reduction using the principal component analysis (PCA) algorithm for all three experiments. The two principal components have been used to generate the t-SNE scatter plots shown in Fig. 2.

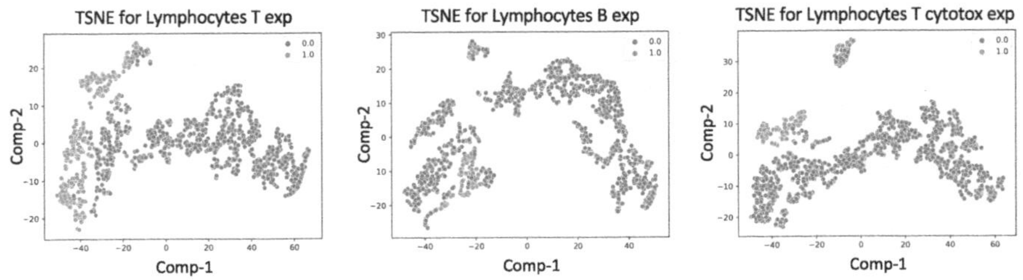


**Fig. 2.** Obtained t-SNE scatterplots considering a subset of 1000 cells for each performed experiment.

In each of the three classification experiments, we exploited all five supervised classification algorithms. Classification has been performed through a repeated nested cross-validation pipeline with five repetitions and five folds. For each repetition, cross-validation has been split into two loops: an inner loop for the hyper-parameters tuning and an outer loop for the performance calculation, avoiding any sort of data peeking. In particular, the hyper-parameters of each estimator have been initialised to an array of values and then tuned through a process of grid search that exhaustively considered all parameter combinations. To avoid dependency on nested cross-validation pipeline from random effects, we decided to perform every experiment five times changing the seed of the random number generator: by averaging the performances across all folds of all repetitions, the effects of particularly bad or particularly good random results have been minimised. For each fold and each repetition, we computed balanced accuracy, Recall, F1-score and the Matthew's correlation coefficient (MCC) since we are particularly interested in the performances of the algorithms for the identification of the positive class (1), and area under the curve (AUC) of the receiver operating characteristic (ROC). The analysis has been run in the two following sequential phases:

1. All available data have been used for training and validation by nested cross-validation pipeline giving average performances as output. In this way, we evaluated the average behaviour of each algorithm through the repetitions of the training phase, changing training data at each run. If an algorithm gives similar results every time, we can conclude that it is robust for the classification task under evaluation;
2. The dataset has been split into two groups (70% and 30%). In this case, we use 70% of data to perform a 5-fold cross-validation training. Then we took the best model from the cross-validation and tested it on the remaining 30% of the dataset.

Data analysis was built using Pandas and *scikit-learn* Python libraries.

## 3 Results and Discussion

The results of the two phases are reported and discussed in the following.

### 3.1 Phase 1 Results

In this paragraph, we deal with mean performances obtained by averaging results from every fold and every repetition of the repeated nested cross-validation, for each of the algorithms evaluated over the validation set. Figure 3 shows the ROC curves, considering standard deviation too. Same results are summarised in tabular form in Table 2 (above), while Table 2 (below) reports the corresponding balanced accuracy indicator, averaged over the same experiments. Since all the experiments provided average values above 0.9, we concluded that the classification algorithms were robust and we proceeded with Phase 2.

**Table 2.** Average results in terms of ROC AUC (above) and balanced accuracy (below) for validation sets (mean ± standard deviation).

| Experiment | SVM | XGB | DT | LR | MLP |
|---|---|---|---|---|---|
| TL | $0.938 \pm 0.005$ | $0.983 \pm 0.001$ | $0.986 \pm 0.002$ | $0.980 \pm 0.000$ | $0.959 \pm 0.001$ |
| BL | $0.927 \pm 0.005$ | $0.970 \pm 0.001$ | $0.944 \pm 0.006$ | $0.974 \pm 0.001$ | $0.869 \pm 0.037$ |
| CTL | $0.988 \pm 0.001$ | $0.982 \pm 0.001$ | $0.975 \pm 0.004$ | $0.968 \pm 0.000$ | $0.911 \pm 0.033$ |
| **Experiment** | **SVM** | **XGB** | **DT** | **LR** | **MLP** |
| TL | $0.938 \pm 0.026$ | $0.982 \pm 0.003$ | $0.985 \pm 0.002$ | $0.980 \pm 0.004$ | $0.962 \pm 0.005$ |
| BL | $0.927 \pm 0.036$ | $0.970 \pm 0.007$ | $0.945 \pm 0.006$ | $0.973 \pm 0.009$ | $0.916 \pm 0.051$ |
| CTL | $0.989 \pm 0.019$ | $0.982 \pm 0.006$ | $0.979 \pm 0.007$ | $0.968 \pm 0.004$ | $0.864 \pm 0.041$ |

**Table 3.** Results in terms of ROC AUC for the test set.

| Experiment | SVM | XGB | DT | LR | MLP |
|---|---|---|---|---|---|
| T Lymphocytes | 0.929 | 0.984 | 0.990 | 0.952 | 0.929 |
| B Lymphocytes | 0.983 | 0.940 | 0.820 | 0.978 | 0.920 |
| Cytotoxic T Lymphocytes | 0.999 | 0.978 | 0.978 | 0.988 | 0.962 |

### 3.2 Phase 2 Results

Table 3 shows ROC AUC values, while Table 4 reports the balanced accuracy (BA), Recall, F1-score, and MCC – obtained on the test set (30% of dataset), for the three algorithms trained on the remaining 70% of data.

Even though the balanced accuracy suggests promising results for each classification task, a more specific evaluation of the classification of positive samples shows some decrease in performances, especially for the DT.

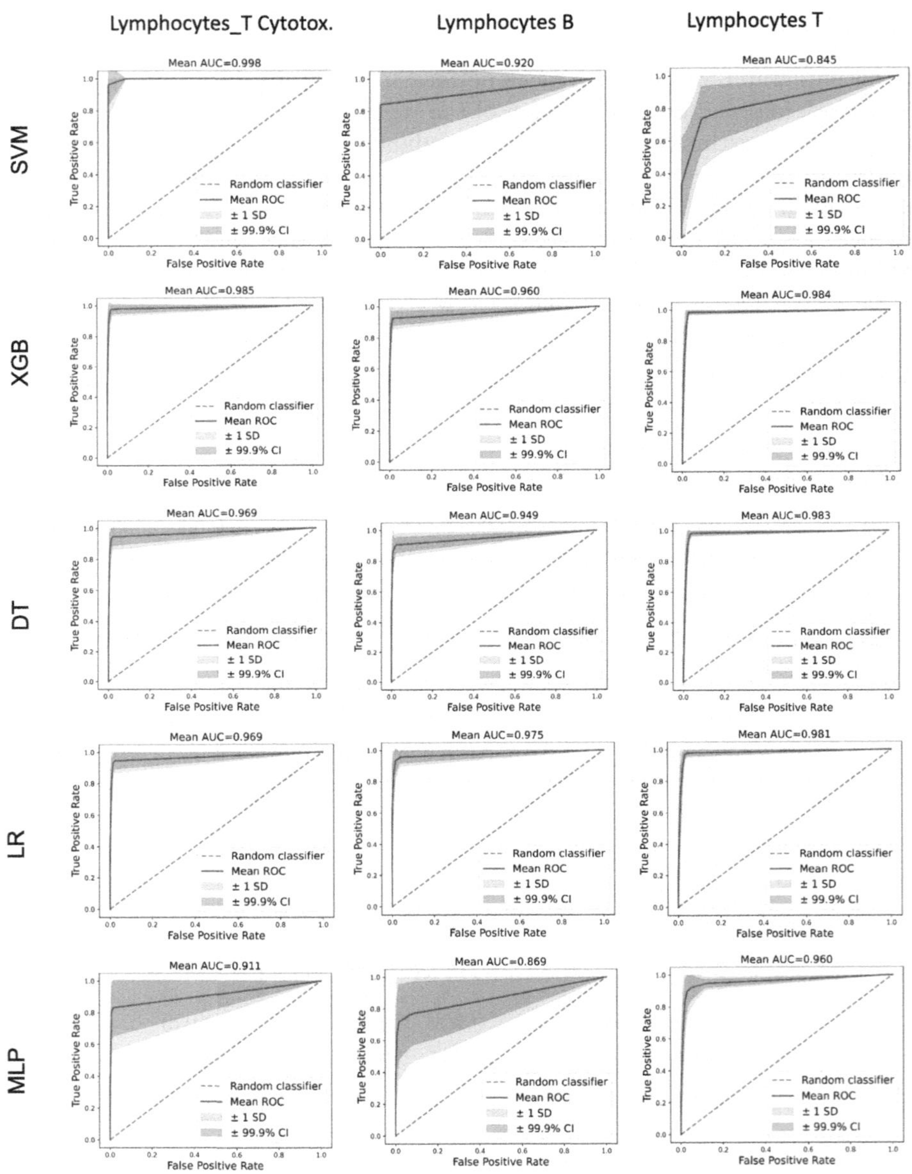


**Fig. 3.** Mean ROC AUC for each algorithm for the classification of total Lymphocytes T, B, T cytotoxic.

### 3.3 Post-hoc Explainability

According to our data analysis, the ML models performed very well in classifying TL, BL, and CTL. However, their explainability is crucial to help clinicians in making decisions [11]. SHAP [9] plots are commonly used to explain the output

**Table 4.** Balanced accuracy – F1 score – Recall – Matthew's correlation coefficient for test set.

| Model | Metric/Experiment | TL | BL | CTL |
|---|---|---|---|---|
| **SVM** | BA | 0.970 | 0.970 | 0.999 |
| | F1 | 0.923 | 0.857 | 0.990 |
| | R | 0.857 | 0.989 | 0.990 |
| | MCC | 0.909 | 0.851 | 0.998 |
| **XGB** | BA | 0.982 | 0.990 | 0.981 |
| | F1 | 0.966 | 0.936 | 0.921 |
| | R | 0.988 | 0.880 | 0.972 |
| | MCC | 0.954 | 0.933 | 0.912 |
| **DT** | BA | 0.985 | 0.973 | 0.982 |
| | F1 | 0.972 | 0.780 | 0.921 |
| | R | 0.990 | 0.640 | 0.972 |
| | MCC | 0.962 | 0.788 | 0.912 |
| **LR** | BA | 0.981 | 0.993 | 0.978 |
| | F1 | 0.966 | 0.960 | 0.904 |
| | R | 0.987 | 0.960 | 0.916 |
| | MCC | 0.955 | 0.956 | 0.892 |
| **MLP** | BA | 0.954 | 0.987 | 0.981 |
| | F1 | 0.918 | 0.913 | 0.911 |
| | R | 0.977 | 0.840 | 0.861 |
| | MCC | 0.890 | 0.910 | 0.903 |

of ML models by attributing the prediction outcome to different features. While SHAP plots can be applied to various models, however, we selected only models that performed better and had complex model behaviour.

With this in view, we exploited SHAP to interpret the classification performance of the DT i.e. model that resulted to be the most robust across all the folds and repetitions of the TL experiment, XGBoost due to its complex ensemble architecture, and MLP for its complex neural network architecture.

Figures 4, 5, and 6 show the plots of SHAP for all the three classification tasks conducted using the XGBoost, DT, and MLP models, respectively. The SHAP-

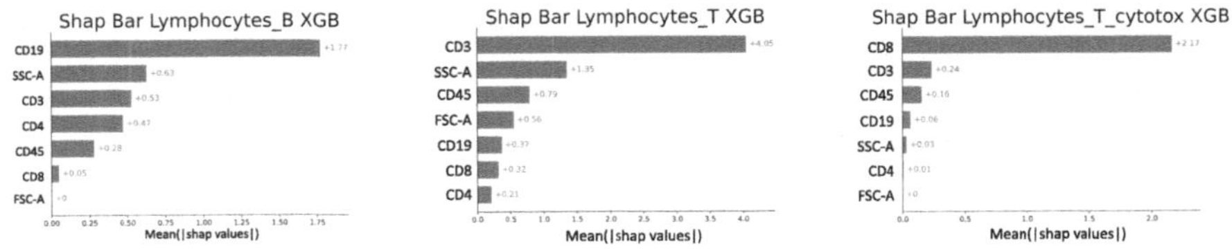


**Fig. 4.** Bar plots of SHAP for XGBoost model to explain the feature contribution towards gating classification.

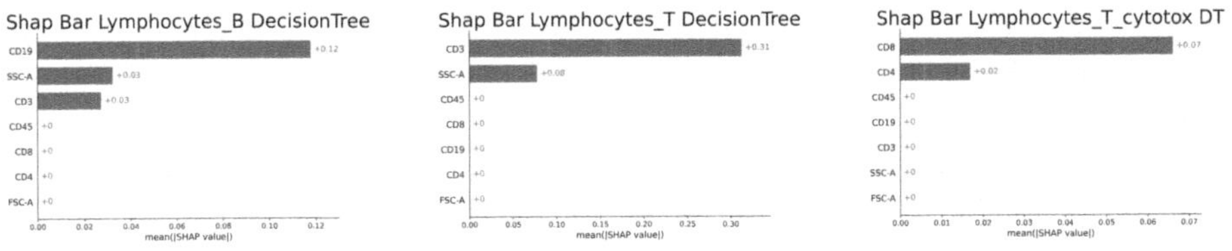


**Fig. 5.** Bar plots of SHAP for DT model to explain the feature contribution towards gating classification.

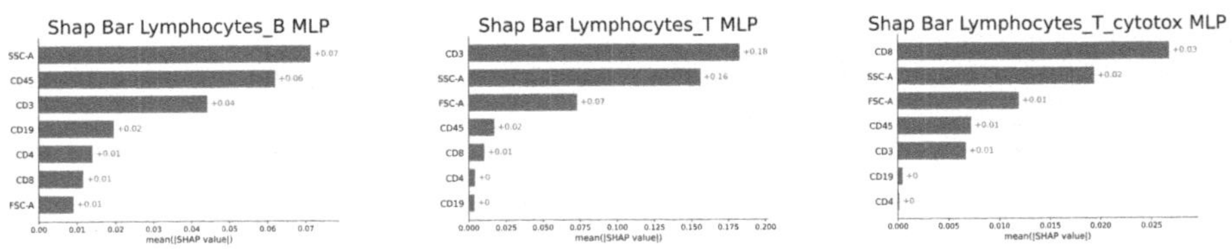


**Fig. 6.** Bar plots of SHAP for MLP model to explain the feature contribution towards gating classification.

based scores for all the events are generated and depicted in order of relevance, with the one at the top being the most relevant event. Each bar represents an event and its contribution to the model classification. On the $x$-axis, the SHAP mean value is presented (calculated by the sum of Shapley values of each event).

Our post-hoc explainability analysis is compared with the manual gating process (with human experts). We have observed that the events outlined as the most contributing to classification outcomes in the SHAP plots are the same utilised in the sequential multiple gating processes for cell identification by domain experts. For example, in Fig. 4, the rightmost plot, the most contributing events in the classification of CTL are $CD8$, $CD3$, and $CD45$. The manual gating process for classifying CTL cells described in [8] is as follows: the first step is to find a gate between $SSC$ and $CD45+$ events, the second gate is performed between $CD3+$ and $CD8+$ events to obtain CTL cells. The same events $CD3+$ and $CD8+$ of the second gate have also shown the highest importance in SHAP plots. Thus, we can reason that events that contributed most to the post-hoc explainability complement the performance of the black box.

Similarly for CTL, DT (see Fig. 5) and MLP-based SHAP scores (see Fig. 6) list CD8 and CD45 in the top three most contributing features in the classification.

## 4 Conclusion

The first aim of this paper was to compare the results of different ML algorithms for the supervised classification of TL, BL, and CTL from flow cytometry data. Mostly, all experiments, with few exceptions, provided high values (above 0.85) of the considered, which led us to the conclusion that ML approaches can provide useful insights in the flow cytometry data analysis domain. In particular, results show that, as concerns the three classification tasks that are discussed in the paper, SVM, LR, and XGBoost are the classification models with higher performances, particularly when considering the Recall and F1-score metrics, while XGBoost and DT the most robust across all the folds and repetitions.

Moreover, since interpretability is a critical priority in this domain, we observed that SHAP-based feature scores consistently supported the manual gating process for classification across all three tasks, indicating the reliability of post-hoc explainability for the clinicians. The ability to provide transparent insights into biologically relevant features enhances trust in model predictions, which is particularly important for clinical decision-making.

To conclude, the pipeline and the overall methodological approach presented in this paper provide valuable insights, suggesting that flow cytometry gating can be automated. However, further evaluation on different datasets is needed to assess the model's robustness and generalisability.

## References

1. Weber, L.M., Nowicka, M., Soneson, C., Robinson, M.D.: diffcyt: differential discovery in high-dimensional cytometry via high-resolution clustering. Commun. Biol. **2**(1), 183 (2019)
2. Verschoor, C.P., Lelic, A., Bramson, J.L., Bowdish, D.: An introduction to automated flow cytometry gating tools and their implementation. Front. Immunol. **6**, 380 (2015)
3. Abdelaal, T., van Unen, V., Höllt, T., Koning, F., Reinders, M., Mahfouz, A.: Predicting cell populations in single cell mass cytometry data. Cytometry Part A **95**(7), 769–781 (2019)
4. Burton, R.J., Ahmed, R., Cuff, S.M., Baker, S., Artemiou, A., Eberl, M.: Cytopy: an autonomous cytometry analysis framework. PLOS Comput. Biol. **17**(6), e1009071 (2021)
5. Hu, Z., Bhattacharya, S., Butte, A.J.: Application of machine learning for cytometry data. Front. Immunol. **12**, 5703 (2022)
6. Gutierrez-Rodrigues, F., et al.: Differential diagnosis of bone marrow failure syndromes guided by machine learning. Blood J. blood–2022017518 (2022)
7. Monaghan, S.A., et al.: A machine learning approach to the classification of acute leukemias and distinction from nonneoplastic cytopenias using flow cytometry data. Am. J. Clin. Pathol. **157**(4), 546–553 (2022)

8. Suffian, M., et al.: Machine learning for automated gating of flow cytometry data. In: CEUR Workshop Proceedings, vol. 3307, pp. 47–56. Sun SITE Central Europe, RWTH Aachen University (2022)
9. Lundberg, S.M., Lee, S.-I.: A unified approach to interpreting model predictions, pp. 4768–4777 (2017)
10. Staats, J., Divekar, A., Philip McCoy, J., Maecker, H.T.: Guidelines for gating flow cytometry data for immunological assays. Immunophenotyping: Methods Protocols 81–104 (2019)
11. Ali, S., et al.: Explainable artificial intelligence (XAI): what we know and what is left to attain trustworthy artificial intelligence. Inf. Fusion 101805 (2023)

# Pre-trained Models Based on Primary Sequence to Classify Antibody Binding to Protein and Non-protein Targets with 80% Accuracy

Sara Joubbi[1,2](✉), Giuseppe Maccari[2], Dario Cardamone[2], Giorgio Ciano[2], Alessio Micheli[1], Paolo Milazzo[1], and Duccio Medini[2]

[1] Department of Computer Science, University of Pisa, Pisa, Italy
sara.joubbi@phd.unipi.it
[2] Data Science for Health Lab, Toscana Life Sciences Foundation, Siena, Italy

**Abstract.** Monoclonal antibodies provide targeted treatment options for various diseases. In infectious diseases, techniques like reverse vaccinology 2.0 can extract potent monoclonal antibodies from human donors. However, there is no control over the specific target to which the antibody binds. Therefore, it is crucial to thoroughly understand the characteristics of the antibody sequence to extract target information and prioritize potential candidates for further *in-vitro* analysis. Addressing this challenge requires integrating both *in-vitro* and *in-silico* approaches. This study aims to explore machine learning algorithms and various augmentation techniques to determine whether monoclonal antibodies bind to proteins or non-protein targets solely based on their sequence. We used oversampling, SMOTE, ESM1-b, and hallucination techniques to enhance sequence information. To the best of our knowledge, this research is the first attempt to apply machine learning methodologies to this specific classification task, resulting in an accuracy rate of 80%, and an MCC of 61%. These findings pave the way for further exploration and introduce a new approach to streamline and enhance antibody development processes.

**Keywords:** Antibodies · Machine Learning · Target classification · Protein target · Non-protein target · AntiBERTy · ProtBERT

## 1 Introduction

Antibodies, also known as immunoglobulins (Ig), are essential proteins in the immune response against pathogens [19]. An antibody is composed of two heavy chains (H) and two light chains (L). These chains consist of a crystallizable fragment ($F_c$) and two antigen-binding fragments ($F_{ab}$). The $F_{ab}$ region contains variable segments (VH and VL), each of which has three hypervariable loops known as complementarity-determining regions (CDRs), as shown in Fig. 1 [13].


M. Vettoretti et al. (Eds.): CIBB 2023, LNBI 14513, pp. 237–254, 2025.
https://doi.org/10.1007/978-3-031-90714-2_18

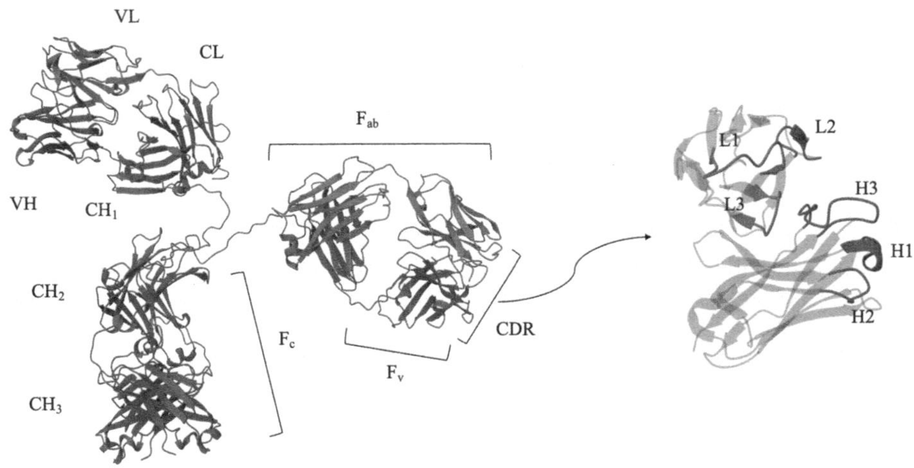


**Fig. 1.** Antibodies are composed of two heavy chains and two light chains, each with variable domains that form a paratope responsible for binding to the target. The paratope consists of six variable loops: three on the heavy chain (H1, H2, and H3) and three on the light chain (L1, L2, and L3). The heavy chains have three constant domains, while the light chains have one. The antibody structure is derived from PDB 1IGT, with a specific focus on the $F_v$ fragment from PDB 1DQL.

Monoclonal antibodies (mAbs), engineered for their high specificity and affinity towards particular targets, offer a promising therapeutic option across various diseases, including cancer, autoimmune diseases, and infectious diseases, with over 100 approved mAbs and 1,000 undergoing clinical studies [13,14]. While vaccines are important for generating immunity to specific diseases and techniques for developing them improve over time, mAbs are emerging as a viable alternative to achieve a similar outcome.

Recent advancements in high-throughput technologies for human immunology and structural biology have enabled the development of innovative techniques for antibody discovery and vaccine development. Reverse vaccinology 2.0 is one such advanced approach, utilizing high-throughput screening of human donor blood to identify neutralizing antibodies against specific pathogens [21]. By focusing on antibodies produced by the human immune system, this method aims to develop more effective vaccines and therapeutics. Despite its potential, experimental methodologies face limitations that can limit their ability to identify potent mAbs with broad protection. A key limitation is the lack of control over the selection of targets for the identified antibodies [13]. This issue is especially significant when variable immunodominant epitopes are present, which can skew the antibody response and limit the generation of mAbs directed against less immunogenic but more conserved epitopes [12,25]. In some cases, targeting specific molecules is necessary to ensure broad protection against particular pathogens and to prevent viral escape [6,23,24]. Consequently, experimental screening would benefit from an initial target identification step to prioritize antibodies that bind to a specific target of interest. Furthermore, protein epitopes often elicit stronger immune responses than non-protein ones, such as polysaccharides, because proteins present more complex and diverse epitopes that are more readily recognized by the immune system. In addition, polysaccharides are less immunogenic than most proteins [1].

An example of this is *S. pneumoniae*, which has over 100 known capsular serotypes. It is not feasible to identify antibodies that provide broad coverage against all of these capsular polysaccharides [4]. In contrast, surface proteins like PhtD and PspA are more conserved and make better targets for antibody therapy [9]. Therefore, antibody binding to protein targets should be further analyzed in laboratories instead of screening antibodies against non-protein targets for the case of *S. pneumoniae*.

To address these challenges, it is essential to integrate *in-vitro* methodologies with *in-silico* approaches to prioritize antibodies based on their targets. Machine learning (ML) algorithms have emerged as practical solutions to reduce costs, expedite processes, and enhance wet lab techniques. Several ML methods have been developed, starting from antibody sequence and structure [13]. These algorithms rely on widely available protein sequences, which can be analyzed using natural language processing (NLP) techniques to encode and use this information for protein [8] or antibody [22] predictions, thus reducing the need for extensive labeled data in downstream applications.

Effective *in-silico* methods, particularly ML, depend on abundant high-quality data. Next-generation sequencing (NGS) technologies have enabled extensive characterization of antibody repertoires, resulting in publicly accessible sequencing data repositories [16]. Although these repositories are valuable for training and validating ML models, challenges such as imbalanced datasets persist and can impact model performance. Data augmentation methods aim to overcome this issue by artificially expanding datasets, thereby enhancing the robustness and generalization of ML models used in biological tasks [18].

In this paper, our objective is to determine whether an antibody sequence binds specifically to a protein target or a non-protein target using only the information within the antibody sequence using state-of-the-art methodologies. We propose several ML methods and augmentation techniques to tackle this task, to prioritize potential candidates for subsequent *in-vitro* characterization. The main contribution of this paper is to demonstrate the applicability of ML methods to a relevant task in mAbs development, which, as far as we know, has not been previously explored. Our study provides valuable insights and presents promising preliminary results that could pave the way for further investigations.

## 2 Materials and Methods

### 2.1 Dataset and Pre-processing

The data used in this study were obtained from the Structural Antibody Database (SAbDab) [7], which includes publicly available antibody structures from the Protein Data Bank (PDB) along with their corresponding experimental information. The dataset consists of 14,340 antibody sequences that bind to various targets, such as proteins, peptides, carbohydrates, nucleic acids, haptens, and combinations (e.g. peptide and proteins). Each entry in the antibody data includes a heavy chain (VH) sequence, a light chain (VL) sequence, and the target type.

We excluded unknown or undefined target antibodies, single-chain antibodies, and redundant sequences from the dataset. We then divided the data into two categories: antibodies that bind to a protein target (2,924 sequences, 90.69%) and antibodies that bind to a non-protein target (315 sequences, 9.31%), as shown in Fig. 2. We focused on the non-protein class, which had fewer samples, and randomly split this dataset into 70% (215 sequences) for training, and 15% (50 sequences) each for validation and testing. To maintain a balanced sample distribution between the two classes in the validation and test sets, we randomly selected 50 sequences from the protein class for each set. Consequently, the validation and test set each contained 100 sequences, while the training set included 3,039 sequences (2,824 protein sequences and 215 non-protein sequences), as summarized in Table 1. The validation set was used for model selection, hyperparameter optimization, and early stopping. For the classification task, Protein was designated as class 0 and Non-protein as class 1.

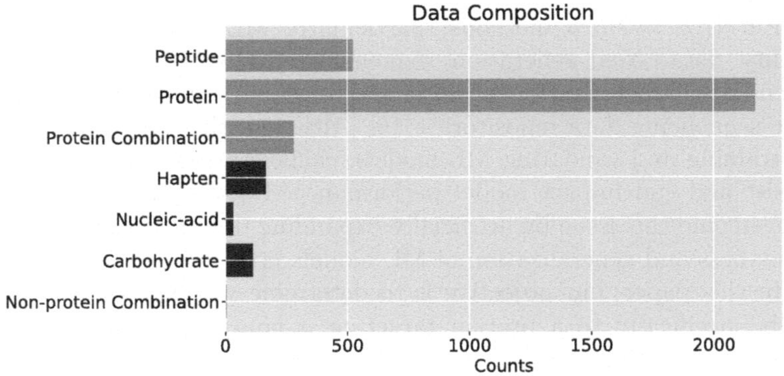


dataset

**Fig. 2.** Data composition of the class Protein and Non-Protein. Class Protein (green) is significantly larger than class Non-protein (magenta). (Color figure online)

**Table 1.** Dataset split for the Protein and Non-protein classes.

| Dataset | Protein | Non-protein | Total |
|---|---|---|---|
| Training | 2,824 (96%) | 215 (70%) | 3,039 |
| Validation | 50 (2%) | 50 (15%) | 100 |
| Test | 50 (2%) | 50 (15%) | 100 |

## 2.2 Data-Augmentation to Maintain the Biological Meaning of the Sequence

In addressing the protein vs. non-protein classification task, we faced the challenge of working with an imbalanced dataset, which often compromises the performance of the underrepresented class. To overcome this issue, we implemented data augmentation. However, generating new sequences that preserved the intended biological meaning (i.e., the target type) posed a significant challenge. We explored different augmentation techniques, as shown in Fig. 3.

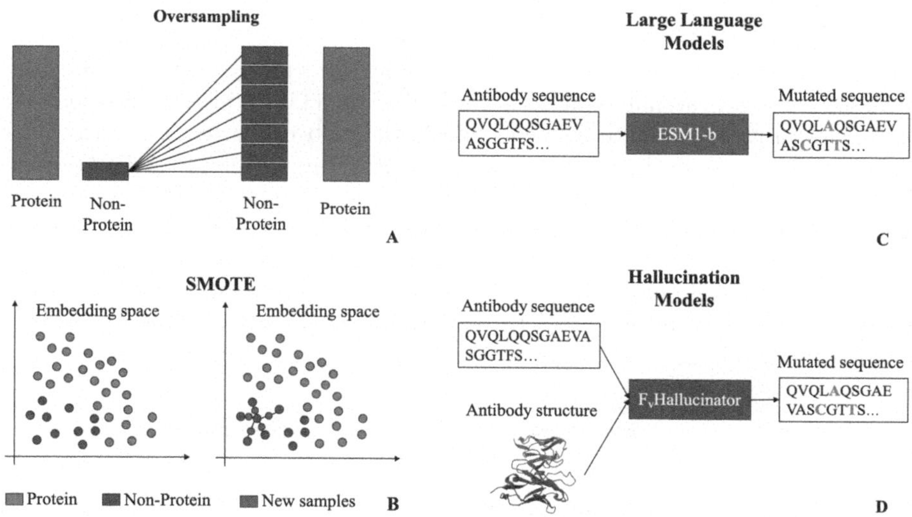


**Fig. 3.** Different augmentation techniques used in this study. A) Oversampling: repeating the non-protein sequence 13 times; B) SMOTE: sampling in the embedding space based on the k-nearest neighbors; C) LLMs: ESM1-b to define biological mutation for Ab sequences; D) Hallucination: use of FvHallucinator to mutate the input sequence while maintaining the same structure.

1. **Oversampling**: The most straightforward way to deal with an imbalanced dataset is by duplicating examples in the minority class (as illustrated in Fig. 3A), even though these examples do not provide any new information to the model.
2. **SMOTE**: A more sophisticated approach is the Synthetic Minority Oversampling Technique (SMOTE) [3], as shown in Fig. 3B. SMOTE generates new instances by randomly selecting an instance from the minority class and its k-nearest neighbors, and then creating synthetic instances along the line segments connecting them.
3. **Large language models**: Another technique involves using a protein language model to modify the amino acid sequence (Fig. 3C). ESM1-b [2] is a

protein language model trained on 86 billion amino acids from 250 million protein sequences using an unsupervised masked language modeling (MLM) objective. ESM1-b was employed to generate mutated sequences.

4. **Hallucination**: Lastly, the hallucination technique was used to mutate protein sequences while preserving their structure and biological significance (Fig. 3D). This method utilizes a ML model to create 3D protein structures from random sequences and predict distances between alpha carbons. The FvHallucinator approach [17] uses a reference structure to generate libraries of Fv sequences through hallucination.

### 2.3 Methods

The classification model consists of an encoder for each sequence, where the embedding are concatenated and classified. The model is divided in three blocks, as shown in Fig. 4: the input sequence block, the embedding block and the classification block. We evaluated two different embedding blocks: ProtBERT [8], a pre-trained 30-layer transformer model trained on Uniref100 (216M sequences) that produces 1024-dimensional embeddings, and AntiBERTy [22], an 8-layer transformer model pre-trained on 558M natural antibody sequences that produce 512-dimensional embeddings. The encoder weights remained fixed during training. By concatenating both VH and VL, we obtain a 2048-dimensional vector for ProtBERT and a 1024-dimensional vector for AntiBERTy.

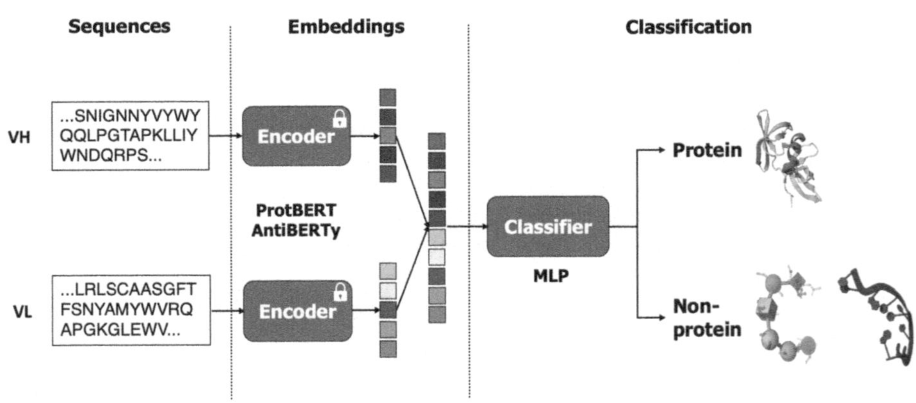


**Fig. 4.** Abtarget architecture. The VH and VL sequences are encoded using PrtoBERT or AntiBERTy. The embeddings are then concatenated and classified using an MLP.

Within the classification block, we used the Multilayer Perceptron (MLP) [11], a feedforward neural network. Before selecting this architecture, we conducted a comparative analysis of the MLP with other well-known ML techniques, including Least Absolute Shrinkage and Selection Operator (LASSO), Support Vector Machine (SVM), k-Nearest Neighbors (KNN), and Random Forest (RF). Moreover, to address the imbalanced dataset issue, we implemented an alternative approach known as one-class classification. This method involves training the model solely on samples from the positive class, supplemented by a limited number of examples from the negative class. The prevailing method for one-class classification is OCSVM (One-Class Support Vector Machine) [10].

### 2.4 Experiments

We conducted the following training set experiments while keeping 100 sequences for both the validation and test sets (refer to Table 2 for the training composition):

1. *Single*: we randomly undersampled 215 sequences for the protein class, training the model with a balanced dataset;
2. *Ensemble (Bagging)*: to create a balanced dataset while exploring all protein classes, we divided the 2,824 sequences into 20 subsamples of 215 sequences, with a 1/3 overlap. The dataset for the non-protein class remained the same for each model. In the final ensemble, we excluded the worst-performing model on the validation set to have an odd number of models. We used voting criteria for the final classification;
3. *Imbalanced*: we trained the model using the entire training set (2,824 sequences of the protein class and 215 sequences of the non-protein class). In this case, we weighted the loss as follows: 1 for the protein class and 13.13 (2,824/215) for the non-protein class;
   - *Imbalanced lr*: we trained the model using the same approach as the Imbalanced experiment. We used the reduceLROnPlateau learning rate scheduler, which reduces the learning rate when the metrics are not improving;
4. *Oversampling*: the model was trained using a dataset of 5,648 sequences, comprising 2,824 sequences of the protein class and 2,824 sequences of the non-protein class. To obtain the 2,824 sequences of the non-protein class, we repeated the original training set of 215 sequences 13 times;
5. *SMOTE sequences*: the model was trained using a dataset of 5,648 sequences; data augmentation was performed on the non-protein sequences using SMOTE;
6. *ESM1-b*: the model was trained with a dataset of 5,648 sequences. The 2,824 sequences of the non-protein class consist of a combination of the original sequences and the ESM1 mutated sequences;
7. *FvHallucinator*: the model was trained using the sequences of the non-protein class produced using FvHallucinator, augmented 13 times.

We maintained the weighting on the loss for all augmentation techniques.

## 3 Results

In this section, the results of the comparison between the two encoders, ProtBERT and AntiBERTy, as described in the methodology section Subsect. 2.3, are

**Table 2.** Training composition for the Protein and Non-protein classes for the different experiments.

| Experiment | Training data | | | Description |
|---|---|---|---|---|
| | Protein | Non-Protein | Total | |
| Single | 215 | 215 | 430 | Balanced dataset. |
| Ensemble | 20 × 215 | 20 × 215 | 20 × 430 | 20 models from 20 subsamples of 215 sequences of the protein class, with a 1/3 overlap. Result with voting criteria. |
| Imbalanced and Imbalanced lr | 2,824 | 215 | 3,039 | Weighted Loss |
| Oversampling, SMOTE, ESM1-b and FvHallucinator | 2,824 | 2,824 | 5,648 | Data augmentation with different techniques |

presented. In Subsect. 3.1, a preliminary study was conducted on the different techniques using a hold-out training, validation, and test dataset. Subsequently, in Subsect. 3.2, a combination of the different encoders was examined. Finally, in Subsect. 3.3, the bootstrap method was used to evaluate the robustness of the experiments. We chose to use the Matthew Correlation Coefficient (MCC) instead of the Area under the ROC Curve (AUC) because it provides a comprehensive evaluation of classification performance, particularly when working with imbalanced datasets [5].

### 3.1 AntiBERTy and ProtBERT Encoders

We conducted a comparison of the classification methods and found that the MLP stands out as the most effective approach, particularly when dealing with imbalanced datasets meanwhile the other methods tend to drop the performances, as demonstrated in Appendix Table 4. Hence, we conducted the experiments outlined in Subsect. 2.4 using the MLP as a classification block. Among the different experiments, the imbalanced dataset split yielded the highest performance for AntiBERTy with the MLP classifier. Conversely, for ProtBERT with the MLP classifier, the oversampling technique exhibited the highest performance. AntiBERTy consistently outperformed ProtBERT in various experiments, achieving higher F1 scores, accuracy, and MCC. This confirms the encoder's specificity for the antibody-related task and highlights the advantage of using a smaller model [15]. The most effective model was AntiBERTy + MLP with the imbalanced dataset, as shown in Fig. 6.

### 3.2 Combination of AntiBERTy and ProtBERT

Given the complementary performance observed between AntiBERTy and ProtBERT when using the MLP, we included both models in each experiment to assess potential performance improvements. This combined experiment is different from the Ensemble experiment described in Subsect. 2.4. As shown in Fig. 5, we combined the predictions of the ProtBERT + MLP model, trained on a specific experiment (e.g., imbalanced dataset), with those of the AntiBERTy + MLP model, also trained on the same experiment. The predictions were then aggregated, and the position with the highest probability score was chosen.

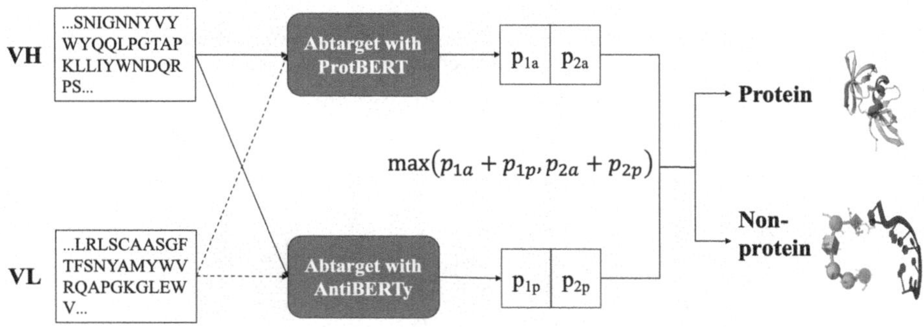


**Fig. 5.** The architecture used for the integration of the two models involves encoding and classifying the VH and VL sequences using the PrtoBERT + MLP model and the AntiBERTy + MLP model respectively. The classification outputs from both models are then aggregated for each class, and the final class is determined based on the class with the highest score.

When analyzing the performance of different models and combinations, clear patterns emerge, as shown in Fig. 6. In scenarios like the Single and Ensemble experiments, the performance surpasses that of the individual models. However, in other experiments, the performance is lower than AntiBERTy but higher than ProtBERT. Notably, ESM1-b stands out as an exception, showing a significant drop in performance when combined with the other encoders. In conclusion, integrating two models with different encoders does not offer a clear advantage in the classification tasks.

### 3.3 Bootstrap on the Different Experiments

The bootstrap technique was used to ensure consistency in arranging the training, validation, and test sets and to assess the reliability of our methodology. This involved resampling the original dataset with replacement to generate multiple new datasets for training and evaluating ML models. We analyzed the results of 20 iterations. The ensemble experiment was not carried out as it is equivalent to performing a bootstrap with 20 repetitions for a single model. Next, we conducted a statistical analysis to determine if the two encoders had a significant difference in the bootstrap results. Using the non-parametric Kruskal-Wallis test, we compared each experiment (see Appendix Table 7) and the overall results

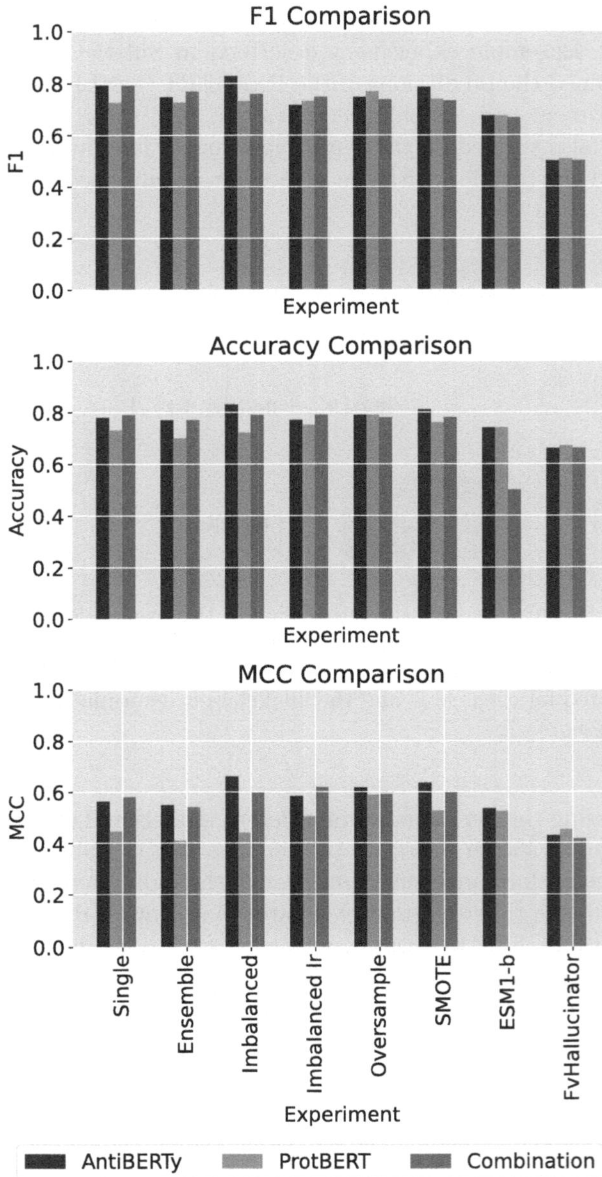


**Fig. 6.** Results of the experiments on different models using MLP classifier and combining AntiBERTy and ProtBERT.

(see Appendix Table 8). The p-values consistently indicated statistical significance ($p < 0.05$) for each metric. However, there was no statistical difference for Oversampling and SMOTE augmentation.

The bootstrap results, presented in Table 3, show that AntiBERTy performed better than ProtBERT across various metrics such as F1 score, accuracy, and MCC. Refer to Appendix Table 6 for precision and recall results. Interestingly, AntiBERTy with an imbalanced dataset and ProtBERT with oversampling augmentation achieved similar scores. However, augmentation techniques did not significantly enhance AntiBERTy's performance beyond that of the imbalanced dataset. Among all augmentation techniques, SMOTE demonstrated the most promising results. Both oversampling and SMOTE effectively improved ProtBERT's performance. It is worth noting that using Large Language Models (LLMs) and the hallucination technique for data augmentation led to a performance decrease.

**Table 3.** Results of the bootstrap on the different experiments in terms of mean and standard deviation.

| Encoder | Experiment | F1 | Accuracy | MCC |
|---|---|---|---|---|
| AntiBERTy | Single | 0.79 ± 0.04 | 0.79 ± 0.04 | 0.58 ± 0.08 |
| | **Imbalanced** | **0.77 ± 0.04** | **0.80 ± 0.03** | **0.61 ± 0.06** |
| | Imbalanced lr | 0.73 ± 0.05 | 0.77 ± 0.03 | 0.57 ± 0.06 |
| | Oversampling | 0.72 ± 0.06 | 0.77 ± 0.04 | 0.59 ± 0.07 |
| | SMOTE | 0.75 ± 0.06 | 0.79 ± 0.04 | 0.60 ± 0.07 |
| | ESM1-b | 0.69 ± 0.06 | 0.75 ± 0.04 | 0.54 ± 0.07 |
| | FvHallucinator | 0.58 ± 0.09 | 0.70 ± 0.04 | 0.49 ± 0.06 |
| ProtBERT | Single | 0.72 ± 0.05 | 0.72 ± 0.04 | 0.44 ± 0.09 |
| | Imbalanced | 0.69 ± 0.08 | 0.74 ± 0.05 | 0.50 ± 0.10 |
| | Imbalanced lr | 0.69 ± 0.07 | 0.74 ± 0.04 | 0.50 ± 0.07 |
| | **Oversampling** | **0.77 ± 0.05** | **0.79 ± 0.04** | **0.60 ± 0.07** |
| | SMOTE | 0.76 ± 0.05 | 0.78 ± 0.04 | 0.58 ± 0.07 |
| | ESM1-b | 0.57 ± 0.08 | 0.69 ± 0.04 | 0.43 ± 0.08 |
| | FvHallucinator | 0.34 ± 0.11 | 0.60 ± 0.04 | 0.31 ± 0.09 |

## 4 Conclusions

Understanding the target type is crucial for selecting antibodies for further characterization and boosting vaccine development. This study employs machine learning to a new task: classify antibody targets based solely on their sequence. We demonstrate that it is possible to differentiate between antibodies that bind to protein or non-protein targets by analyzing their sequences. To assess the model's performance, we conducted bootstrap resampling 20 times and consistently obtained similar results compared to the previous random split. We then used the Kruskal-Wallis test to determine if there was a statistically significant difference between the two encoders. The results showed a $p - value < 0.05$ for each metric. It is worth noting that AntiBERTy, despite having an imbalanced dataset, and ProtBERT, with oversampling augmentation, achieved the best results.

One possible explanation for the disparity in performance between AntiBERTy and ProtBERT could be attributed to their inherent characteristics. AntiBERTy, with its specialized training on antibody-related data, enables the MLP to achieve superior performance with fewer instances. Conversely, ProtBERT's training on proteins may necessitate a larger dataset to accurately distinguish variations in the embedding space. The relatively lower performance of ESM and FvHallucinator could be attributed to the inherent similarity of the generated sequences, which might introduce noise instead of offering additional information.

These experiments demonstrate the use of pre-trained models to predict the target type of antibodies, using sequence data and dealing with imbalanced class distributions. With an accuracy rate of 80%, we are making progress in identifying potential antibody candidates, which can save time and cost in subsequent *in-vitro* characterization. Although our results are promising, several aspects require further investigation. The limitation caused by the lack of data and class imbalance remains significant, and our future direction is to address this by augmenting labeled data from literature sources for non-protein classes. Additionally, we plan to explore the transition from binary to multi-class classification, as this could greatly enhance the model's effectiveness in antibody characterization and allow for finer differentiation between different target subclasses. However, it is important to note that splitting the subclass datasets, considering the available data (see Fig. 2), could worsen the issue of class imbalance.

To investigate the behavior of the model, we can improve our methodology by incorporating eXplainable Artificial Intelligence (XAI). This will enhance interpretability and trustworthiness of the model.

**Acknowledgments.** This work is partially supported by PNRR ECS00000017 Tuscany Health Ecosystem - Spoke 6 "Precision medicine & personalized healthcare", funded by the European Commission under the NextGeneration EU program. We would like to thank the University of Pisa Data Center for providing the necessary hardware resources for this project. We also want to express our gratitude to Pietro Liò and Domenico Tortorella for their help and support throughout this project.

**Disclosure of Interests.** The authors have no competing interests to declare that are relevant to the content of this article.

**Availability of Data and Software Code.** Our software code is available at the following URL: https://github.com/dasch-lab/abtarget.

# Appendix

## Classification Head Comparison

We conducted a preliminary performance comparison among various classifiers: MLP, LASSO, SVM, KNN, RF and OCSVM. The evaluation was carried out using both balanced and imbalanced data splits (see Table 4). The hyperparameters of each method were tuned using the validation set.

The MLP classifier is composed of two fully connected layers with a LeakyReLU activation function between them. For ProtBERT, the dimensions of the first layer are 2048 to 1024, while for AntiBERTy, they are 1024 to 512. The second layer outputs a dimension of 2 (the number of classes). The MLP was trained for 50 epochs, using cross-entropy loss, a batch size of 16, and the Adam optimizer with an initial learning rate of $3 \times 10^{-5}$.

For the **LASSO** experiments, we selected regularization parameters of 0.7 for AntiBERTy with the single dataset, 0.33 for AntiBERTy with the imbalanced dataset, and 0.1 for ProtBERT with both the single and imbalanced datasets. In the **SVM** experiments, the linear kernel produced the best results with a regularization term of 10 for AntiBERTy in the single experiment, 3 for the imbalanced experiment, 5 for ProtBERT in the single experiment, and 10 for the imbalanced dataset. For the **RF** algorithm, we implemented 100 estimators using the entropy criterion for AntiBERTy in both the single and imbalanced datasets. As for ProtBERT, we employed 200 estimators for the single dataset and 50 for the imbalanced dataset, using the entropy criterion. In the **KNN** experiments, we selected k = 12 for the single experiment and k = 5 for the imbalanced dataset with AntiBERTy, while for ProtBERT, we chose k = 8 for the single dataset and k = 5 for the imbalanced dataset, using a 'distance' weight in all scenarios. The **OCSVM** experiments employed the 'rbf' kernel with the gamma parameter set to 'auto' for the AntiBERTy single and imbalanced experiments. For the PrtoBERT single experiments, we opted for a 'sigmoid' kernel, whereas for the ProtBERT imbalanced experiments, a 'poly' kernel with a degree of 5 was selected. The nu parameter was set to 0.5 for the balanced dataset and 0.07 for the imbalanced dataset, which consisted of 265 non-protein sequences compared to 3139 training sequences. The results indicate that the MLP is the only method that either maintains or enhances performance with imbalanced data, demonstrating its efficacy in effectively learning from more samples.

**Table 4.** Comparison between MLP, RF, KNN, SVM, LASSO, and OCSVM on a random split as defined in Table 1.

| Encoder | Classifier | Experiment | Precision | Recall | F1 | Accuracy | MCC |
|---|---|---|---|---|---|---|---|
| AntiBERTy | MLP | Single | 0.75 | 0.84 | 0.79 | 0.78 | 0.56 |
| | | **Imbalanced** | **0.82** | **0.84** | **0.83** | **0.83** | **0.66** |
| | OCSVM | Single | 0.41 | 0.30 | 0.35 | 0.44 | -0.13 |
| | | Imbalanced | 0.25 | 0.02 | 0.04 | 0.48 | -0.10 |
| | LASSO | Single | 0.75 | 0.76 | 0.75 | 0.75 | 0.50 |
| | | Imbalanced | 0.96 | 0.44 | 0.60 | 0.71 | 0.50 |
| | SVM | Single | 0.76 | 0.82 | 0.79 | 0.78 | 0.56 |
| | | Imbalanced | 0.93 | 0.50 | 0.65 | 0.73 | 0.52 |
| | RF | Single | 0.73 | 0.76 | 0.75 | 0.74 | 0.48 |
| | | Imbalanced | 0.94 | 0.30 | 0.45 | 0.64 | 0.38 |
| | KNN | Single | 0.69 | 0.82 | 0.75 | 0.73 | 0.45 |
| | | Imbalanced | 0.91 | 0.42 | 0.58 | 0.69 | 0.45 |
| ProtBERT | MLP | Single | 0.73 | 0.73 | 0.73 | 0.73 | 0.45 |
| | | Imbalanced | 0.76 | 0.70 | 0.73 | 0.72 | 0.44 |
| | OCSVM | Single | 0.44 | 0.40 | 0.42 | 0.45 | -0.10 |
| | | Imbalanced | 0.54 | 0.92 | 0.68 | 0.57 | 0.20 |
| | LASSO | Single | 0.71 | 0.74 | 0.73 | 0.72 | 0.44 |
| | | Imbalanced | 1.0 | 0.34 | 0.51 | 0.67 | 0.45 |
| | SVM | Single | 0.71 | 0.78 | 0.74 | 0.73 | 0.46 |
| | | Imbalanced | 1.0 | 0.14 | 0.25 | 0.57 | 0.27 |
| | RF | Single | 0.78 | 0.72 | 0.72 | 0.76 | 0.52 |
| | | Imbalanced | 1.0 | 0.32 | 0.48 | 0.66 | 0.44 |
| | KNN | Single | 0.71 | 0.78 | 0.74 | 0.73 | 0.46 |
| | | Imbalanced | 0.96 | 0.44 | 0.60 | 0.71 | 0.50 |

## Environment Setting

The hardware and software resources used for model development and experimentation are presented in Table 5.

## Bootstrap Full Results

Additional information regarding the bootstrap evaluation is presented in Table 6, which includes also precision, recall.

## Statistical Analysis

Statistical analysis was conducted to ascertain whether there exists a significant difference between the two encoders. The outcomes of the Kruskal-Wallis test are

**Table 5.** Development environments and requirements.

| System | Ubuntu 20.04.5 LTS |
|---|---|
| CPU | AMD EPYC 7413 24-Core Processor |
| RAM | 16×4GB; 2.67MT/s |
| GPU | NVIDIA A100-SXM-80GB |
| CUDA version | 11.5 |
| Programming language | Python 3.7.15 |
| Deep learning framework | Pytorch [20] (Torch 1.12.1, torchvision 0.13.1) |

**Table 6.** Results of the bootstrap on the different experiments in terms of mean and standard deviation for the MLP.

| Encoder | Exp. | Precision | Recall | F1 | Accuracy | MCC |
|---|---|---|---|---|---|---|
| AntiBERTy | Single | 0.81 ± 0.08 | 0.78 ± 0.05 | 0.79 ± 0.04 | 0.79 ± 0.04 | 0.58 ± 0.08 |
| | Imb. | 0.65 ± 0.01 | 0.88 ± 0.01 | 0.77 ± 0.04 | 0.80 ± 0.03 | 0.61 ± 0.06 |
| | Imb. lr | 0.63 ± 0.01 | 0.88 ± 0.01 | 0.73 ± 0.05 | 0.77 ± 0.03 | 0.57 ± 0.06 |
| | Over. | 0.60 ± 0.01 | 0.92 ± 0.00 | 0.72 ± 0.06 | 0.77 ± 0.04 | 0.59 ± 0.07 |
| | SMOTE | 0.64 ± 0.01 | 0.91 ± 0.00 | 0.75 ± 0.06 | 0.79 ± 0.04 | 0.60 ± 0.07 |
| | ESM1-b | 0.56 ± 0.01 | 0.91 ± 0.01 | 0.69 ± 0.06 | 0.75 ± 0.04 | 0.54 ± 0.07 |
| | FvH | 0.42 ± 0.01 | 0.96 ± 0.01 | 0.58 ± 0.09 | 0.70 ± 0.04 | 0.49 ± 0.06 |
| ProtBERT | Single | 0.72 ± 0.08 | 0.72 ± 0.03 | 0.72 ± 0.05 | 0.72 ± 0.04 | 0.44 ± 0.09 |
| | Imb. | 0.62 ± 0.11 | 0.82 ± 0.05 | 0.69 ± 0.08 | 0.74 ± 0.05 | 0.50 ± 0.10 |
| | Imb. lr | 0.58 ± 0.10 | 0.84 ± 0.05 | 0.69 ± 0.07 | 0.74 ± 0.04 | 0.50 ± 0.07 |
| | Over. | 0.70 ± 0.01 | 0.86 ± 0.00 | 0.77 ± 0.05 | 0.79 ± 0.04 | 0.60 ± 0.07 |
| | SMOTE | 0.68 ± 0.01 | 0.85 ± 0.00 | 0.76 ± 0.05 | 0.78 ± 0.04 | 0.58 ± 0.07 |
| | ESM1-b | 0.43 ± 0.01 | 0.88 ± 0.01 | 0.57 ± 0.08 | 0.69 ± 0.04 | 0.43 ± 0.08 |
| | FvH | 0.21 ± 0.01 | 0.96 ± 0.01 | 0.34 ± 0.11 | 0.60 ± 0.04 | 0.31 ± 0.09 |

presented in Table 7, while the comprehensive results can be found in Table 8. The obtained p-values consistently revealed statistical significance ($p < 0.05$) for each metric. Nevertheless, no statistical difference was observed for Oversampling and SMOTE augmentation.

**Table 7.** Results of the Kruskal-Wallis test for the two encoders for different experiments. $P - values > 0.05$ are shown in bold font.

| Experiment | p-value F1 | p-value accuracy | p-value MCC |
|---|---|---|---|
| Single | 6.99e−05 | 1.67e−05 | 1.59e−05 |
| Imbalanced | 7.92e−04 | 7.91e−05 | 3.93e−04 |
| Imbalanced lr | **0.05** | 0.01 | 0.01 |
| Oversampling | 0.01 | **0.10** | **0.65** |
| SMOTE | **0.83** | **0.42** | **0.19** |
| ESM1-b | 3.09e−05 | 6.96e−05 | 1.09e−04 |
| FvHallucinator | 6.38e−07 | 5.39e−07 | 7.33e−07 |

**Table 8.** Results of the Kruskal-Wallis test for the two encoders across all experiments.

| Metric | p-value |
|---|---|
| F1 | 3.22e−04 |
| Accuracy | 2.99e−08 |
| MCC | 3.62e−11 |

## List of Abbreviations

| | |
|---|---|
| AUC-ROC | Area Under the ROC Curve |
| CDR | Complementary-Determining Regions |
| $F_{ab}$ | Antigen-Binding Fragment |
| $F_c$ | Crystallizable Fragment |
| FvH | FvHallucinator |
| H | Heavy chain |
| Ig | Immunoglobulins |
| Imb | Imbalanced |
| Imb lr | Imbalanced learning rate |
| KNN | k-Nearest Neighbors |
| L | Light chain |
| LASSO | Least Absolute Shrinkage and Selection Operator |
| LLM | Large Language Model |
| lr | Learning Rate |
| mAb | Monoclonal Antibodies |
| MCC | Matthew Correlation Coefficient |
| ML | Machine Learning |
| MLM | Masked Language Modeling |
| MLP | Multilayer Perceptron |
| NGS | Next-Generation Sequencing |
| NLP | Natural Language Processing |
| OCSVM | One-Class Support Vector Machine |

| | |
|---|---|
| Over. | Oversampling |
| RF | Random Forest |
| SAbDab | Structural Antibody Database |
| SMOTE | Synthetic Minority Oversampling Technique |
| SVM | Support Vector Machine |
| VH | Variable Heavy |
| VL | Variable Light |
| XAI | eXplainable Artificial Intelligence |

## References


1. Ada, G., Isaacs, D.: Carbohydrate-protein conjugate vaccines. Clin. Microbiol. Infect. **9**(2), 79–85 (2003). https://doi.org/10.1046/j.1469-0691.2003.00530.x
2. Bhattacharya, N., et al.: Interpreting Potts and transformer protein models through the lens of simplified attention. In: Pacific Symposium on Biocomputing 2022, pp. 34–45. World Scientific (2021). https://doi.org/10.1142/9789811250477_0004
3. Chawla, N.V., Bowyer, K.W., Hall, L.O., Kegelmeyer, W.P.: Smote: synthetic minority over-sampling technique. J. Artif. Intell. Res. **16**, 321–357 (2002). https://doi.org/10.1613/jair.953
4. Chen, Z., et al.: Human monoclonal antibodies isolated from a primary pneumococcal conjugate vaccinee demonstrates the expansion of an antigen-driven hypermutated memory b cell response. BMC Infect. Dis. **18**, 1–11 (2018). https://doi.org/10.1186/s12879-018-3517-7
5. Chicco, D., Jurman, G.: The Matthews correlation coefficient (MCC) should replace the ROC AUC as the standard metric for assessing binary classification. BioData Min. **16**(1), 4 (2023). https://doi.org/10.1186/s13040-023-00322-4
6. Dingens, A.S., Arenz, D., Weight, H., Overbaugh, J., Bloom, J.D.: An antigenic atlas of HIV-1 escape from broadly neutralizing antibodies distinguishes functional and structural epitopes. Immunity **50**(2), 520–532 (2019). https://doi.org/10.1016/j.immuni.2018.12.017
7. Dunbar, J., Krawczyk, K., Leem, J., Baker, T., Fuchs, A., Georges, G., et al.: Sabdab: the structural antibody database. Nucleic Acids Res. **42**(D1), D1140–D1146 (2014). https://doi.org/10.1093/nar/gkt1043
8. Elnaggar, A., Heinzinger, M., Dallago, C., Rehawi, G., Wang, Y., Jones, L., et al.: Prottrans: toward understanding the language of life through self-supervised learning. IEEE Trans. Pattern Anal. Mach. Intell. **44**(10), 7112–7127 (2021). https://doi.org/10.1109/TPAMI.2021.3095381
9. Gingerich, A.D., Royer, F., McCormick, A.L., Scasny, A., Vidal, J.E., Mousa, J.J.: Synergistic protection against secondary pneumococcal infection by human monoclonal antibodies targeting distinct epitopes. J. Immunol. **210**(1), 50–60 (2023). https://doi.org/10.4049/jimmunol.2200349
10. Gôlo, M., de Souza, M.C., Rossi, R.G., Rezende, S.O., Nogueira, B.M., Marcacini, R.M.: One-class learning for fake news detection through multimodal variational autoencoders. Eng. Appl. Artif. Intell. **122**, 106088 (2023). https://doi.org/10.1016/j.engappai.2023.106088

11. Haykin, S.: Neural Networks: A Comprehensive Foundation. Prentice Hall PTR (1994)
12. Jain, T., Sun, T., Durand, S., Hall, A., Houston, N.R., Nett, J.H., et al.: Biophysical properties of the clinical-stage antibody landscape. Proc. Natl. Acad. Sci. **114**(5), 944–949 (2017). https://doi.org/10.1073/pnas.1616408114
13. Joubbi, S., et al.: Antibody design using deep learning: from sequence and structure design to affinity maturation. Briefings Bioinform. **25**(4) (2024). https://doi.org/10.1093/bib/bbae307
14. Kaplon, H., Crescioli, S., Chenoweth, A., Visweswaraiah, J., Reichert, J.M.: Antibodies to watch in 2023. In: MAbs, vol. 15, p. 2153410. Taylor & Francis (2023). https://doi.org/10.1080/19420862.2022.2153410
15. Kenlay, H., Dreyer, F.A., Kovaltsuk, A., Miketa, D., Pires, D., Deane, C.M.: Large scale paired antibody language models. arXiv preprint arXiv:2403.17889 (2024)
16. Kovaltsuk, A., Leem, J., Kelm, S., Snowden, J., Deane, C.M., Krawczyk, K.: Observed antibody space: a resource for data mining next-generation sequencing of antibody repertoires. J. Immunol. **201**(8), 2502–2509 (2018). https://doi.org/10.4049/jimmunol.1800708
17. Mahajan, S.P., Ruffolo, J.A., Frick, R., Gray, J.J.: Hallucinating structure-conditioned antibody libraries for target-specific binders. Front. Immunol. **13**, 999034 (2022). https://doi.org/10.3389/fimmu.2022.999034
18. Maharana, K., Mondal, S., Nemade, B.: A review: data pre-processing and data augmentation techniques. Glob. Transitions Proc. **3**(1), 91–99 (2022). https://doi.org/10.1016/j.gltp.2022.04.020
19. Norman, R.A., Ambrosetti, F., Bonvin, A.M., Colwell, L.J., Kelm, S., Kumar, S., et al.: Computational approaches to therapeutic antibody design: established methods and emerging trends. Brief. Bioinform. **21**(5), 1549–1567 (2020). https://doi.org/10.1093/bib/bbz095
20. Paszke, A., et al.: Pytorch: an imperative style, high-performance deep learning library. In: Advances in Neural Information Processing Systems, vol. 32 (2019). https://doi.org/10.48550/arXiv.1912.01703
21. Rappuoli, R., Bottomley, M.J., D'Oro, U., Finco, O., De Gregorio, E.: Reverse vaccinology 2.0: human immunology instructs vaccine antigen design. J. Exp. Med. **213**(4), 469–481 (2016). https://doi.org/10.1084/jem.20151960
22. Ruffolo, J.A., Gray, J.J., Sulam, J.: Deciphering antibody affinity maturation with language models and weakly supervised learning. arXiv preprint arXiv:2112.07782 (2021)
23. Wu, N.C., Wilson, I.A.: Influenza hemagglutinin structures and antibody recognition. Cold Spring Harb. Perspect. Med. **10**(8), a038778 (2020). https://doi.org/10.1101/cshperspect.a038778
24. Xiang, Y., et al.: Superimmunity by pan-sarbecovirus nanobodies. Cell Reports **39**(13) (2022). https://doi.org/10.1016/j.celrep.2022.111004
25. Zhou, J., Le, C.Q., Zhang, Y., Wells, J.A.: A general approach for selection of epitope-directed binders to proteins. Proc. Natl. Acad. Sci. **121**(19), e2317307121 (2024). https://doi.org/10.1073/pnas.2317307121

# Inferring Breast Cancer Subtype Associations Using an Original Omics Integration Based on Non-negative Matrix Tri-Factorization

Camila Riccio-Rengifo[1](✉), Silvia Cascianelli[2], Gaia Ceddia[2], and Marco Masseroli[2]

[1] Omics Science Research Institute (iÓMICAS), Pontificia Universidad Javeriana Cali, Cali, Colombia
camila.riccio@javerianacali.edu.co
[2] Politecnico di Milano, Milan, Italy
{silvia.cascianelli,gaia.ceddia,marco.masseroli}@polimi.it

**Abstract.** In the oncogenomics domain, there is an increasing interest and need for computational methodologies that harness multidimensional patient characterizations. These methods aim to decipher the growing availability and heterogeneity of omics and annotation data at our disposal. In this study, we propose a two-phase workflow designed for data integration and knowledge extraction, applied to the hot clinical topic of breast cancer subtyping and employing Non-negative Matrix Tri-Factorization (NMTF). This technique decomposes a non-negative data matrix into three lower-rank matrices for exploration and pattern discovery. Our NMTF-based innovative strategy jointly analysed association matrices of a multi-partite network including different omics data of breast cancer patients and their corresponding subtypes, and found a network reconstruction able to maximize correct patient-subtype predictions. Based on this optimized reconstruction, the workflow second stage builds subtype-specific subnetworks and latent representations of all their nodes to establish associations also between subtypes and involved omics data. Our approach can delve into latent patterns and intricate relationships spanning various input data within a multidimensional framework. Thus, it not only sheds fresh light on BRCA subtypes but also offers adaptability for analogous classification efforts on similar clinical issues.

**Keywords:** Omics data integration · Cancer subtyping · Non-negative Matrix Tri-Factorization · Latent representations

---

**Supplementary Information** The online version contains supplementary material available at https://doi.org/10.1007/978-3-031-90714-2_19.


M. Vettoretti et al. (Eds.): CIBB 2023, LNBI 14513, pp. 255–272, 2025.
https://doi.org/10.1007/978-3-031-90714-2_19

## 1 Introduction

The growing abundance of different omics data and clinical annotations, especially for cancer patients, is prompting the advancement of computational methodologies capable of integrating heterogeneous types of information to derive original insights and ultimately lead to new knowledge. Non-negative Matrix Tri-Factorization (NMTF) [1] is an effective solution to merge multiple data types into a unified network representation. Furthermore, its iterative factorization process can be enhanced by incorporating an extended graph topology with multiple datasets and association matrices, either binary or weighted. Consequently, NMTF is promising for multidimensional omics-based analyses, revealing latent connections among data of different types and their relevance to the task at hand. Indeed, NMTF has already been employed in various biomedical applications requiring the integration of different information into a unified predictive model, as evidenced by previous studies as [2,3,28].

Cancer stratification, which concerns the identification/prediction of the relevant tumor subtypes associated with patient samples, is pivotal for enhancing clinical management. Similarly to other tumors, breast cancer (BRCA) can be stratified into distinct intrinsic subtypes, namely Luminal A (LumA), Luminal B (LumB), Basal, HER2, and Normal-like, initially identified through unsupervised analysis of microarray expression data from a thousand of protein-coding genes [4,5]. BRCA subtyping through the exploration of biological characteristics and molecular differences among subtypes allowed for enhancing prognostic, therapeutic, and recurrence predictions for patients. Traditionally, BRCA patients are categorized into subtypes using the PAM50 test [6], relying on 50 genes and their correlation with pre-calculated centroids. Recently, machine learning-based approaches have been employed to address the BRCA subtyping challenge, leading to the exploration of broader gene spaces and even multi-omics spaces [7,8]. However, even when starting from a wide multi-omics context, these methods need to pinpoint and concentrate on a subset of predictive features to accomplish the classification task, rather than performing extensive and comprehensive data integration and exploration, as NMTF can do. Furthermore, the tri-factorization is able to capture non-linear interactions, differently from linear techniques like Singular Value Decomposition or Low-rank approximations, and offers a richer and more interpretable representation of the integrated data interactions than standard two-factor Non-negative Matrix Factorization (NMF). Indeed, the outer factors of each tri-factorization are shared between multiple interconnected data matrix decompositions.

This work introduces a novel computational workflow aimed at uncovering insights into cancer stratification by integrating patient subtype annotations with their mRNA and microRNA expression profiles, along with the co-expression values of these latter. This original solution eventually led to accomplishing the predictive subtyping task while inferring new knowledge about these omics' roles in different subtypes, and the corresponding subtype-specific relationships. A comprehensive patient-centric network linking these elements was built, and all the association matrices (weighted with continuous or discrete values) were analyzed using NMTF to predict missing subtype annotations for a randomly selected patient subset, used to assess predictive capabilities.

Notably, the reconstruction obtained at the NMTF iteration that maximizes accuracy in subtype predictions was leveraged to identify also associations of mRNA and microRNA genes with each subtype. The exploration of mRNA (henceforth simply called genes) and microRNA (miRNA) relationships with BRCA subtypes, which so far had been primarily focused on identifying signatures exhibiting differential expression behaviours (including up- and down-regulated genes and miRNAs) [9–13], was therefore improved. The so-inferred associations offer insights into the involvement of genes and miRNAs and their connections with various BRCA subtypes, uncovering both novel links and significant correspondences with established state-of-the-art knowledge. This strength is demonstrated here in the context of BRCA subtyping, but could be similarly used to study other cancer types or heterogenous disease contexts.

## 2 Materials and Methods

### 2.1 Data Source

We extracted and utilized publicly available BRCA data from The Cancer Genome Atlas (TCGA) [14] [15] to construct a comprehensive patient-centric network encompassing subtype annotations, mRNA (simply referred to as genes) expression quantification, miRNA expression, and co-expression values within and between these omics. Specifically, we retrieved data from 817 TCGA BRCA patients, which included gene and miRNA expression profiles, and were jointly analysed with the PAM50 test in [16] to obtain the subtype annotation for each patient. From this dataset, we identified 415 patients with LumA, 176 with LumB, 136 with Basal, 65 with Her2, and 25 with Normal subtypes. Initially, the corresponding expression profiles comprised 25,150 genes and 1,580 miRNAs (not null for all patients) and underwent separate preprocessing steps. Raw expression read counts were first normalized by calculating reads per million (RPM) for each gene or miRNA $x_i$ in each sample $s_j$ as:

$$RPM = \frac{\#\ reads\ mapped\ to\ x_i}{total\ \#\ reads\ for\ sample\ s_j} * 10^6; \tag{1}$$

then, a $log_2$ transformation was applied to RPM values. To streamline subsequent analyses and reduce computational complexity, we retained only the most highly expressed genes/miRNAs. For both genes and miRNAs separately, we computed Z-scores from the RPM values based on their mean and standard deviation across patients, selecting genes/miRNAs with an average Z-score exceeding the 75th percentile of all the corresponding Z-scores. Ultimately, for each patient, we preserved a set of 6,286 genes and a set of 395 miRNAs. These sets underwent min-max normalization separately, adding also an $\epsilon$ value to scale all quantities between $\epsilon$ and 1.

### 2.2 BRCA Network Configuration and NMTF Analysis

In this work, the integration of BRCA data and their relationships was achieved through a multi-partite network consisting of four node types: 1) subtypes, 2) patients, 3) genes, and 4) miRNAs, along with their corresponding interconnections (subtype-patient, patient-gene, patient-miRNA, and gene-miRNA), and further gene-gene and miRNA-miRNA inner connections. This configuration resulted in the generation of the following association matrices:

- $R_{12}$: A $n_1 \times n_2$ binary matrix representing subtype annotations of patients, where each column represents a patient with a single 1-value, and $R'_{12}$ is masked version for performing the patient subtyping task.
- $R_{23}$ and $R_{24}$: Continuous matrices of sizes $n_2 \times n_3$ and $n_2 \times n_4$ respectively, containing normalized RPM values of genes and miRNAs for each patient.
- $R_{34}$: A $n_3 \times n_4$ continuous co-expression matrix computed based on absolute values of Pearson correlation between each gene-miRNA pair, considering their expression values across all patients.

Note that $n_1 = 5$ corresponds to the number of intrinsic subtypes, $n_2 = 817$ is the number of patients, $n_3 = 6,286$ represents the number of genes, and $n_4 = 395$ is the number of miRNAs considered in this study. Additionally, for gene and miRNA nodes, we incorporated intra-class edges within the network. These edges are placed where intra-class co-expression values exceed a threshold required to give the corresponding network a scale-free topology. Such self-loops are described using weighted adjacency matrices for genes ($W_3$) and miRNAs ($W_4$). The full network configuration is illustrated in Fig. 1.

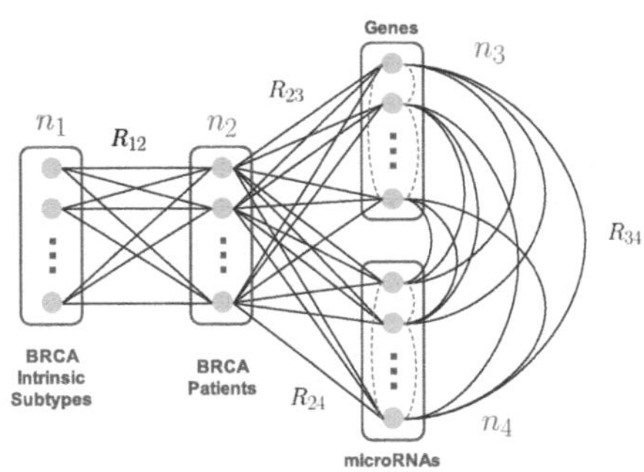


**Fig. 1.** Schematic representation of the multi-partite network that integrates four BRCA data types and their relationships. Labels on the edges are the name of the corresponding association matrices.

### 2.3 Non-negative Matrix Tri-Factorization

Non-negative Matrix Tri-Factorization (NMTF) is a linear algebra algorithm designed to factorize any input matrix of non-negative elements into three matrices, referred to as latent factors or components [1]. One key advantage of NMTF is its ability to extend and link the decomposition of association matrices in a multi-partite network as the one under analysis in this study. For further details related to the iterative approach of the Non-negative Matrix Tri-Factorization and the used update rules please refer to the Appendix. The NMTF decomposition of the association matrices obtained from our BRCA network configuration is detailed in Eqs. (2–5). Note that, the association matrices are interconnected through the factor matrices $G_2$, $G_3$, and $G_4$ in the NMTF decomposition. Specifically, $G_2$ is involved in the decomposition of the association matrices $R_{12}$, $R_{23}$, and $R_{24}$, $G_3$ contributes to the decomposition of $R_{23}$ and $R_{34}$, while $G_4$ contributes to the decomposition of $R_{24}$ and $R_{34}$. This interconnectedness facilitates the integration of information across subtypes, patients, genes, and miRNAs, allowing for a comprehensive analysis of the dataset.

$$R_{12} \approx G_1 H_{12} G_2^T \tag{2}$$

$$R_{23} \approx G_2 H_{23} G_3^T \tag{3}$$

$$R_{24} \approx G_2 H_{24} G_4^T \tag{4}$$

$$R_{34} \approx G_3 H_{34} G_4^T \tag{5}$$

The first step of the NMTF algorithm involves initializing the factor matrices. For the multi-partite network depicted in Fig. 1, these matrices include $G_1$, $G_2$, $G_3$, $G_4$, $H_{12}$, $H_{23}$, $H_{24}$, and $H_{34}$. To streamline this process, it is sufficient to initialize the four $G$ matrices first. Subsequently, the $H$ matrices can be computed by solving the corresponding Eqs. (2–5) for each $H$ matrix, subject to the constraint $G_i G_i^T = G_i^T G_i = I$ for $i \in \{1, 2, 3, 4\}$, replacing $R_{12}$ by $R'_{12}$ in Eq. 2.

The Singular Value Decomposition (SVD) strategy was employed for the initialization due to its widespread use, deterministic nature, and its ability to reduce the number of iterations required for convergence [17]. SVD is a matrix tri-factorization technique from linear algebra that decomposes matrix $R_{AB}$ into the product of three matrices: $R_{AB} = U_{AB} S_{AB} V_{AB}^T$, where $U_{AB}$ and $V_{AB}$ are orthogonal and $S_{AB}$ is a diagonal matrix with non-negative real numbers along its diagonal, referred to as singular values. These are arranged in descending order, with larger singular values capturing more variance in the data. Meanwhile, the columns of $U_{AB}$ and $V_{AB}$ represent the left and right singular vectors of $R_{AB}$, respectively. To align with the dimensions of the NMTF decomposition and capture the dominant patterns in the data, the factor matrices $G_i$ with $i \in \{1, 2, 3\}$ were initialized using the first $k_i$ columns of the corresponding $U$ matrix, while $G_4$ was initialized using the first $k_4$ columns of the $V$ matrix. This approach ensures that each $G$ matrix is initialized from the Singular Value Decomposition (SVD) of a different $R$ matrix, thereby incorporating information from all association matrices into the NMTF decomposition.

## 3 Workflow for Breast Cancer Subtyping

A computational workflow is proposed for predicting BRCA subtype assignments, leveraging NMTF and machine learning techniques. This workflow was tailored to the network configuration outlined in Fig. 1 and consists of two consecutive stages, illustrated in Fig. 2. The first stage involves patient subtyping, while the second stage focuses on predicting subtype associations for genes and microRNAs. Each stage is elucidated in the subsequent subsections.

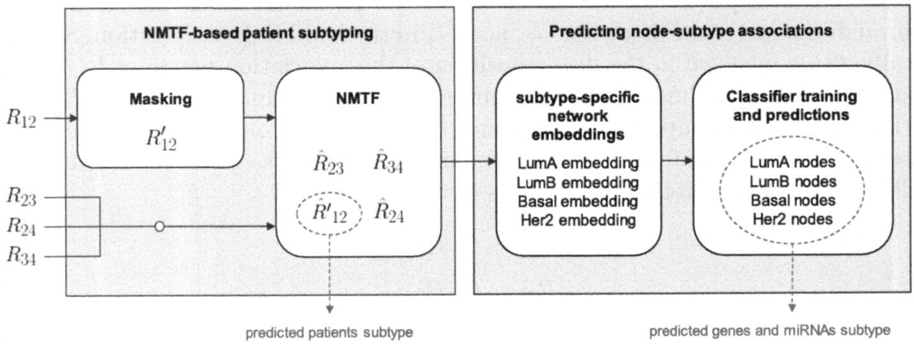


**Fig. 2.** Proposed Computational Workflow for BRCA Subtype Prediction Using NMTF and Machine Learning Techniques.

### 3.1 First Stage: NMTF-Based Patient Subtyping

Exploring the task of BRCA patient subtyping in the first step of our workflow involved masking the $R_{12}$ matrix obtaining the $R'_{12}$ matrix and conducting an iterative analysis using NMTF. To this aim, we devised a binary mask $M$, mirroring the dimensions of $R_{12}$, to randomly select 40% of patients while maintaining the same subtype proportions as the entire population. This 40% subset of masked patients is evenly divided between a validation set and a test set, while the remaining patients/columns of $R_{12}$ are preserved in the new matrix $R'_{12}$, which is built as follows:

$$R'_{12}[\cdot, j] = \begin{cases} R_{12}[\cdot, j], & \text{if } M[\cdot, j] = 0 \\ 0.5, & if M[\cdot, j] = 1 \end{cases} \tag{6}$$

Figure 3 provides an illustration of $R'_{12}$ and the sampling process.

In our network configuration we substituted $R_{12}$ with the masked matrix $R'_{12}$. Then, considering $R'_{12}$, we applied the aforementioned Singular Value Decomposition strategy to initialize all the $G$ matrices and proceed with the subsequent NMTF decomposition. Our objective extended beyond simply achieving a satisfactory approximation of all matrices through NMTF. Indeed, we aimed to assess whether the actual subtype associations of masked patients could manifest as predominant values in $\hat{R}'_{12}$ during the NMTF iterations devoted to reconstructing the masked $R'_{12}$ matrix. To study this, we set a time horizon of 5,000 NMTF iterations (after several tuning efforts), and at intervals of every 10 iterations we

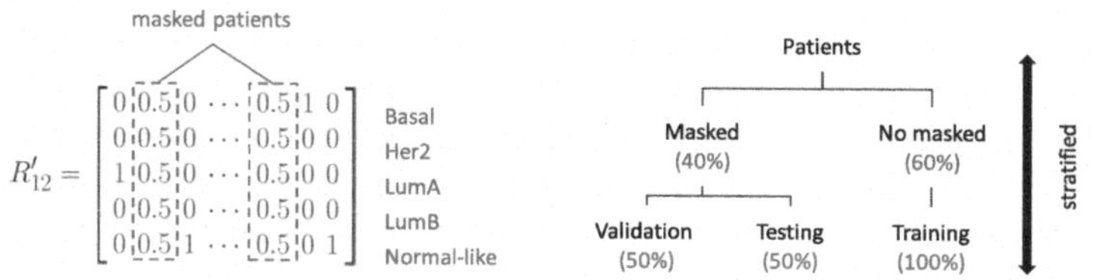


**Fig. 3.** Illustration of the masking process applied to the $R_{12}$ matrix.

investigated the reconstructed $\hat{R}'_{12}$ to identify any indication of the real subtype for each masked patient, instead of the simpler expected convergence toward the initialized value of 0.5 without any class preference.

Specifically, given $\hat{R}'_{12}$, our focus was on the masked patients of the validation set: leveraging their original subtype assignments from $R_{12}$ as references, we assessed the Average Precision Score (APS). This metric was calculated as the area under the Precision-Recall curve for the validation set only while varying the $\delta$ value, which is required as a threshold to binarize the continuous values in $\hat{R}'_{12}$ from 0 to 1. The so-obtained APS provided insights into the predictive performance of our NMTF-based solution. Ultimately, we retained the *best iteration* yielding for $\hat{R}'_{12}$ the maximum APS on the validation set.

Each masked patient was accordingly associated with an exclusive subtype based on the highest value observed for that patient in $\hat{R}'_{12}$ at the *best iteration*. This *best iteration* was fine-tuned using the validation set, but the corresponding $\hat{R}'_{12}$ was employed to predict subtypes for all the masked patients, irrespective of whether they belonged to the validation or test set. Then, focusing on the test set only, comprising patients not previously considered, we could obtain an unbiased validation of the predictive capability of our NMTF-based solution. Indeed, also utilizing the predicted subtypes of the validation set for quantifying the subtyping efficacy of our approach would have been inappropriate, given the calculation of APS and *best iteration* on this particular set.

### 3.2 Second Stage: Subtype Associations for Genes and MicroRNAs

In the subsequent stage of our workflow, the comprehensive network that we derived from NMTF at the *best iteration* was used to build four distinct sub-networks corresponding to LumA, LumB, Basal, and Her2 subtypes, each one encompassing all the genes and microRNAs but only the patients annotated or predicted to belong to the respective subtype. Within each sub-network, we employed the Node2Vec technique [18] to obtain an embedding, i.e., a latent representation of reduced dimensionality, for any node (patient, gene, or microRNA). The overall network topology was, therefore, embedded based on node similarities and relationships detected using random walks. Node2Vec embeddings provided homogeneous and concise representations of all nodes within each sub-network, facilitating the setting of a binary classification task to predict subtype associations not only for patients but also for genes and microRNAs.

To establish associations between each node and its corresponding subtype, for each class-specific subnetwork we allocated 90% of patient-node embeddings into a training set. We also added a comparable and evenly-distributed volume of negative examples, which were simply real patient embeddings from the remaining three sub-networks. For each class-specific binary task, several classifiers underwent training and tuning using 5-fold cross-validation and grid search methodology. These included ensemble techniques such as Decision Tree classifier (DTC), Random Forest classifier (RFC), Gradient Boosting (GB), and Extreme Gradient Boosting (XGB), together with Logistic Regression (LR) and Support Vector Machines (SVM). However, Logistic Regression and Support Vector Machines achieved disappointing results in at least one subtype classification task and were consequently excluded from subsequent comparisons. Despite each model could assign subtype associations to patients as well as genes and microRNAs, the evaluation of correctness could be performed only based on the available patient predictions. Classifiers were assessed on both cross-validation and testing to identify the most promising model or models to explore also gene and microRNA associations with each subtype. Subsequently, considering the so-selected models, we delved deeper into recognising genes and microRNAs linked exclusively to one of the subtypes.

## 4 Results

This section provides a thorough analysis of our workflow performance in predicting patient subtypes and their associations with genes and miRNAs. The analysis is divided into two subsections: "Subtype Predictions for Patient" and "Subtype Associations for Genes and miRNAs". The corresponding code is freely available on GitHub at https://github.com/DEIB-GECO/NMTF_for_subtyping. In the first subsection, we discuss the performance metrics of our patient subtyping approach based on Non-negative Matrix Tri-Factorization on randomly selected test sets of patients. In the second subsection, we detail Node2Vec-based node embeddings and binary classifiers used to estimate the association with specific breast cancer subtypes also for each gene and miRNA under consideration. Initially, we focus on the XGB classifier, which demonstrates promising performance across all subtypes. Then, we extend our analysis to an ensemble of classifiers, including DTC, RFC, GB and XGB. We show their computational performance and delve into the predicted associations between genes/miRNAs and breast cancer subtypes. Furthermore, we present the validation of the involved genes and miRNAs against known interactions documented in miRTarBase [19], thereby affirming the biological relevance of our predictions.

### 4.1 Subtype Predictions for Patients

The performance metrics summarising the results of our NMTF-based subtyping approach on randomly selected test sets of patients are presented in Table 1. These values were collected from 10 different masks generated in as many runs of our workflow's first stage. Overall, the results demonstrate high and consistent performance, indicating the efficacy of our approach even in such an imbalanced class scenario. Notably, the best iteration, which achieved the maximum APS on the validation set, remained quite stable across the 10 runs ($349 \pm 62.6$), despite each run involving up to 5,000 NMTF iterations. This stability underscores the reliability of our workflow in performing the subtyping task. The results in Table 1 showcase promising accuracy, precision, recall, and F1-score across subtypes, with slight variations observed between different runs and with a decrease in the recall of LumB cases, that are partially misclassified as LumA. Notably, only a few samples were predicted as Normal-like, even reducing the small initial dimensionality of this controversial class: lacking prognostic ortherapeutic relevance, Normal-like patients were excluded from further analyses.

The specificity, negative predictive value (NPV), and Matthews correlation coefficient (MCC) metrics also exhibit favourable performance, especially when considering the Basal and Her2 subtypes. These findings provide confidence in the effectiveness of our NMTF-based patient subtyping approach. Given the minimal differences among the ten runs, we just selected and focused on the run with the highest performance in testing to extract the entire network at its best iteration, based on the validation set (iteration 450, with APS = 0.849). Subsequently, we proceeded with the second stage of our workflow, leveraging this extracted network to predict subtype associations also for genes and miRNAs.

**Table 1.** Results for patient subtyping from running the first stage of our workflow ten times using different random masks.

| | accuracy | | precision | | recall | | F1-score | |
|---|---|---|---|---|---|---|---|---|
| | average ± std | best run | average ± std | best run | average ± std | best run | average ± std | best run |
| LumA | 0.828 ± 0.025 | 0.858 | 0.807 ± 0.042 | 0.837 | 0.879 ± 0.038 | 0.894 | 0.840 ± 0.015 | 0.865 |
| LumB | 0.842 ± 0.023 | 0.871 | 0.601 ± 0.213 | 0.726 | 0.532 ± 0.192 | 0.648 | 0.564 ± 0.200 | 0.685 |
| Basal | 0.982 ± 0.002 | 0.983 | 0.923 ± 0.010 | 0.930 | 0.974 ± 0.013 | 0.971 | 0.948 ± 0.007 | 0.950 |
| Her2 | 0.968 ± 0.005 | 0.968 | 0.802 ± 0.029 | 0.800 | 0.802 ± 0.029 | 0.800 | 0.802 ± 0.029 | 0.800 |
| Normal | 0.971 ± 0.002 | 0.969 | 0.663 ± 0.237 | 0.500 | 0.116 ± 0.039 | 0.160 | 0.193 ± 0.061 | 0.242 |
| Overall | 0.918 ± 0.009 | 0.929 | 0.799 ± 0.023 | 0.824 | 0.791 ± 0.026 | 0.824 | 0.795 ± 0.024 | 0.825 |

| | specificity | | NPV | | MCC | |
|---|---|---|---|---|---|---|
| | average ± std | best run | average ± std | best run | average ± std | best run |
| LumA | 0.777 ± 0.082 | 0.821 | 0.865 ± 0.036 | 0.882 | 0.664 ± 0.036 | 0.717 |
| LumB | 0.927 ± 0.028 | 0.933 | 0.881 ± 0.035 | 0.906 | 0.481 ± 0.172 | 0.606 |
| Basal | 0.984 ± 0.002 | 0.985 | 0.995 ± 0.003 | 0.994 | 0.938 ± 0.008 | 0.940 |
| Her2 | 0.983 ± 0.002 | 0.983 | 0.983 ± 0.002 | 0.983 | 0.784 ± 0.032 | 0.783 |
| Normal | 0.998 ± 0.002 | 0.995 | 0.973 ± 0.001 | 0.974 | 0.262 ± 0.077 | 0.271 |
| Overall | 0.950 ± 0.006 | 0.957 | 0.948 ± 0.006 | 0.956 | 0.744 ± 0.030 | 0.781 |

### 4.2 Subtype Associations for Genes and MiRNAs

In the second stage of our workflow, we made use of Node2Vec to obtain latent representations for all the nodes in subtype-specific subnetworks. Any of these subnetworks is then used in combination with binary classifiers to estimate the association of each node with the subtype under exam, as already described in Sect. 3.2. Accordingly, given each classification, we could devise predicted associations between genes/miRNAs and BRCA subtypes, extract the exclusive ones for each specific subtype, and compare them both with associations and interactions known at the state-of-the-art. First, we considered ensemble methods using the same model type differently tuned and trained for each subtype. All these assessed ensemble classifiers reached quite comparable overall performances in cross-validation and testing when considering both positive and negative cases for a given subtype, with no clear distinction of the best approach. Yet, for 3 out of 4 subtypes, the Extreme Gradient Boosting was the best in recognizing the class of interest, in terms of recall over its 10% of patients left out for testing. Accordingly, the next subsection is focused on XGB. However, as an alternative, we also explored ensemble strategies making use of different types of classification models, each one optimized for the specific subtype of interest. The corresponding results are discussed after the end of the next subsection.

### 4.3 XGB Estimated Associations

**Computational Performance.** Table 2 summarizes results of key performance metrics, including precision, recall, and F1-score, for each subtype, along with the number of testing patients considered. The performance metrics provide valuable insights into the XGB model's ability to classify patients into different subtypes based on the embeddings of the best network reconstructed with NMTF.

**Table 2.** Patients prediction performance with XGB classifier.

| subtype | testing patients | recall | precision | F1-score |
|---|---|---|---|---|
| LumA | 44 (/443) | 0.68 | 0.53 | 0.60 |
| LumB | 16 (/157) | 0.69 | 0.55 | 0.61 |
| Basal | 14 (/142) | 0.71 | 0.56 | 0.63 |
| Her2 | 7 (/67) | 0.57 | 0.57 | 0.57 |

Overall, the XGB classifier demonstrated good performance across subtypes, with relatively high recall values indicating its effectiveness in correctly identifying patients belonging to each subtype. Note that the XGB classifier exhibits particularly the best performance in predicting the Basal subtype, as evidenced by the highest recall (0.71) and F1-score (0.63) among all subtypes.

**Gene/miRNA Subtype Associations.** Table 3 reports the number of predicted gene and miRNA associations to each subtype obtained from the XGB classifier. Considering that some associations may overlap among different subtypes, it details their overall number as well as the number of exclusive ones. To validate our findings as well as uncover novel connections, we considered genes and miRNAs exclusively associated with one subtype. For gene predictions, we compared the exclusive sets with those reported in the Molecular Signature Database (MSigDB) [9], representing up- and down-regulated genes for each subtype [20]. miRNA predictions were instead compared with those identified as differentially expressed in previous studies in the literature [10–13,21]. Although the matches with known associations were limited, these were largely significant according to Fisher's tests with Benjamini/Hochberg correction, also for p-value $< 0.005$, as suggested by Benjamin et al. [22] for more stringent significance thresholds, considering the union of exclusive genes/miRNAs for each subtype as the background set. The newly identified associations would be worthy of further investigation to comprehend their subtype-specific roles. For instance, the SNRPD1 gene, already implicated in cell proliferation in cancer [23], emerged as linked to the Basal subtype, known as the most aggressive BRCA subtype. Similarly, the hsa-mir-342 miRNA, associated with the Her2 subtype, could act as a potential tumor growth inhibitor for this class, as suggested in [24].

**Table 3.** Results of our workflow's second stage: XGB subtype associations for genes and miRNAs specifying all those *predicted* but possibly overlapping among subtypes, those *exclusive* of a single subtype and those already *reported* at the state of the art.

| subtype | genes (6,286) | | | miRNAs (395) | | |
|---|---|---|---|---|---|---|
| | predicted | exclusive | reported | predicted | exclusive | reported |
| LumA | 4,008 | 602 | 3 (/8) (37.5%)* | 244 | 43 | 3 (/6) (50.1%)* |
| LumB | 2,960 | 319 | 10 (/42) (23.8%) ** | 194 | 21 | 2 (/16) (12.5%)* |
| Basal | 2,995 | 306 | 19 (/99) (19.2%) ** | 191 | 22 | 1 (/15) (6.7%) |
| Her2 | 2,825 | 281 | 3 (/12) (25.0%)* | 171 | 16 | 1 (/15) (6.7%) |

Fisher's test: *p-value $< 0.05$; ** p-value $< 0.005$

**Confirmation of MiRTarBase Known Interactions.** To validate the biological significance of our predicted associations for genes and miRNAs linked to the same subtype, we evaluated their interactions based on those at the state-of-the-art available in miRTarBase version 9.0 [19]. Gene-miRNA pairs associated with the same BRCA subtype should indeed demonstrate interplay, reflecting joint involvement in the regulatory networks of the specific subtype.

Here, we examined all potential gene-miRNA pairs predicted within the same subtype by the XGB classifier. Subsequently, we identified exclusive pairs for each subtype and compared them with the interactions catalogued in miRTarBase, restricting our analysis to the 31,725 interactions involving genes and miRNAs considered in the initial network construction. The outcomes of this analysis are summarized in Table 4. Interestingly, we observed a significant number of reported known interactions within the exclusive gene-miRNA pairs for each subtype, as confirmed by Fisher's tests with Benjamini/Hochberg correction and p-value $< 0.005$, as suggested by Benjamin et al. [22] for more stringent significance thresholds. This suggests a concordance between our predicted associations and those documented in miRTarBase, reinforcing the biological plausibility of our subtype predictions. These findings underscore the potential utility of our associations in identifying biologically relevant gene-miRNA interactions, related only to specific cancer subtypes, facilitating further investigations into the molecular mechanisms underlying breast cancer heterogeneity.

**Table 4.** Number of gene-miRNA pairs predicted within the same subtype by the XGB classifier, and reported within miRTarBase.

| subtype | predicted | exclusive | reported (31,725) |
|---|---|---|---|
| LumA | 977,952 | 467,914 | 6,509** |
| LumB | 574,240 | 211,669 | 2,529** |
| Basal | 572,045 | 214,161 | 2,370** |
| Her2 | 483,075 | 168,816 | 2,254* |

* for p-value $< 0.05$; ** for p-value $< 0.005$

### 4.4 Multi-classifier Estimated Associations

In this subsection, we expand our analysis to incorporate a multi-classifier solution to estimate the associations between patients, genes, miRNAs, and breast cancer subtypes. By leveraging the diverse strengths of different ensemble methods (e.g., XGB, DTC, RFC and GBC), we aimed to enhance the robustness and accuracy of our classification results for each subtype.

**Computational Performance.** The computational performance of multiple ensemble methods in estimating associations between patient nodes and BRCA subtypes are similar to the ones of the individual XGB classifier. Table 5 provides a comprehensive overview of these results considering key performance metrics such as precision, recall, and F1-score for each breast cancer subtype. This analysis considers the average and standard deviation across results obtained from the four classifiers (DTC, RFC, GBC, and XGB) and highlights the classifier with the best performance for each metric and subtype.

Notably, any ensemble method demonstrates consistent performance across all subtypes, with Basal exhibiting the highest average recall and average F1-score values, followed closely by LumA subtype predictions. Similarly, for all

**Table 5.** Patients prediction performance with ensemble classifiers.

| subtype | testing patients | precision | best precision | recall | best recall | F1-score | best F1-score |
|---|---|---|---|---|---|---|---|
| LumA | 44 (/443) | $0.54 \pm 0.17$ | 0.58 (DTC) | $0.65 \pm 0.15$ | 0.89 (RFC) | $0.59 \pm 0.08$ | 0.69 (DTC) |
| LumB | 16 (/157) | $0.50 \pm 0.09$ | 0.55 (XGB) | $0.56 \pm 0.07$ | 0.69 (XGB) | $0.53 \pm 0.06$ | 0.61 (XGB) |
| Basal | 14 (/142) | $0.56 \pm 0.03$ | 0.60 (DTC) | $0.66 \pm 0.03$ | 0.71 (XGB) | $0.61 \pm 0.02$ | 0.63 (XGB) |
| Her2 | 7 (/67) | $0.56 \pm 0.16$ | 0.67 (DTC) | $0.51 \pm 0.06$ | 0.57 (XGB) | $0.48 \pm 0.14$ | 0.62 (RFC) |

the ensemble models except for XGB, LumB appeared as the class of more difficult recognition. These results emphasize the significance of exploring multiple classifiers to assess the robustness and reliability of predictions for different breast cancer subtypes. This coherence not only confirmed the suitability of the obtained node latent representation for the subsequent classification task but allowed us to select for each subtype the best patient classification model and assume it as the best in tracing also subtype associations for genes and miRNAs in subsequent analysis. Overall, XGB still achieved the best performance over the majority of the subtypes except for LumA, where the RFC exhibited superior capabilities. Indeed, for the LumA subtype, RFC reached the highest recall (0.89) and precision and F1-score comparable with the best ones obtained using DTC. Therefore, in our multi-classifier solution, we selected the Random Forest as the best classifier for the LumA subtype, together with different Extreme Gradient Boosting models for all the other subtypes.

**Gene/miRNA Known Subtype Associations.** We then focused on investigating the genes and miRNA subtype associations using the RFC for LumA class, while retaining the also previously used XGB classifiers for the other subtypes. Table 6 presents the results of this analysis, detailing the number of predicted genes and miRNAs within each subtype, along with the exclusive ones and their comparison with known reported associations. For LumA, the RFC predicted 5,455 genes and 337 miRNAs, of which 827 genes and 55 miRNAs were exclusive to this subtype. Although the same XGB models were used to predict associations for the other subtypes as in the previous analysis, the higher number of predicted associations with LumA led to a reduction in exclusive associations for the other subtypes. Consequently, while these associations remained significantly enriched compared to those reported in the literature (with the exception of the Her2 class), the overlaps and Fisher's test results were less promising than those obtained using only XGB-based classifiers. Thus, the improved recall for the LumA subtype achieved by the RFC model did not result in better subtype-specific associations for exclusive genes and miRNAs-at least not according to the validation performed in this study, based on current knowledge.

**Confirmation of MiRTarBase Known Interactions.** Table 7 presents the results of validating gene-miRNA pairs, associated with the same breast cancer subtype by the multi-classifier solution, against miRTarBase associations. For

**Table 6.** Results of our workflow's second stage using subtype associations for genes and miRNAs specifying all those *predicted* but possibly overlapping among subtypes, those *exclusive* of a single subtype and those already *reported* at state of the art. Specifically, subtype associations are predicted by the RFC for the LumA class and by XGB classifiers for all the other subtypes.

| subtype | genes (6,286) | | | miRNAs (395) | | |
|---|---|---|---|---|---|---|
| | predicted | exclusive | reported | predicted | exclusive | reported |
| LumA | 5,455 | 827 | 4 (/8) (50%)* | 337 | 55 | 6 (/9) (66.7%)** |
| LumB | 2,960 | 102 | 3 (/30) (10%)* | 194 | 8 | 2 (/12) (16.7%)** |
| Basal | 2,995 | 102 | 7 (/83) (8.4%)** | 191 | 7 | 1 (/15) (6.7%)* |
| Her2 | 2,825 | 118 | 0 (/8) (0.0%) | 171 | 6 | 1 (/13) (7.7%)* |

Fisher's tests: * p-value < 0.05, ** for p-value < 0.005

LumA subtype predictions, out of a total of 250,079 predicted pairs, 228,031 were exclusive to LumA. Among these exclusive pairs, 2,297 were reported also in miRTarBase, indicating again a significant overlap between our predicted associations and those documented in the database. Also for the other subtypes, Fisher's tests with p-value $< 0.005$, indicated strong concordance between the predicted associations and those documented in miRTarBase.

**Table 7.** Number of gene-miRNA pairs predicted within the same subtype by the ensemble of classifiers, and validated within miRTarBase.

| subtype | predicted | exclusive | reported (31,725) |
|---|---|---|---|
| LumA | 250,079 | 228,031 | 2,297** |
| LumB | 73,343 | 60,979 | 681** |
| Basal | 72,762 | 60,750 | 854** |
| Her2 | 52,440 | 44,040 | 455** |

Fisher's tests: * p-value < 0.05, ** p-value < 0.005

### 4.5 Comparison of Gene/miRNA Resulting Interactions and Enrichment Analyses

XGB algorithm is known for its efficiency and effectiveness in handling large datasets and complex relationships between variables. Here, the XGB classifiers emerged not only as powerful machine learning solutions for assigning each node (patient, gene or miRNA) with subtype associations but also demonstrated the ability to capture interaction mechanisms among different gene and miRNA nodes exclusively associated with the same subtype. In the context of breast cancer patient subtyping, the XGB classifiers reached the most promising performance for almost all the classes, especially in terms of recall, considering the ability to correctly identify patients belonging to each subtype. Furthermore, in the context of predicting gene and miRNA associations with different subtypes, these models found significant overlapping with knowledge at the state-of-art, when focusing on the emerged subtype-specific genes and miRNAs either singularly or considering their mutual interactions. On the other hand, though the multi-classifier solution combined the strengths of multiple ensemble algorithms to improve the recall of LumA predictions, the analysis of the so-predicted gene and miRNA associations with all the breast cancer subtypes highlighted slightly worse validation results, with the only exception of finding more miRNA associations with LumA subtype confirmed at the state-of-the-art. Also, the multi-classifier results determined a drop of the subtype-specific gene-miRNA interacting pairs and, among them, of the known ones confirmed by miRTarBase.

Thus, the XGB-based classification emerged overall as the most encouraging solution to guide further investigations of exclusive genes and miRNAs associated with different breast cancer subtypes, aiming to better capture biologically relevant roles and interplay mechanisms peculiar to each of these heterogeneous subtypes. Accordingly, to support the biological interpretability of these newly identified associations, we also performed enrichment analyses subtype-by-subtype using g:Profiler [25]. Gene enrichment suggested that the unique biological processes and functions associated with each breast cancer subtype could guide the development of more targeted therapeutic strategies. For instance, protein binding and kinase activity in Her2-positive patients could be worthy of further investigation, while inspecting intracellular processes might be more relevant for Basal subtype patients. The enrichment of miRNAs in specific pathways from KEGG [26] and WikiPathways [27] databases, provided valuable insights into the molecular mechanisms involving miRNAs in different breast cancer subtypes. For instance, the Basal subtype was linked to pathways controlling cell proliferation and survival, HER2 to growth factor signalling, LumA to hormone response, and LumB to cell cycle regulation. Further details about enrichment analysis results are reported as Supplementary information.

## 5 Discussion and Conclusion

In this work, we designed and implemented a two-stage computational framework to integrate omics data and derive insights for a classification task of interest. Specifically, we focused on BRCA subtyping application to demonstrate the predictive capability of our solution in estimating subtype associations not only for different patients, but also when examining network nodes representing genes or miRNAs. This joint objective allowed us to maximize the classification performance based on the available, known, subtype-patients associations and infer knowledge about the subtype-related genes, miRNAs and interactions, especially those exclusively peculiar to one specific subtype.

When employing NMTF in the first stage of the workflow, we achieved promising prediction of BRCA subtypes for masked patients, arising inherently from the network iterative reconstruction. Particularly, the NMTF-based patient subtyping approach demonstrated high and consistent performance across breast cancer subtypes, despite being in a scenario with imbalanced classes. Furthermore, the stability of our approach across multiple runs and random subsets of masked patients underscores its reliability and effectiveness in accurately classifying breast cancer patients into distinct subtypes. Then, in the second stage of the workflow, leveraging subtype-specific subnetworks and network-based embeddings, we found relevant associations between each subtype and distinct mRNA and miRNA genes. This exploration both uncovered novel connections and retrieved statistically significant confirmations of associations and interactions already known at the state-of-the-art.

In the end, the most reliable subtype assignments for genes and miRNAs, considered both singularly or within interacting pairs, appeared to be those obtained using XGB classifiers for all the classes. Indeed, our analysis of predicted gene and miRNA associations with breast cancer subtypes revealed significant overlaps with associations reported in literature and databases, validating the biological relevance of our predictions. The identification of exclusive associations for each subtype highlights potential molecular mechanisms underlying breast cancer heterogeneity and offers interesting insights. Our findings reinforced the biological relevance of the identified miRNAs-subtype associations and pointed out their potential roles in subtype-specific oncogenic processes, which could be studied for developing targeted therapies or understanding resistance mechanisms in breast cancer treatment. Further investigations on gene/miRNA-subtype associations based on our findings could indeed enhance personalized medicine, ultimately leading to improve clinical decision-making processes.

Thus, our proposed solution facilitated the discovery of potential miRNA and gene markers for BRCA subtypes, holding promise for enhancing BRCA patient stratification and personalized treatment. Future research directions may include validating predicted associations experimentally, exploring additional omics data types, and translating research findings into actionable clinical applications. Furthermore, through the introduction of a novel hybrid predictive approach encompassing NMTF application and machine learning principles, like the utilization of training and testing phases, this study enriches the scientific literature by presenting a valuable alternative to straightforward classification methods, able to deal jointly with data integration, predictive tasks and knowledge inference.

## References

1. Ding, C., Li, T., Peng, W., Park, H.: Orthogonal nonnegative matrix tri-factorizations for clustering. In: Proceedings of the 12th ACM SIGKDD International Conference on Knowledge Discovery and Data Mining, pp. 126–135 (2006)
2. Vitali, F., Marini, S., Pala, D., et al.: Patient similarity by joint matrix trifactorization to identify subgroups in acute myeloid leukemia. JAMIA Open **1**(1), 75–86 (2018)
3. Ceddia, G., Pinoli, P., Ceri, S., Masseroli, M.: Matrix factorization-based technique for drug repurposing predictions. IEEE J. Biomed. Health Inform. **24**(11), 3162–3172 (2020)

4. Perou, C.M., et al.: Molecular portraits of human breast tumours. Nature **406**(6797), 747–752 (2000)
5. Sørlie, T., et al.: Gene expression patterns of breast carcinomas distinguish tumor subclasses with clinical implications. Proc. Natl. Acad. Sci. U.S.A. **98**(19), 10869–10874 (2001)
6. Parker, J.S., Mullins, M., Cheang, M.C., et al.: Supervised risk predictor of breast cancer based on intrinsic subtypes. J. Clin. Oncol. **27**(8), 1160–1167 (2009)
7. Cascianelli, S., Molineris, I., Isella, C., et al.: Machine learning for RNA sequencing-based intrinsic subtyping of breast cancer. Sci. Rep. **10**(1), 1–13 (2020)
8. Cristovao, F., Cascianelli, S., Canakoglu, A., et al.: Investigating deep learning based breast cancer subtyping using pan-cancer and multi-omic data. IEEE/ACM Trans. Comput. Biol. Bioinf. **19**(1), 121–134 (2020)
9. Liberzon, A., Subramanian, A., Pinchback, R., et al.: Molecular signatures database (MSigDB) 3.0. Bioinformatics **27**(12), 1739–1740 (2011)
10. Arun, R.P., Cahill, H.F., Marcato, P.: Breast cancer subtype-specific miRNAs: networks, impacts, and the potential for intervention. Biomedicines **10**(3), 651 (2022)
11. Haakensen, V.D., Nygaard, V., Greger, L., et al.: Subtype-specific micro-RNA expression signatures in breast cancer progression. Int. J. Cancer **139**(5), 1117–1128 (2016)
12. Søkilde, R., Persson, H., Ehinger, A., et al.: Refinement of breast cancer molecular classification by miRNA expression profiles. BMC Genomics **20**, 1–12 (2019)
13. Peng, F., Zhang, Y., Wang, R., et al.: Identification of differentially expressed miRNAs in individual breast cancer patient and application in personalized medicine. Oncogenesis **5**(2), e194–e194 (2016)
14. The Cancer Genome Atlas Network et al.: Comprehensive molecular portraits of human breast tumours. Nature **490**(7418), 61–70 (2012)
15. Pallotta, S., Cascianelli, S., Masseroli, M.: RGMQL: scalable and interoperable computing of heterogeneous omics big data and metadata in R/bioconductor. BMC Bioinform. **23**(1), 123 (2022)
16. Ciriello, G., et al.: Comprehensive molecular portraits of invasive lobular breast cancer. Cell **163**(2), 506–519 (2015)
17. Qiao, H.: New SVD based initialization strategy for non-negative matrix factorization. Pattern Recogn. Lett. **63**, 71–77 (2015)
18. Grover, A., Leskovec, J.: node2vec: scalable feature learning for networks. In: Proceedings of the 22nd ACM SIGKDD International Conference on Knowledge Discovery and Data Mining, pp. 855–864 (2016)
19. Huang, H.Y., et al.: mirtarbase update 2022: an informative resource for experimentally validated mirna-target interactions. Nucleic Acids Res. **50**(D1), D222–D230 (2022)
20. Smid, M., Wang, Y., Zhang, Y., et al.: Subtypes of breast cancer show preferential site of relapse. Can. Res. **68**(9), 3108–3114 (2008)
21. Triantafyllou, A., Dovrolis, N., Zografos, E., et al.: Circulating miRNA expression profiling in breast cancer molecular subtypes: applying machine learning analysis in bioinformatics. Cancer Diagnosis Prognosis **2**(6), 739 (2022)
22. Benjamin, D.J., et al.: Redefine statistical significance. Nat. Hum. Behav. **2**(1), 6–10 (2018)
23. Dai, X., Yu, L., Chen, X., Zhang, J.: SNRPD1 confers diagnostic and therapeutic values on breast cancers through cell cycle regulation. Cancer Cell Int. **21**, 1–16 (2021)

24. Lindholm, E.M., Leivonen, S.K., Undlien, E., et al.: mir-342-5p as a potential regulator of HER2 breast cancer cell growth. MicroRNA **8**(2), 155–165 (2019)
25. Kolberg, L., Raudvere, U., Kuzmin, I., Adler, P., Vilo, J., Peterson, H.: g: Profiler-interoperable web service for functional enrichment analysis and gene identifier mapping (2023 update). Nucleic Acids Res. **51**(W1), W207–W212 (2023)
26. Kanehisa, M., Goto, S.: Kegg: kyoto encyclopedia of genes and genomes. Nucleic Acids Res. **28**(1), 27–30 (2000)
27. Agrawal, A., et al.: Wikipathways 2024: next generation pathway database. Nucleic Acids Res. **52**(D1), D679–D689 (2024)
28. Gu, Q., Zhou, J.: Co-clustering on manifolds. In: Proceedings of the 15th ACM SIGKDD International Conference on Knowledge Discovery and Data Mining, pp. 359–368 (2009)

# Screening the Bioactivity of the P450 Enzyme by Spiking Neural Networks

Mauro Nascimben(✉) and Lia Rimondini

Department of Health Sciences, Università del Piemonte Orientale, 28100 Novara, Italy
mauro.nascimben@uniupo.it, lia.rimondini@med.uniupo.it

**Abstract.** The current study investigated the potential usage of spiking neural networks to score P450 bioactivity, predicting the enzyme's ability to catalyze xenobiotics. Numerical experiments focused on enzyme P450 3A4 and 2C9 isoforms to evaluate optimal network configurations and molecular fingerprint length in forecasting compounds where P450 was active or not. Reacher structural representations of the molecules improved the SNN outcomes employing leaky-integrate-and-fire neuronal models. In addition, this neuron performed slightly better than the one that included synaptic decay. Overall, the accuracies obtained in a benchmark library of compounds were in line with other machine learning techniques encountered in the previous literature, which further supports the applicability of spiking neural networks for quantitative structure-activity analysis. The SNN hyperparameter evaluation found substantial stability across simulations. In the future, the application of selected SNN configurations in conjunction with neuromorphic hardware could be considered as a high-performance computing alternative for virtual chemoinformatics screening to significantly improve energy efficiency and achieve accelerated calculations.

**Keywords:** spiking neural network · virtual screening · QSAR · P450 · molecular fingerprints

## 1 Introduction

The availability of vast libraries of compounds facilitates the computer-assisted identification of drugs or other active molecules as an alternative to running hit-picking on biochemical assays from the laboratory. For example, chemoinformatic approaches focusing on examining the chemical structure of the molecules to predict a target activity are taking advantage of machine learning (ML) algorithms as virtual screening tools to scan these extensive datasets [19,22,24,38]. The benefits of virtual screening molecular databases are both in terms of costs and time reduction compared to experimental trials; however, chemical databases, like those containing molecular structures, are often overwhelming in terms of the number of compounds and information characterizing the compound itself, and analyzing this high-dimensional data requires significant computational power [18]. In the


M. Vettoretti et al. (Eds.): CIBB 2023, LNBI 14513, pp. 273–292, 2025.
https://doi.org/10.1007/978-3-031-90714-2_20

chemoinformatics discipline of virtual screening, which is the process of searching through sizable chemical libraries for possible compounds that exhibit a peculiar bioactivity, building and refining machine learning models on chemical data sets might necessitate a substantial investment in computer power. Consequently, the problem of the high computational cost of predictions arises because, once trained, applying these models (especially in real-time applications or for large data sets) can be computationally demanding. Furthermore, chemoinformatics models must be detailed and complex in order to achieve high accuracy in predicting molecular properties or interactions, which often necessitates running multiple simulations or cross-validations, which adds to the workload. Chemical data frequently needs extensive preparation, such as computationally intensive normalization, modification, and cleaning, before analysis can begin. The resource intensity of chemoinformatics jobs is further increased by the need for access to cloud computing or high-performance computing (HPC) clusters, which are costly to maintain and run. Other than with quantum computers, neuromorphic hardware might be an exploitable alternative in chemoinformatics to achieve energy-efficient computing [1]. Also, parallelism is intrinsic to neuromorphic systems, since each neuron in the network can function independently and process data alongside other neurons at the same time. This advantage, coupled with the ability to scale processing capabilities by simply adding more neuromorphic cores, could lead to systems that handle increasingly complex problems efficiently. Conventional HPC systems face the von Neumann bottleneck, where the distinct separation of memory and processing components results in delays and energy waste due to the constant back-and-forth transfer of data between these two elements; in contrast, neuromorphic hardware, by combining memory (synaptic weights) and processing units (neurons) into a single integrated system, can significantly minimize the need for data movement. This feature could be interesting for chemoinformatics computing where the data representations could be sparse, and also in the view of building computers blending traditional and neuromorphic components, creating hybrid systems that leverage the strengths of both approaches for solving complex and non-linear problems [33]. Spiking Neural Networks (SNNs) are considered the software counterpart of neuromorphic hardware: both SNNs and neuromorphic hardware are designed to mimic the functioning of the human brain more closely than traditional artificial neural networks and conventional computing hardware [5]. At present, the research on SNN for virtual screening and quantitative structure-activity prediction could be considered in its initial stage, with only a few attempts available in the literature [13,23,25].

### 1.1 Spiking Neural Networks as Virtual Screening Tools

Spiking Neural Networks are artificial neural networks (ANN) with energy-efficient features when employed on neuromorphic hardware, commonly built with fewer neurons and faster pulse-based information propagation than traditional ANN. In [25], SNNs were tested in several public-domain databases of molecules for toxicity detection. In that investigation, SNN achieved consistent performance compared to the state-of-the-art literature on the same data types. Another work verified SNNs for bioaccumulation prediction by examining the performance of different neuronal models and comparing the results with the present body of knowledge [26], conjointly performing numerical experiments on a multiclass dataset. The effectiveness of binding SNNs to neuromorphic hardware was demonstrated on a chemoinformatics task by [23], which achieved impressive speed on 3D molecular similarity search. Although another manuscript describes the application of SNN for in silico chemistry for covalent docking [13], their application is still in the embryonic phase; thus, additional evidence might intensify the usage of SNN for the virtual screening of compounds. Usually, databases of molecules encode the compounds' structure in convenient representations for small and large molecules. Small organic molecules databases could encode the substructural format into vectors of alphanumeric characters called the simplified molecular input line entry system (SMILES), which could be further transformed into bit strings called molecular fingerprints (MF). Alternative linear notations include SMARTS, InChi, and InChikeys [39]. These character codes could preserve 2D or 3D information about compounds, with the latter being preferred as the methodology to represent more complex molecules such as peptides [10]. By transforming molecular structures into binary molecular fingerprints, which are vectors of 0 s and 1 s, the input matrix of the SNN is the molecular structure itself, assumed equivalent to a spike train, the natural input of SNNs, which are binary sequences. The use of binary molecular fingerprints aligns with one of the key features of SNNs, the concept of "sparsity". Under this assumption, the binarization of the molecular structure is conceptually equivalent to a binary spike train, the inherent input of SNNs. Similarly to spike trains, where 0 s and 1 s signify whether the presynaptic neuron is sending information or not over time, in molecular fingerprints, the binary digits represent the presence or absence of specific structural features, patterns, or substructures within the molecule. This parallels neuroscience, where the precise timing of spikes relative to each other (the distance between spikes) can encode information, such as sensory inputs, motor commands, or other neural data. Similarly, in molecular fingerprints, the structural information of the compound is encoded in precise positions, denoting specific features or substructures of the molecule, such as functional groups, atom types or bonding patterns.

### 1.2 P450 Bioactivity

Cytochromes P450 are a group of biocatalysts with crucial roles in the metabolism of fatty acids and foreign chemicals (aka xenobiotics). In addition, they are involved in the steroidogenesis and hepatic clearance of numerous compounds [3]. There are numerous isoforms of cytochrome P450 enzymes, each with its own specific substrate preferences and functions. Notable cytochrome P450 isoforms are:

- 1A2 isoform which is involved in the metabolism of caffeine, as well as various environmental toxins and carcinogens found in cigarette smoke and grilled foods.

- 2D6 plays a crucial role in the metabolism of many drugs, including antidepressants, antipsychotics, beta-blockers, and opioids.
- 3A4 is one of the most abundant cytochrome P450 enzymes in the liver and is involved in the metabolism of a wide range of drugs, including statins, calcium channel blockers, benzodiazepines, and many others.
- 2C9 and 2C19 are isoforms important for the metabolism of various drugs, including nonsteroidal anti-inflammatory drugs, anticoagulants, and some antidepressants.
- 2E1 is primarily involved in the metabolism of alcohol and certain environmental toxins.

Due to the wide range of metabolic interactions mediated by P450, several libraries collecting its bioactivity could be retrieved from public repositories to assist drug development: the database employed in the present study was derived from the collection initially created by [43]. Among the sub-types of P450 enzymes, the 3A4 and 2C9 isoforms have relevant implications for toxicity and drug absorption; therefore, [28] organized a reduced dataset from [43] to study the interplay between P450 and drug-like compounds.

The literature of machine learning methodologies and datasets targeting compounds binding to cytochrome P450 is heterogeneous. A considerable amount of chemoinformatics works derive predictions from functional or functional plus structural descriptors of the molecules rather than its structure only [45]. Other academic studies found that combining the outputs of multiple models could reach a more accurate and reliable final prediction (consensus mechanism), as in [28,37,44]. Consunsus could improve generalization by controlling overfitting, and robustness combining the strengths of multiple models; however, if the models in the ensemble are too similar (highly correlated), their errors may also be correlated, which reduces the benefit of consensus. Another strategy to approach P450 inhibition forecasting could capitalize on merging distinct data sources as in [35], where a random forest classifier have exhibited encouraging outcomes in forecasting the interactions between small molecules and cytochrome P450 enzymes by fusing P450 datasets from PubChem. In this case, keeping a low similarity between molecules coming from disparate libraries could ensure a good generalization of the model.

### 1.3 Aim of the Study

The current investigation aims at testing the ability of SNNs to score P450 bioactivity: numerical experiments will evaluate various SNN configurations in terms of hyperparameters, input complexity, and neuronal model, and reveal further insights on the application of SNN in chemoinformatics. After the necessary knowledge is developed in this novel field, a complete hardware and software-based SNN framework could be established for virtual molecular screening.

## 2 Methods

The public-domain database of molecules in [28] comprised the P450 bioactivity information of 9122 compounds encoded in SMILES format. A few preprocessing steps (i.e., canonicalization by removing duplicate structures and disconnected entities) were already performed by the curators of the compound database. Further preprocessing steps using the RDkit Python package [20] were:

- Standardizing functional groups to ensure that molecules are represented consistently.
- Detection and standardization of aromatic systems so that all aromatic rings are represented uniformly
- Eventual removal of salts, solvents, or counterions that are not part of the actual molecule of interest, by keeping the core structure to focus the analysis on relevant compounds' features only.
- Checking 3D arrangement of atoms to guarantee that stereocenters are correctly defined and consistent, ensuring that chiral centers, E/Z isomers, and other stereochemical features are coherently represented.
- Converting the molecule into its neutral form, and assuring that atoms have correct valences and that charges are properly assigned.

Afterward, five types of MF were created from the SMILES: a 166 bits long MAACS MF and radius 2 Morgan MF with lengths 256, 512, 1024, and 2048 bits (aka circular fingerprints). Morgan Fingerprints are widely used in cheminformatics to represent molecular structures. They are based on the concept of circular neighborhoods around each atom and capture molecular topology in a very efficient way: Morgan fingerprints capture the environment of atoms in a circular manner, where each atom is iteratively connected to its neighbors within a specific radius, and the radius controls how many bonds away from the central atom should be included. Each circular environment is hashed into a bitstring, meaning that the molecular substructures around an atom are transformed into unique integer identifiers; identifiers are used to create the binary vector, which represents the fingerprint of the molecule. MACCS (Molecular ACCess System) are a type of molecular fingerprint that uses a predefined set of structural keys to describe molecules. Instead of dynamically generating patterns, like Morgan fingerprints, MACCS fingerprints are based on a fixed dictionary of substructures. MACCS fingerprints are based on a set of 166 predefined structural keys, which correspond to specific molecular substructures or functional groups: the keys include things like aromatic rings, halogens, carbonyl groups, and others. Because each MACCS key corresponds to a specific feature, each bit represents whether or not the corresponding structural key is present in the molecule. Since each bit corresponds to a known chemical substructure, MACCS fingerprints are more interpretable than many other types of fingerprints; however, due to its fixed set of predefined keys, MACCS fingerprints are less sensitive to small structural differences than Morgan fingerprints, which can be both an advantage and a limitation depending on the context. To summarize, Morgan fingerprints offer more customization (e.g., radius, bit length) and capture more specific details about a molecule's structure compared to MACCS, which uses a fixed dictionary of features. Class imbalance in the binary labels was addressed by undersampling the majority class; the updated dataset shaped 3626 instances for each class. A similarity indicator was calculated to judge the degree of heterogeneity of the molecules constituting the MAACS dataset. The Jaccard-Tanimoto index, as a closeness measure of two molecular fingerprints, is commonly used to quantify the analogy of two molecules structurally. It is calculated as the ratio of the size of the intersection of two binary vectors (i.e., molecular fingerprints) to the size of their union. The coefficient ranges from 0 to 1: a value close to 0 means the two molecules share no common features (completely dissimilar), whereas an index near 1 means the two molecules are identical in terms of their structural features (perfect resemblance). Higher values indicate a greater degree of similarity between the two molecules based on their fingerprints. Other than easing similarity interpretation, its calculation is computationally efficient, especially when comparing large databases of molecular fingerprints. The Jaccard-Tanimoto coefficient was 0.364, a similarity below the suggested threshold of 0.8 [15]; a low affinity ensured the SNN would not learn the same family of molecules affecting the model's generalization.

The SNNs were coded in Python 3.9 under the Torch library [8] and tested using nested cross-validation (CV, Fig. 1). This kind of CV was a convenient choice to keep hyperparameter tuning (inner loop) and model evaluation (outer loop) independent; it has been demonstrated that nested CV approaches produce robust and unbiased performance estimates regardless of sample size [42]. During CV the number of active or inactive entries was proportional for both the P450 isoforms. After CV, the final prediction was made on a test set (20% of the total data, as suggested in [12]) of unseen MF. Furthermore, to evaluate the reproducibility of the outcomes and control the generalization error, the procedure was repeated ten times. The SNNs architectures had a single hidden layer, with a number of input neurons equal to MF length and two output neurons. In the output layer, the neuron with the highest firing rate was selected to establish the predicted class (bioactive compound or not). The primary neuronal model was the leaky integrate-and-fire (LIF), but its extended version with synaptic conductance (SYN) has also been tested [11]. The LIF model is computationally efficient and simple, making it useful for large-scale simulations of neural networks; despite its simplicity, the LIF model captures essential features of neuronal dynamics, such as integration of inputs, leakage, and spiking. It essentially describes the membrane potential $V(t)$ of a neuron over time which evolves according to the differential equation:

$$\tau_m \frac{dV(t)}{dt} = -\big(V(t) - V_{rest}\big) + R_m I(t) \tag{1}$$

where:

- $\tau_m$: is the membrane time constant.
- $V_{rest}$: is the resting potential of the neuron.
- $R_m$: is the membrane resistance.

– $I(t)$: is the input current.

The model is "leaky" because the membrane potential decays over time towards the resting potential $V_{rest}$ if no input current is applied. This decay reflects the passive loss of charge across the cell membrane, modeled by the $-\big(V(t) - V_{rest}\big)$ term. When the membrane potential $V(t)$ reaches a threshold value, the neuron "fires" an action potential (also called spike, usually represented as a 1 in a binary vector). After firing, the membrane potential is reset to a lower value, often back to $V_{rest}$: the resetting models the refractory period during which the neuron cannot fire again immediately. The SYN neuron model extends the LIF model by including synaptic conductances that vary over time. The membrane potential $V(t)$ is now subject to synaptic inputs through conductance changes:

$$\tau_m \frac{dV(t)}{dt} = -\big(V(t) - V_{rest}\big) + \sum_i g_i(t)\big(E_i - V(t)\big) \tag{2}$$

Synaptic conductance $g_i(t)$ changes dynamically in response to presynaptic spikes. When a presynaptic neuron fires, the synaptic conductance increases and then decays over time, usually following an exponential function. Excitatory and inhibitory synapses have different conductance values $g_i$ and reversal potentials $E_i$. The SYN model captures the effects of synaptic dynamics on the neuron's membrane potential, making it more biologically realistic, allowing for the interaction of excitatory and inhibitory synaptic inputs, which can dynamically influence the membrane potential and the likelihood of the neuron firing. The SYN model is more biologically realistic but computationally demanding than the basic LIF model because it requires the simulation of time-varying synaptic conductances and their effects on membrane potential.

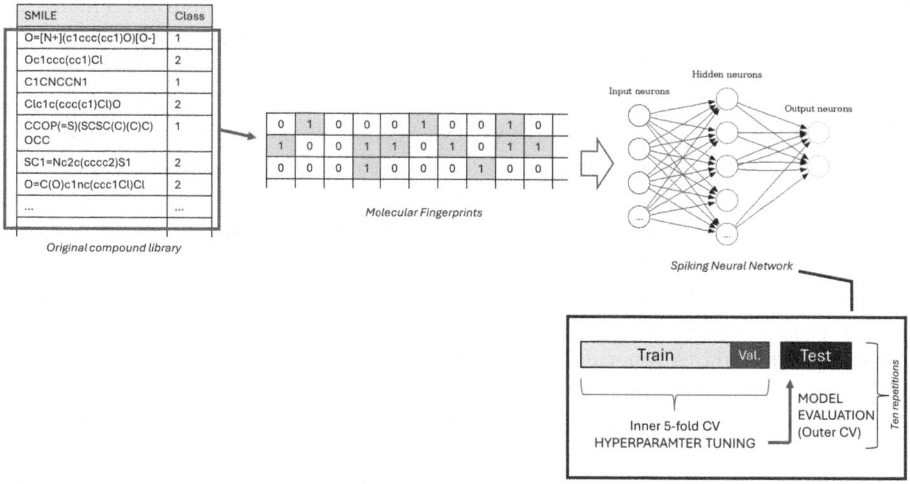


**Fig. 1.** Overview of the experimental procedure during the numerical experiments.

When fed to the SNNs, the MFs were divided into mini-batches of 64 examples to avoid the so-called "generalization gap", the difference between the performance on the training data and unseen test data [31]. Training on the entire dataset at once can lead to overfitting because the model might become too specialized to the specific patterns of the training data; dividing the input dataset into minibatches ensures that the model sees a slightly different view of the data at each step, promoting better generalization to new data because the algorithm learns to generalize from smaller, more varied samples of the data. In practice, instead of always following the gradient based on the entire dataset (which might lead to getting stuck in local minima or saddle points), minibatches provide a less consistent gradient estimate, allowing the model to potentially escape from these traps. The gradients computed using minibatches are noisy approximations of the true gradient which would be computed using the entire dataset. This noise acts as a form of regularization, preventing the model from fitting too closely to the training data, and encourages the model to learn more robust features offering enhanced generalization to unseen data. Also, when updating weights based on minibatch gradients, the model tends to converge to a more general solution because the noise from different minibatches averages out over time smoothing the learning process, and leading to better generalization. Additionally, frequent updates to the model parameters compared to batch gradient descent, where the parameters are updated only after processing the entire dataset, often lead to faster convergence, as the model continuously adjusts its parameters based on new information. Moreover, subdividing the data into manageable chunks, allow large datasets to be trained even on limited hardware.

The differentiability issue of the SNN discrete inputs to compute backpropagation was solved through a sigmoid surrogate gradient function as in [27]. In spiking neural networks, neurons communicate through discrete spikes, which makes the use of standard backpropagation challenging due to the non-differentiable nature of spike generation. Considering the LIF neuron, The membrane potential V(t) evolves over time based on incoming spikes and decays towards the resting potential. The neuron generates a spike when V(t) crosses a certain voltage threshold. The firing mechanism is described by the Heaviside step function which is non-differentiable.Instead of directly using the non-differentiable unit step function during backpropagation, it could be replaced with a smooth, differentiable function to approximate it (in this case, the sigmoid). During the forward pass, the neuron still spikes according to the original step function, but during the backward pass, the gradient is computed using the derivative of the sigmoid function. Hyperparameter tuning involved neuron-specific parameters such as the membrane decay rate, synaptic neurons' conductance current or more general network values. These latter parameters comprised the surrogate gradient slope, number of hidden neurons, optimizers (from Python Torch library 1.9.0) and related parameters, number of epochs, and activation of gradient clipping. Hyperparameter detection was conducted by random search sampling a regular grid of possible values. The SNN architecture and methodologies applied in the current work resembled those of the articles in [25,26], where additional details on the mathematical formulations could be found. The output layer of two neurons produced a binary vector simulating a spike train: class label was determined by the neuron of the two displaying the higher number of spiking events.

Models were evaluated through balanced accuracy (BA, Eq. 3), a metric used to evaluate the performance of a binary classifiers, specifically designed to address the limitations of traditional accuracy in situations where the dataset is imbalanced (i.e., one class significantly outnumbers the other).

$$BA = \frac{Sensitivity + Specificity}{2} \quad (3)$$

$$Sensitivity = \frac{True\ positives}{True\ positives + False\ Negatives} \quad (4)$$

$$Specificity = \frac{True\ negatives}{True\ negatives + False\ Positives} \quad (5)$$

Unlike simple accuracy, which can be misleading in such cases, balanced accuracy provides a more meaningful measure of classifier performance by giving equal weight to the true positive rate (sensitivity, Eq. 4) and the true negative rate (specificity, Eq. 5). By giving equal importance to the performance on both the positive and negative classes, balanced accuracy ensures that the model's ability to correctly identify both classes is reflected in the metric, making it a fairer evaluation criterion when class distributions are skewed. Also, this is true when comparing models targeting different datasets extracted from the literature because it avoids the bias that can arise in measurements affected by class imbalances. Essentially, balanced accuracy incorporates both the true positive rate (minimizing false negatives) and the true negative rate (minimizing false positives), making it sensitive to both Type I (false positive) and Type II (false negative) errors. The reason of relying on BA was the enhanced interpretability and simplicity connected with having a single number capable of judging all numerical experiments performed, facilitating the comparisons with other studies during meta-analysis.

## 3 Results

The numerical experiments aimed to assess the performance of LIF and SYN neural models in predicting the bioactivity of P450, evaluating the accuracy of the predictions. The models considered and reported in the following paragraphs were only those with a difference between validation and test balanced accuracy (BA) below 2% to avoid the possibility of overfitting. Also, the computations determined the optimal length of MF that produced the best outcomes. The experiments were complemented with a meta-analysis of previous literature in the field to arrive at conclusive results. The purpose of this meta-analysis was to contextualize the findings of the numerical experiments and provide a broader picture of the subject matter. Lastly, a description of the SNN hyperparameters was included to understand which was more critical in determining the outcomes.

### 3.1 Comparison Between Neural Models

The outcomes on MAACS MF included the standard deviations between different runs (STD). The impact of distinct neuronal models was reported in Table 1: the BA values are close to each other on validation (outer CV) and test sets, ensuring an adequate overfitting control. Also, the BAs of the two neuron types are lightly different.

**Table 1.** The SNN were tested with different biological neuron models; table includes validation and test set mean accuracies. ID is the numerical simulation identification number.

| Neuron | ID | Val. BA MEAN | Test BA MEAN | Test BA STD |
|---|---|---|---|---|
| LIF | 11740 | 79.91% | 78.62% | ±1.20% |
| SYN | 44947 | 75.46% | 75.65% | ±2.17% |

### 3.2 MF Length Evaluation

Another numerical experiment tested the relation between the complexity of MF and P450 bioactivity prediction. The goal was to understand if more refined molecular representations might provide better scores in solving this task (Table 2).

**Table 2.** The SNN outcomes when the input was a MF longer than MAACS (length in bits); table includes validation and test set accuracies.

| MF length | ID | Val. BA MEAN | Test BA MEAN | Test BA STD |
|---|---|---|---|---|
| 256 | 88305 | 80.46% | 79.23% | ±0.80% |
| 512 | 17877 | 78.24% | 80.19% | ±0.94% |
| 1024 | 70999 | 78.43% | 80.05% | ±0.95% |
| 2048 | 54766 | 82.81% | 81.52% | ±0.88% |

During this experiment, the SNNs were coded using the LIF neuronal model. The best accuracies pertain to the MF with 2048 bits in length, and in general, all values are higher than those obtained with MAACS MF.

### 3.3 Meta-Analysis on P450-Related Data

In the literature, it could be possible to find articles testing machine learning algorithms to P450 datasets. Differences between repositories arise from the type of MF utilized, the scoring metric, differences in data partition, and network training organization. Several papers extract functional descriptors from a database of molecules, thus excluding the possibility of comparisons with the present investigation, which employs only structural information. Additionally, the most recent publications focus on applying structural MF and functional features to larger datasets assembled by merging different databases to enhance overall accuracy [30]. Another limitation is the presence of six possible isoform-based databases regarding P450, further cutting down the number of compatible works.

The [28] study achieved a 78% non-error rate applying N-nearest neighbors [41], a custom classifier inspired by K-nearest neighbors catching the similarity between molecules on the extended-connectivity fingerprints (ECFP). These results were obtained on a similar dataset as in the current study, whereas other outcomes reported in the following paragraphs involved other collections of molecules.

In [7], the authors tested MAACS and FP4 fingerprints on P450 3A4 isoform, getting 75.5% accuracy with SVM classifier, 70.5% with decision trees, 72.6% with K-nearest neighbors, and 63.7% with Naïve Bayes.

The work of [4] on P450 3A4 isoform on a different dataset made of 4470 proprietary compounds, reached 82% accuracy from Barnard Chemical Information fingerprint of 4096 bits and SVM as a classifier.

Authors of [16] developed an in-silico model targeting ECFP and FCFP (functional connectivity fingerprint) MF training K-nearest neighbors on a dataset of 1382 compounds. Results were validated on an external set of 14 compounds: the 2D6 isoform was predicted correctly on 57.14% of the samples, whereas on 3A4, correctness was 71.42%.

In [2], the authors compared several classifiers to the FP-GNN deep learning method [6] on five P450 isoforms. The FP-GNN architecture require the concatenation of three MF types, additionally forwarding also molecular graphs to the neural netwrok for enhancing the predictive power of this chemoinformatic framework. However, the study also performed comparisons on other classifiers (Naive Bayes, Random Forest, Support Vector Machine, extreme gradient boosting, and deep neural networks) employing MFs only, evaluating datasets of 10284 (isoform 3A4) and 11711 (2D6 isoform) training examples. In general, the accuracies on 2D6 are lower than those on 3A4 confirming the trend found in the other literature sources (Table 3 reports only those on 3A4).

The LIF models with the lowest test set BA (on MAACS MF) and highest BA (2048 bits MF) were visually compared to the outcomes collected in Table 3. Figure 2 shows the performance of the SNNs compared to other machine learning methods for P450 bioactivity scoring from molecules' structures. The lower horizontal red line is the BA of "11740", whereas the upper one is the best BA found on the test set employing 2048 bits long MF (experiment "54766"). The accuracies from previous models were shown as bars for each classifier. Previous models did not perform better than the SNNs; findings show a substantial similarity between the SVM classifier on 4096 bits MF (from [4]) and LIF-based SNN on 2048 bits MF.

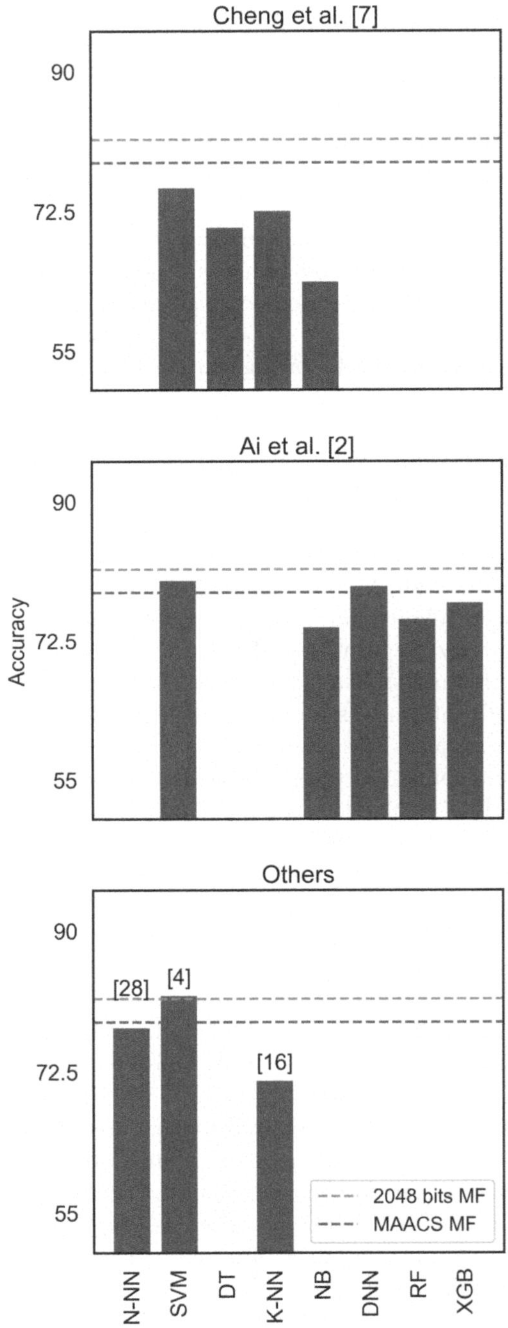


**Fig. 2.** Results of meta-analysis from Table 3: the horizontal lines report the BA of the present investigation on MAACS MF (experiment ID "11740") and 2048 bits MF (experiment ID "54766") compared to other models extracted from previous literature.

**Table 3.** Summary of the meta-analysis comparing other sources from literature.

| Ref. | Classifier | Accuracy | MF |
|---|---|---|---|
| [28] | N-NN | 78% (non-error rate) | ECFP |
| [7] | SVM | 75.50% | MAACS and FP4 |
| [7] | Decision trees | 70.50% | MAACS and FP4 |
| [7] | K-NN | 72.60% | MAACS and FP4 |
| [7] | Naïve Bayes | 63.70% | MAACS and FP4 |
| [4] | SVM | 82% | Barnard C. I. |
| [16] | K-NN | 71.42% | ECFP and FCFP |
| [2] | DNN | 79.5% | Morgan |
| [2] | RF | 75.3% | Morgan |
| [2] | SVM | 80.2% | Morgan |
| [2] | Naïve Bayes | 74.3% | Morgan |
| [2] | XGB | 77.4% | Morgan |

### 3.4 SNN Hyperparameters

Hyperparameter tuning helps to find the combination of hyperparameters that yields the best performance metrics for the models: exploring different values and combinations, one can identify settings that lead to optimal accuracy and gain a deeper understanding of how different hyperparameters affect the behavior and performance of the model [40]. Additionally, searching through the hyperparameter space can find configurations that prevent overfitting and improve the model's ability to generalize to unseen data. Understanding how hyperparameters interact with different data types can help develop more robust and generalizable models and provide insights into the underlying characteristics of the data and the problem domain. Also, hyperparameter values that lead to stable training dynamics should be identified to avoid issues like exploding or vanishing gradients [36].

Although sources in the literature report that hyperparameter tuning for SNN is complex to execute [32], the values found in the current investigation remain relatively stable. The hyperparameters of the best SNN models are included in Table 4.

**Table 4.** SNN Hyperparameters (Num Hidden: number of neurons in the hidden layers, Beta: membrane decay, Grad slope: slope of the surrogate gradient, Opt: Torch optimizer, LR: learning rate, WD: weight decay).

| ID | Neuron | MF | Num Hidden | Beta | Grad Slope | Opt | LR | WD |
|---|---|---|---|---|---|---|---|---|
| 11740 | LIF | MAACS | 1000 | 0.95 | 25 | Adamax | 0.0001 | 0.01 |
| 44947 | SYN | MAACS | 1000 | 0.8 | 75 | Adagrad | 0.01 | 0 |
| 88305 | LIF | 256 | 1000 | 0.95 | 50 | Adagrad | 0.001 | 0 |
| 17877 | LIF | 512 | 2000 | 0.95 | 50 | SGO | 0.01 | 0 |
| 70999 | LIF | 1024 | 1000 | 0.95 | 50 | SGO | 0.001 | 0 |
| 54766 | LIF | 2048 | 1000 | 0.95 | 50 | SGO | 0.001 | 0 |

One piece of information emerging from Table 4 is that the longer MF produced the best accuracies with the stochastic gradient descent optimizer. Analyzing the relationship between multiple parameters involved in the SNN (independent variables) regarding BA on the test set (dependent variable), it turned out that the learning rate was significantly crucial in the model ($R^2 = 0.93$). For this reason, Fig. 3 depicts a part of the hyperparameter space of LIF "11740" analysis, including the Torch optimizers and the associated learning rate over a scatterplot with test versus validation sets BA.

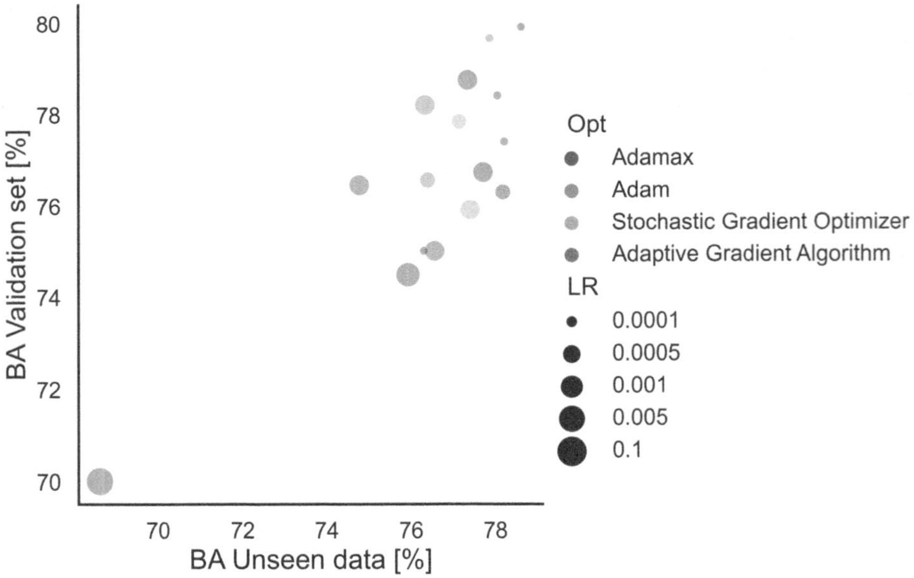


**Fig. 3.** The figure illustrates the Torch optimizer and the associated learning rate for each validation and test set BA pair of LIF-based MAACS MF analysis.

The top-right corner of the scatterplot contains several values from the Adamax optimizer with small learning rates. Other hyperparameters that might have some relevance in the models but did not pass the standard significance threshold were Beta ($p = 0.064$) and weight decay ($p = 0.098$). Regarding the Torch optimizers, which are responsible for updating the model parameters during the training process to minimize the loss function, Fig. 4 displays the relation between MF length and the different optimizers employed in the numerical experiments. Accuracies related to stochastic gradient optimizer (abbreviated as SGO) are slightly better than others.

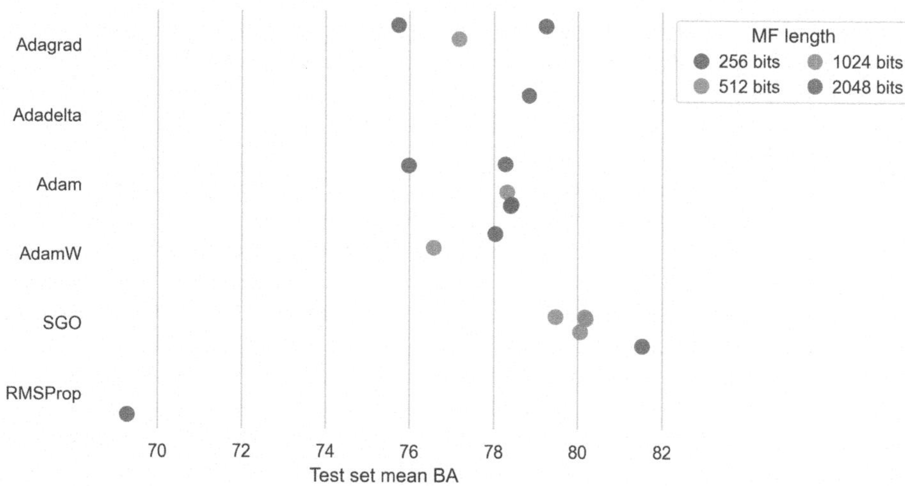


**Fig. 4.** Relation between BA scores on Test set, the Torch optimizer and MF length.

## 4 Discussion and Conclusions

In general, chemoinformatics virtual screening plays a significant role in the early stages of drug discovery and development, particularly in identifying potential drug candidates. The impact on drug safety, persistence, and bioactivation in patients is profound, influencing the overall success of a drug in clinical settings. Virtual screening helps in predicting the toxicity and adverse effects of compounds before they reach clinical trials: by simulating interactions between drug candidates and biological targets or off-targets, chemoinformatics can identify potential safety concerns such as toxicity, mutagenicity, or interactions with proteins that might lead to side effects. Additionally, chemoinformatics can assess the pharmacokinetic properties of compounds, including absorption, distribution, metabolism, and excretion (ADME); in particular, virtual screening can predict how long a drug will stay active in the body, its bioavailability, and its ability to reach the target site in adequate concentrations. Another aspect is that through virtual screening one can predict the potential bioactivation of a drug, where a compound is metabolized into a more reactive or toxic form for identifying prodrugs that become active only after metabolic conversion or for avoiding drugs that might produce harmful metabolites. Finally, early identification of unsuitable compounds through virtual screening saves resources by focusing experimental efforts on the most promising candidates, thus streamlining the drug development pipeline.

The current study predicted if a compound interacts or not (active or inactive) with two isoforms of Cytochromes P450 enzyme. The activity of Cytochrome P450 (CYP 450) enzymes plays a crucial role in drug metabolism: the isoforms mainly involved in drug metabolism are 3A4 and 2C9. The CYP 3A4 isoform interacts with more than half of all clinically used drugs, it is particularly known for metabolizing large and lipophilic molecules, and accounts for approximately 43% of the metabolism of known drugs. The CYP 2C9 primarily metabolizes non-steroidal anti-inflammatory drugs and weakly acidic molecules with a hydrogen bond acceptor. Similarly to CYP 3A4, CYP 2C9 also plays a significant role in drug metabolism

When a P450 enzyme is active, it means that the enzyme is capable of catalyzing the oxidation of substrates, which are often drugs or other xenobiotics (foreign chemicals). Active compounds often have specific structural characteristics that favor binding (substrate binds effectively to the enzyme's active site), and catalysis (enzyme facilitates the addition of an oxygen atom to the substrate, often resulting in increased solubility and easier excretion from the body). An inactive P450 enzyme means that it does not effectively catalyze the oxidation of substrates. For instance, small and flexible molecules tend to be inactive with respect to CYP enzymes. Several factors influence whether a P450 enzyme will be active or inactive: highly lipophilic compounds tend to be more active (lipophilicity), and larger and more complex molecules with specific structural features are more likely to be active.

The prediction of P450 activity attracted several researchers' attention due to this enzyme's key role in xenobiotics metabolism. Besides, commercial models were developed to estimate small molecules' interaction with P450 [34]. Compared to another SNN virtual screening tools previously published [25], where a single neuronal model was employed in all SNNs, the current work showed a comparable performance of the LIF and SYN neurons. The result is in accordance with the conclusions of another research on bioaccumulation [26], where the accuracy of LIF was slightly better than SYN, as in the present work. However, SYN neurons require an additional parameter to be tuned, and after further optimization, the score might increase.

The current study explored different types of molecular fingerprints; indeed, there are structural fingerprints (subdivided into substructure-based fingerprints and path-based fingerprints), topological fingerprints (like Extended Connectivity Fingerprints), pharmacophore fingerprints, and functional fingerprints (numerical vectors). Molecular fingerprints are used to rapidly screen large chemical libraries to identify compounds similar to a known active molecule or to identify potential drug candidates. Researchers usually employ organized benchmarks to evaluate and compare the effectiveness of different algorithms in predicting the biological activity of molecules based on their structural features. Indeed, molecular fingerprints are suitable for large-scale computational analyses, such as virtual screening, because they reduce complex molecular structures to a manageable and searchable format. Usually, compound libraries and benchmark datasets are in SMILES format (Simplified Molecular Input Line Entry System), and do not contain molecular fingerprints. Indeed, SMILES are usually standardized and preprocessed before conversion into the desired fingerprint format. In [25], only MAACS and Morgan MFs of 1024 bits were compared together. In that article, the accuracies were analogous, whereas in the current investigation, longer MF improved the scores. Longer data strings may account for a more detailed molecular description aiding prediction. Indeed, one strategy used by other researchers to improve the performance of P450 bioactivity prediction was to build augmented datasets, including multiple features [21]: for example, MFs are often combined with molecular descriptors to merge structural and functional information. However, the results of the current study focus on SNN targeting structural data only, and augmented schemes could be applied in future studies to test SNNs for further P450 bioactivity determination. Another remark pertains to data complexity because compared to [25], where the Tanimoto index was, on average, 0.263, and the other study on multi-class bioaccumulation in [26] where the Tanimoto index was 0.21, the present dataset has an higher degree of heterogeneity of the compounds.

One potential limitation of existing machine learning models in chemoinformatics is connected with the hardware bottleneck impairing speed and computational power. Indeed, the chemoinformatics research is investigating profitable techniques to overcome this issue; for example exploring stochastic computing as in [9]. Stochastic computing aims at reducing energy consumption through simplified arithmetic operations using random bit streams to represent and process information, while SNNs leverage the sparse and event-driven nature of spike-based communication. A point of contact between SNN and stochastic computing is the presence of probabilistic elements in their formulations: stochastic computing inherently deals with probabilities through random bit streams, while SNNs can use probabilistic inference in neuron models to process information and learning in SNNs (e.g., through spike-timing-dependent plasticity) can be viewed as a form of Bayesian updating [14,17]; intriguingly, stochastic computing could be integrated into SNNs [23]. Future alternatives for HPC are quantum and neuromorphic computers: algorithms targeting these hardware types could be exploited for energy-efficient and faster outcomes [29]. The numerical experiments on P450 and previous works on SNN for molecular toxicity and bioaccumulation provided evidence for applying SNN in chemoinformatics' quantitative structure-activity analysis. Prospective studies might produce an integrated neuromorphic software and hardware-accelerated framework taking advantage of the presented results.

While restraining model evaluation to balanced accuracy metrics facilitated and simplified the comparisons among results and between current outcomes and literature, balanced accuracy averages the recall for each class, emphasizing the model's ability to correctly identify all instances of each class. However, in some scenarios, precision might be important to address, and focusing solely on balanced accuracy could lead to a suboptimal model for those cases.

**Disclosure of Interests.** The authors received funding from European Union's Horizon 2020 research and innovation program under grant agreement No. 860462.

## References

1. Aboumerhi, K., Güemes, A., Liu, H., Tenore, F., Etienne-Cummings, R.: Neuromorphic applications in medicine. J. Neural Eng. **20**(4), 041004 (2023)
2. Ai, D., Cai, H., Zhao, D., Wang, L.: Deepcyps: a deep learning platform for enhanced cytochrome p450 activity prediction. Front. Pharmacol. **14**, 1099093 (2023)
3. Anzenbacher, P., Anzenbacherová, E.: Cytochromes P450 and metabolism of xenobiotics. Cell. Mol. Life Sci. **58**(5), 737–747 (2001). https://doi.org/10.1007/pl00000897
4. Arimoto, R., Prasad, M.A., Gifford, E.M.: Development of CYP3A4 inhibition models: comparisons of machine-learning techniques and molecular descriptors. SLAS Discovery **10**(3), 197–205 (2005)
5. Balaji, A., et al.: Mapping spiking neural networks to neuromorphic hardware. IEEE Trans. Very Large Scale Integr. (VLSI) Syst. **28**(1), 76–86 (2019)
6. Cai, H., Zhang, H., Zhao, D., Wu, J., Wang, L.: Fp-gnn: a versatile deep learning architecture for enhanced molecular property prediction. Briefings Bioinform. **23**(6), bbac408 (2022)
7. Cheng, F., et al.: Classification of cytochrome p450 inhibitors and noninhibitors using combined classifiers. J. Chem. Inf. Model. **51**(5), 996–1011 (2011)
8. Eshraghian, J.K., et al.: Training spiking neural networks using lessons from deep learning. arXiv preprint arXiv:2109.12894 (2021)
9. Frasser, C.F., et al.: Using stochastic computing for virtual screening acceleration. Electronics **10**(23), 2981 (2021)
10. Gao, K., Nguyen, D.D., Sresht, V., Mathiowetz, A.M., Tu, M., Wei, G.W.: Are 2D fingerprints still valuable for drug discovery? Phys. Chem. Chem. Phys. **22**(16), 8373–8390 (2020)
11. Gerstner, W., Kistler, W.M.: Mathematical formulations of hebbian learning. Biol. Cybern. **87**(5), 404–415 (2002)
12. Gholamy, A., Kreinovich, V., Kosheleva, O.: Why 70/30 or 80/20 relation between training and testing sets: a pedagogical explanation. Int. J. Intell. Technol. Appl. Stat. **11**(2), 105–111 (2018)
13. Gu, Z., et al.: Discovery of covalent lead compounds targeting 3cl protease with a lateral interactions spiking neural network. J. Chem. Inf. Model. **64**(8), 3047–3058 (2024)
14. Jang, H., Simeone, O., Gardner, B., Gruning, A.: An introduction to probabilistic spiking neural networks: probabilistic models, learning rules, and applications. IEEE Signal Process. Mag. **36**(6), 64–77 (2019)
15. Jasial, S., Hu, Y., Vogt, M., Bajorath, J.: Activity-relevant similarity values for fingerprints and implications for similarity searching. F1000Research **5**(591), 591 (2016)

16. Jensen, B.F., Vind, C., Padkjær, S.B., Brockhoff, P.B., Refsgaard, H.H.: In silico prediction of cytochrome P450 2D6 and 3A4 inhibition using gaussian kernel weighted k-nearest neighbor and extended connectivity fingerprints, including structural fragment analysis of inhibitors versus noninhibitors. J. Med. Chem. **50**(3), 501–511 (2007)
17. Kasabov, N.: To spike or not to spike: a probabilistic spiking neuron model. Neural Netw. **23**(1), 16–19 (2010)
18. Kell, D.B., Samanta, S., Swainston, N.: Deep learning and generative methods in cheminformatics and chemical biology: navigating small molecule space intelligently. Biochem. J. **477**(23), 4559–4580 (2020)
19. Kimber, T.B., Chen, Y., Volkamer, A.: Deep learning in virtual screening: recent applications and developments. Int. J. Mol. Sci. **22**(9), 4435 (2021)
20. Landrum, G., et al.: Rdkit: a software suite for cheminformatics, computational chemistry, and predictive modeling. Greg Landrum **8** (2013)
21. Lee, J., Basith, S., Cui, M., Kim, B., Choi, S.: In silico prediction of multiple-category classification model for cytochrome P450 inhibitors and non-inhibitors using machine-learning method. SAR QSAR Environ. Res. **28**(10), 863–874 (2017)
22. Mathai, N., Kirchmair, J.: Similarity-based methods and machine learning approaches for target prediction in early drug discovery: performance and scope. Int. J. Mol. Sci. **21**(10), 3585 (2020)
23. Morro, A., et al.: A stochastic spiking neural network for virtual screening. IEEE Trans. Neural Netw. Learn. Syst. **29**(4), 1371–1375 (2017)
24. Murugan, N.A., Priya, G.R., Sastry, G.N., Markidis, S.: Artificial intelligence in virtual screening: models versus experiments. Drug Discovery Today (2022)
25. Nascimben, M., Rimondini, L.: Molecular toxicity virtual screening applying a quantized computational SNN-based framework. Molecules **28**(3), 1342 (2023)
26. Nascimben, M., Spriano, S., Rimondini, L., Venturin, M.: Molecular fingerprint based and machine learning driven QSAR for bioconcentration pathways determination. In: Bretti, G., Natalini, R., Palumbo, P., Preziosi, L. (eds.) Mathematical Models and Computer Simulations for Biomedical Applications, pp. 193–215. Springer Nature Switzerland, Cham (2023)
27. Neftci, E.O., Mostafa, H., Zenke, F.: Surrogate gradient learning in spiking neural networks: bringing the power of gradient-based optimization to spiking neural networks. IEEE Signal Process. Mag. **36**(6), 51–63 (2019)
28. Nembri, S., Grisoni, F., Consonni, V., Todeschini, R.: In silico prediction of cytochrome P450-drug interaction: QSARs for CYP3A4 and CYP2C9. Int. J. Mol. Sci. **17**(6), 914 (2016)
29. Omar, Ö.H., Del Cueto, M., Nematiaram, T., Troisi, A.: High-throughput virtual screening for organic electronics: a comparative study of alternative strategies. J. Mater. Chem. C **9**(39), 13557–13583 (2021)
30. Ouzounis, S., et al.: A robust machine learning framework built upon molecular representations predicts cyp450 inhibition: Toward precision in drug repurposing. OMICS J. Integr. Biol. **27**(7), 305–314 (2023)
31. Oyedotun, O.K., Papadopoulos, K., Aouada, D.: A new perspective for understanding generalization gap of deep neural networks trained with large batch sizes. Appl. Intell. 1–17 (2022)
32. Parsa, M., Mitchell, J.P., Schuman, C.D., Patton, R.M., Potok, T.E., Roy, K.: Bayesian-based hyperparameter optimization for spiking neuromorphic systems. In: 2019 IEEE International Conference on Big Data (Big Data), pp. 4472–4478. IEEE (2019)

33. Pastur-Romay, L.A., Cedrón, F., Pazos, A., Porto-Pazos, A.B.: Deep artificial neural networks and neuromorphic chips for big data analysis: pharmaceutical and bioinformatics applications. Int. J. Mol. Sci. **17**(8), 1313 (2016)
34. Peach, M.L., et al.: Computational tools and resources for metabolism-related property predictions. overview of publicly available (free and commercial) databases and software. Futur. Med. Chem. **4**(15), 1907–1932 (2012)
35. Plonka, W., Stork, C., Šícho, M., Kirchmair, J.: Cyplebrity: machine learning models for the prediction of inhibitors of cytochrome p450 enzymes. Bioorganic Med. Chem. **46**, 116388 (2021)
36. Probst, P., Boulesteix, A.L., Bischl, B.: Tunability: importance of hyperparameters of machine learning algorithms. J. Mach. Learn. Res. **20**(53), 1–32 (2019)
37. Rácz, A., Keserű, G.M.: Large-scale evaluation of cytochrome p450 2c9 mediated drug interaction potential with machine learning-based consensus modeling. J. Comput. Aided Mol. Des. **34**, 831–839 (2020)
38. Rodríguez-Pérez, R., Miljković, F., Bajorath, J.: Machine learning in chemoinformatics and medicinal chemistry. Annu. Rev. Biomed. Data Sci. **5**, 43–65 (2022)
39. Saldívar-González, F.I., Huerta-García, C.S., Medina-Franco, J.L.: Chemoinformatics-based enumeration of chemical libraries: a tutorial. J. Cheminformatics **12**(1), 1–25 (2020). https://doi.org/10.1186/s13321-020-00466-z
40. Simon, S., Kolyada, N., Akiki, C., Potthast, M., Stein, B., Siegmund, N.: Exploring hyperparameter usage and tuning in machine learning research. In: 2023 IEEE/ACM 2nd International Conference on AI Engineering–Software Engineering for AI (CAIN), pp. 68–79. IEEE (2023)
41. Todeschini, R., Ballabio, D., Cassotti, M., Consonni, V.: N3 and BNN: two new similarity based classification methods in comparison with other classifiers. J. Chem. Inf. Model. **55**(11), 2365–2374 (2015)
42. Vabalas, A., Gowen, E., Poliakoff, E., Casson, A.J.: Machine learning algorithm validation with a limited sample size. PLoS ONE **14**(11), e0224365 (2019)
43. Veith, H., et al.: Comprehensive characterization of cytochrome P450 isozyme selectivity across chemical libraries. Nat. Biotechnol. **27**(11), 1050–1055 (2009)
44. Wang, N.N., et al.: Machine learning to predict metabolic drug interactions related to cytochrome p450 isozymes. J. Cheminformatics **14**(1), 23 (2022)
45. Xiong, Y., Qiao, Y., Kihara, D., Zhang, H.Y., Zhu, X., Wei, D.Q.: Survey of machine learning techniques for prediction of the isoform specificity of cytochrome p450 substrates. Curr. Drug Metab. **20**(3), 229–235 (2019)

# Enhancing Functional Interpretability in Gene Expression Analysis Through Biologically-Guided Feature Selection

Sofia Mongardi(✉), Silvia Cascianelli, and Marco Masseroli

Department of Electronics, Information and Bioengineering (DEIB), Politecnico di Milano, Milan, Italy
{sofia.mongardi,silvia.cascianelli,marco.masseroli}@polimi.it

**Abstract.** Reducing the high dimensionality of the original feature space through the use of feature selection algorithms is crucial in gene-expression-based predictive tasks to potentially improve performance and provide a better understanding of each feature's power and biological meaning. Feature selection approaches like LASSO and other embedded techniques select small subsets of relevant features based solely on their quantitative contribution and predictive power, often leading to the selection of features with limited biological relevance. This work aims to provide a wide exploratory analysis of LASSO feature selection to assess the effects of different hyper-parameters on the selection of the most relevant features and their corresponding biological significance. Then, it introduces a new approach that can guide LASSO in the selection of the features by considering their predictive power as well as their biological relevance. With this intention, this work proposes a novel Gene Information Score to estimate each gene's biological relevance and shows its use in enhancing the feature selection.

**Keywords:** Machine learning · feature selection · biological interpretability · genomics · classification

## 1 Introduction

Feature selection algorithms are extensively used in a wide range of prediction tasks to identify the most relevant features and enhance performance [1–3]. Embedded feature selection methods directly integrate the feature selection step into the learning process, simultaneously selecting the most predictive features and optimizing the prediction tasks under study. The Least Absolute Shrinkage and Selection Operator (LASSO) [4] is one of the most popular and used examples of such techniques. LASSO selects features by regularizing the model parameters, i.e., it shrinks the coefficient of each feature, eventually setting to zero those associated with features that are considered not predictive by the model. As a result, LASSO identifies a smaller subset of features having the


M. Vettoretti et al. (Eds.): CIBB 2023, LNBI 14513, pp. 293–307, 2025.
https://doi.org/10.1007/978-3-031-90714-2_21

highest predictive power, reducing risks of overfitting, noise, and the curse of dimensionality.

The selection of this smaller subset of features strictly depends on the regularization hyper-parameter $\lambda$ and on a threshold value, or *tolerance*, which controls the convergence of the prediction optimization algorithm. The $\lambda$ parameter specifies how much to shrink the model coefficients, with higher values of $\lambda$ corresponding to stronger regularization and fewer features selected. The convergence of the prediction optimization function is also key in guiding the feature selection process: when the convergence value is below the tolerance threshold, the optimization stops. Thus, the higher this threshold, the faster the convergence and the more features are selected.

In classification tasks involving gene expression data, it is crucial to decrease the high dimensionality of the original gene/feature space to improve the reliability and interpretability of the results. A small subset of predictive features allows for a more comprehensive understanding of the predictive importance and biological relevance of each feature. Yet, embedded methods like LASSO select features solely based on their quantitative contribution to the predictive task, which can lead to the selection of a feature subset with limited biological significance and poor interpretability. Nonetheless, in the huge initial gene space, multiple feature subsets can have equivalent predictive power.

Accordingly, in this work, we first provide a wide exploratory analysis of LASSO feature selection. Specifically, we tuned and compared LASSO-regularized Logistic regression models using different *tolerance* values, and found high variability of the selected genes in the corresponding feature subsets. Indeed, the different feature subsets achieved comparable predictive performance but displayed strongly unequal biological significance and interpretability power. Therefore, we then propose a first approach to guide LASSO towards a gene expression-based classification that is more functionally interpretable, by not only considering the predictive power of the selected features but also their biological relevance. We evaluate the proposed method in two different application use cases concerning the classification of Breast Cancer (BRCA) patients and Colorectal Cancer (CRC) patients into their corresponding intrinsic subtypes, using publicly available RNA-seq data from The Cancer Genome Atlas (TCGA) [5] project.

In the following, we summarize our exploratory analysis and the obtained evidence. Then, we describe the workflow and the results of our proposal to enhance the LASSO feature selection. To consider the biological relevance of each gene, we define a Gene Information Score, computed considering gene annotations from the Gene Ontology (GO) [6]. This creates opportunities for the potential incorporation of a measure of gene biological relevance directly into an embedded feature selection method to further guide its choices.

# 2 Materials and Methods

## 2.1 Datasets and Pre-processing

In this section, we describe the datasets for the two application use cases considered, which concerned the prediction of cancer-related subtypes of patients based on gene expression data.

**Breast Cancer Subtyping Dataset.** We extracted and used publicly available RNA-Seq profiles of BRCA patients from TCGA [5,7,8]. The dataset included gene expression data for 25,150 genes and 1,053 samples. The gene expression profiles were pre-processed as follows. We first discarded any miRNA gene (1,650), then removed the samples (10) with the top-five expressed genes accounting for at least 20% of the total sample raw counts, and discarded the genes (1,421) that were not expressed in at least 80% of the samples. Genes were considered not expressed if the corresponding raw count was $\leq 4$. We also removed all the non-coding genes (4,155) according to the NCBI Reference Sequence Database annotation [9]. We then normalized the gene raw counts by computing the *reads per million* (RPM) of each gene $g_i$ in each sample $s_j$: $RPM_{ji} = \frac{\#\ reads\ mapped\ to\ g_i}{total\ \#\ reads\ for\ sample\ s_j} * 10^6$, and then a $log_2$ transformation of the normalized RPM values. A total of 17,924 genes were retained after pre-processing.

For all the remaining TCGA samples (1,043), we retrieved their associated BRCA intrinsic subtypes from cBioPortal[1], obtained from the popular PAM50 test [10]. We considered the *Basal* (175), *HER2-enriched* - Her2 (81), *Luminal A* - LumA (543), and *Luminal B* - LumB (207) subtypes for our analyses, for a total of 1,006 patients (see Table 1), discarding the *Normal-like* (37) class whose clinical significance remains controversial.

**Colorectal Cancer Subtyping Dataset.** We extracted and used publicly available raw RNA-seq profiles of Colon adenocarcinoma (COAD) and Rectum adenocarcinoma (READ) patients from TCGA [5,8]. The initial dataset comprised gene expression data of 58,387 genes and 620 primary tumoral samples. We followed the same pre-processing steps described above. We first discarded miRNA genes (1,732), removed the samples with the top-five expressed genes accounting for at least 20% of the total sample raw counts (0), removed the not expressed genes (21,659), and lastly, we retained only the coding ones (17,559). We normalized the gene raw counts by computing the *reads per million* (RPM). After pre-processing, we retrieved each sample's primary subtype based on the Colorectal cancer Intrinsic Subtypes (CRIS) obtained through the Nearest Template Prediction algorithm (NTP) in [11]. We considered all subtypes in our analyses: CRIS A (160), CRIS B (83), CRIS C (165), CRIS D (98), and CRIS E (114), for a total of 620 samples (see Table 2).

[1] https://www.cbioportal.org/.

**Table 1.** Class distributions in the training and testing sets for the BRCA dataset.

| Set | Basal | Her2 | LumA | LumB | Total |
|---|---|---|---|---|---|
| Training | 140 | 65 | 434 | 165 | 804 |
| Test | 35 | 16 | 109 | 42 | 202 |
| All | 175 | 81 | 543 | 207 | 1,006 |
| Percentage (%) | 17.40 | 8.05 | 53.97 | 20.58 | 100.0 |

**Table 2.** Class distributions in the training and testing sets for the CRC dataset.

| Set | CRIS A | CRIS B | CRIS C | CRIS D | CRIS E | Total |
|---|---|---|---|---|---|---|
| Training | 128 | 67 | 132 | 78 | 91 | 496 |
| Test | 32 | 16 | 33 | 20 | 23 | 124 |
| All | 160 | 83 | 165 | 98 | 114 | 620 |
| Percentage (%) | 25.80 | 13.39 | 26.61 | 15.81 | 18.39 | 100.0 |

### 2.2 Logistic Regression with LASSO Regularization

After pre-processing, we used stratified sampling to split the datasets, selecting 80% of the patient samples for training and 20% for testing. Class distributions for the training and testing sets are reported in Table 1 for the BRCA dataset and in Table 2 for the CRC dataset. The training data was normalized using min-max normalization to prevent larger-scale features from dominating the feature selection process and the classification. We normalized the using the min and max statistics that were calculated from the training dataset. A multi-class Logistic regression (LR) with LASSO regularization was chosen as a linear classifier to perform the subtype-prediction tasks. LASSO-regularized LR minimizes the following cost function:

$$L(\theta) = -\left[\sum_{i=1}^{n}\sum_{k=1}^{K}[y^{(i)} = k]\log\frac{\exp(\theta^{(k)\top}x^{(i)})}{\sum_{j=1}^{K}\exp(\theta^{(j)\top}x^{(i)})}\right] + \lambda\sum_{j=1}^{K}||\theta^{(j)}||_1 \quad (1)$$

where $n$ and $K$ denote the number of instances and classes, respectively, $\theta^{(k)}$ is the vector of model parameters for class $k$, $\lambda$ is the LASSO regularization hyper-parameter, and $y^{(i)}, x^{(i)}$ are the class and feature vectors of the i-th sample, respectively. We used the LASSO-regularized LR model implemented in the scikit-learn Python package[2].

We performed 5-fold cross-validation (CV) to tune the parameter $\lambda$ over a range of values (i.e., 0.2, 0.5, 1.0, 2.0, 5.0) that can substantially reduce the huge dimensionality of the original feature space (17,924 for BRCA and 17,559 for CRC), while using the default value (1e-4) for the *tolerance*. The LASSO-regularized LR was then trained with the tuned $\lambda$ value using three different

[2] https://scikit-learn.org/stable/.

*tolerance* values (i.e., 1e-2, 1e-3, 1e-4). For each of these values, we trained the model on 10 different permutations of the training samples to assess the robustness of the results to the order of the training samples in terms of classification performance, both on the validation and final test set, and selected feature subsets. We used the *accuracy* (mean and standard deviation across the 10 permutations) as the performance evaluation metric to assess and compare the different classification models along with the *stability* of the selected feature subsets (their union and intersection across the 10 permutations).

### 2.3 Functional Annotations

To evaluate the biological relevance of the feature subsets selected by the regularized model, we used their annotations with terms of the Gene Ontology, which offers a standardized vocabulary of terms structured as a directed acyclic graph (DAG) to represent term relationships and is divided into three sub-ontology describing biological processes (BP), cellular components (CC) and molecular functions (MF).

We retrieved all GO terms annotated to the genes retained after preprocessing; we used the biomaRt[3] R library, extracting GO annotation terms for a total of 18,324 annotated genes. The biomaRt library only provides the most specific GO annotation terms (i.e., those at the highest depth in the DAG hierarchy, according to the true-path-rule [12]) for each gene. Therefore, to obtain all the GO annotation terms of our genes, we extracted all the ancestors at any depth level on the GO DAG for these specific GO terms, considering all and only the DAG *'is_a'* and *'part_of'* relationships.

### 2.4 Enrichment Analysis

We performed different biological validation analyses to evaluate the biological relevance and interpretability of each selected feature subset. Firstly, we extracted the number of significantly enriched GO terms; we used the Fisher's Exact test [13] to assess, separately for each dataset-specific subtype, the annotation GO terms enriched in the subset of robust subtype-specific genes (from intersection across the 10 permutations) with respect to the entire set of genes in the dataset. The p-values were adjusted for multiple testing using Benjamini-Hochberg false discovery rate (FDR) correction, and the significance level was set to 0.05. We used MalaCards [14], an integrated database of human diseases, related phenotypes and associated genes and annotations covering multiple sources such as OMIM, DrugBank, Pubmed, and GeneCards, to extract Breast Cancer- and Colorectal Cancer-specific gene signatures. Then, we computed the intersection between the union of the robust features (intersection across the 10 permutations) selected across the different subtypes and the disease-specific signatures. From MalaCards, we also retrieved the GO terms related to BRCA (59) and CRC (26) and evaluated the degree of overlap between these BRCA-

[3] https://bioconductor.org/packages/release/bioc/html/biomaRt.html.

and CRC-related terms and the significantly enriched ones across the different BRCA and CRC subtypes, for the different levels of *tolerance*. We additionally extracted cancer-specific and disease-specific pathways from the KEGG database [15], an integrated database storing information on the molecular functions of genes and their associated pathways, and computed the intersection between the robust features selected across the different subtypes and the gene signatures associated with such pathways.

# 3 Classification Results

In this Section, we report and compare the subtyping results of the LASSO-regularized LR when using 4 different *tolerance* values, considering each time 10 different models trained on permutations of all training samples. We also present the enrichment analysis results assessing the biological relevance and interpretability of the feature subsets under comparison.

## 3.1 Breast Cancer Subtyping

The results of all the different examined multi-class LASSO-regularized LR models are shown in Table 3. Higher *tolerance* values led to the selection of larger feature sets, which, however, appear not to be robust to permutations of the training samples (higher difference between number of union and intersection features). By decreasing the *tolerance* value, we were able to retrieve more stable, yet smaller feature subsets. Moreover, it seems that larger feature subsets did not yield a significant difference in the predictive performance. Indeed, all classifiers achieved comparable performances on the test set, regardless of the different *tolerance* values.

**Table 3.** Classification performances and number of features selected as **Union**, **Intersection**, or **Not robust** across 10 permutations on the BRCA dataset.

| Tolerance | CV[a] accuracy | Test accuracy | Union | Intersection | Not robust[b] |
|---|---|---|---|---|---|
| 1e-2 | 0.90290 (± 0.00087) | 0.94059 (± 0.0) | 3,677 | 3,018 | 659 |
| 1e-3 | 0.90687 (± 0.00037) | 0.94059 (± 0.0) | 726 | 711 | 15 |
| 1e-4 | 0.90425 (± 0.0) | 0.93069 (± 0.0) | 195 | 194 | 1 |
| PM[c] | 0.90800 (± 0.0) | 0.93069 (± 0.0) | 225 | 224 | 1 |

[a] Cross-validation (CV).
[b] Genes selected as features not in all 10 permutations.
[c] Proposed method to biologically guide LASSO.

Conversely, results from the enrichment analyses (see Tables 4 and 5) show that the robust feature subsets across the 10 subsets obtained with a *tolerance* value of 1e-2 retain much more known biological information compared to feature subsets obtained with a *tolerance* of 1e-4. Particularly, for the former case the number of significant GO terms for each BRCA subtype is much larger, as well as the degree of overlap between the selected features and the Malacards and KEGG pathway gene signatures. Furthermore, the number of GO statistically significant terms related to BRCA according to MalaCards is also higher (17 vs. 7).

**Table 4.** Enrichment analysis results for the BRCA dataset.

| Tolerance Subtype | 1e-2 | | 1e-4 | | Proposed Method | |
|---|---|---|---|---|---|---|
| | N | GO terms | N | GO terms | N | GO terms |
| Basal | 950 | 8,710 (87, 1.0%) | 30 | 1,564 (124, 8.0%) | 23 | 1,791 (365, 20.4%) |
| Her2 | 983 | 9,067 (93, 1.0%) | 51 | 2,376 (0) | 47 | 2,879 (488, 17.0%) |
| LumA | 1,173 | 9,739 (360, 3.7%) | 76 | 2,747 (170, 6.2%) | 83 | 3,471 (365, 10.5%) |
| LumB | 1,019 | 9,387 (168, 1.8%) | 57 | 2,267 (30, 1.3%) | 93 | 3,762 (162, 4.3%) |

N: size of the selected feature subset.
GO terms: number of GO annotation terms retrieved (and the corresponding number and percentage of significantly enriched terms, with FDR p-value $< 0.05$).

**Table 5.** Additional enrichment analysis results for the BRCA dataset.

| Tolerance | Malacards BRCA genes | KEGG BRCA pathways | KEGG Cancer pathways | Malacards BRCA GO terms |
|---|---|---|---|---|
| 1e-2 | 259 | 69 | 237 | 17 |
| 1e-4 | 37 | 4 | 18 | 7 |
| PM[a] | 34 | 9 | 27 | 10 |

Number of matches between the robust selected features and the Malacards and KEGG gene signatures and the Malacards BRCA-associated GO terms.
[a] Proposed method to biologically guide LASSO.

### 3.2 Colorectal Cancer Subtyping

The results of the different classification experiments are reported in Table 4. Larger feature sets were selected for higher *tolerance* values. As observed before for the BRCA dataset, by decreasing the *tolerance* value, we favored the selection of smaller feature subsets that also appeared to be robust to permutations of the training samples. Comparable performances were achieved on the final test set, suggesting that larger feature sets were not able to significantly increase the classification performance.

As for the previous application case on the BRCA dataset, results from the enrichment analyses (see Tables 7 and 8) suggest that the larger robust feature

**Table 6.** Classification performances and number of features selected as **Union**, **Intersection**, or **Not robust** across 10 permutations on the CRC dataset.

| Tolerance | CV[a] accuracy | Test accuracy | Union | Intersection | Not robust[b] |
|---|---|---|---|---|---|
| 1e-2 | 0.79061 (± 0.00151) | 0.83871 (± 0.0) | 4,888 | 4,296 | 592 |
| 1e-3 | 0.79630 (± 0.0) | 0.85484 (± 0.0) | 1,075 | 1,057 | 18 |
| 1e-4 | 0.76665 (± 0.00092) | 0.83871 (± 0.0) | 371 | 369 | 2 |
| PM[c] | 0.75804 (± 0.0) | 0.79032 (± 0.0) | 390 | 389 | 1 |

[a] Cross-validation (CV).

[b] Genes selected as features not in all 10 permutations.

[c] Proposed method to biologically guide LASSO.

subsets (higher *tolerance*) include genes providing much more known biological information with respect to the smaller feature subsets obtained with a *tolerance* of 1e-4. Indeed, a higher number of significant GO terms for each CRC-specific subtype is obtained for a larger *tolerance* value. The number of selected genes overlapping with Malacards and KEGG pathway gene signatures is also higher in this scenario.

**Table 7.** Enrichment analysis results for the CRC dataset.

| Tolerance Subtype | 1e-2 | | 1e-4 | | Proposed Method | |
|---|---|---|---|---|---|---|
| | N | GO terms | N | GO terms | N | GO terms |
| CRIS A | 1,160 | 9,756 (37, 0.4%) | 67 | 2,304 (15, 0.7%) | 76 | 3,482 (170, 4.9%) |
| CRIS B | 1,289 | 10,421 (74, 0.7%) | 81 | 2,067 (2, 0.1%) | 86 | 3,565 (132, 3.7%) |
| CRIS C | 1,224 | 10,113 (3, 0.03%) | 74 | 2,494 (7, 1.4%) | 88 | 3,838 (108, 2,8%) |
| CRIS D | 1,300 | 10,212 (97, 0.9%) | 92 | 3,029 (0) | 91 | 3,718 (112, 3.0%) |
| CRIS E | 1,241 | 10,223 (42, 0.4%) | 86 | 2,416 (0) | 89 | 3,593 (93, 2.6%) |

N: size of the selected feature subset.

GO terms: number of GO annotation terms retrieved (and the corresponding number and percentage of significantly enriched terms, with FDR p-value $< 0.05$).

## 4 Proposed Method

In this Section, we introduce our proposed approach to guide LASSO in the selection of features by also considering their biological relevance, estimated with our *Gene Information Score*, also described here below. Lastly, we discuss its results in terms of classification performance as well as enrichment analyses to assess the biological relevance and interpretability of the selected feature subsets.

**Table 8.** Additional enrichment analysis results for the BRCA dataset.

| Tolerance | Malacards CRC genes | KEGG CRC pathways | KEGG Cancer pathways | Malacards CRC GO terms |
|---|---|---|---|---|
| 1e-2 | 286 | 24 | 340 | 0 |
| 1e-4 | 28 | 1 | 28 | 0 |
| PM[a] | 41 | 4 | 47 | 2 |

Number of matches between the robust selected features and the Malacards and KEGG gene signatures and the Malacards CRC-associated GO terms.
[a] Proposed method to biologically guide LASSO.

### 4.1 Calculation of Gene Information Score

To compute an information score for each gene in the dataset, we took advantage of the GO annotations extracted as illustrated in Sect. 2.3. We built an annotation matrix with genes as rows, GO terms as columns, and binary values (1/0) to indicate whether a gene is annotated with a given GO term or not. Starting from this annotation matrix, we built a weighted annotation matrix as proposed in [16]. To this aim, for each GO term, we considered its depth and number of descendants in the GO DAG to compute the following measure of importance, named *structure-based Information Content* ($IC_{struct}$), as proposed in [17]:

$$IC_{struct}(t_j) = \frac{depth(t_j)}{max\_depth} \times \left(1 - \frac{(log(desc(t_j)) + 1)}{log(total\_terms)}\right) \tag{2}$$

where $depth(t_j)$ refers to the maximum depth of the term $t_j$ in the GO hierarchy, $max\_depth$ is the maximum depth of the GO DAG, and $desc(t_j)$ is the number of descendants of term $t_j$ in the hierarchy. Notice that for each term, the $IC_{struct}$ is here computed considering only the GO sub-ontology the term belongs to, and ignoring inter-sub-ontology relationships. Also, the term $max\_depth$ is computed with respect to each separate GO sub-ontology. Each entry in the weighted annotation matrix is obtained by replacing the one-values of the binary annotation matrix with the $IC_{struct}$ of the corresponding term. Based on such weighted annotation matrix $W$, we defined the *Gene Information Score* ($GIS$) as:

$$GIS(g_i) = \frac{\sum_{j:W_{g_i,t_j}>0} W_{g_i,t_j}}{\sum_{j:W_{g_i,t_j}>0} 1} \tag{3}$$

where $W_{g_i,t_j}$ is the weight for the term $t_j$ associated with the $i$-th gene $g_i$ in the matrix W. Through this formula, we obtain *GIS* values in the range [0,1], where a value of $GIS(g_i) = 0$ indicates that no prior biological information could be extracted for the gene $g_i$ in the considered knowledge bases. The higher the *GIS* for a given gene, the higher the number of terms associated with it.

### 4.2 Experiments and Results

Here, we present the findings obtained by applying our proposed approach to the two application use cases considered.

**Breast Cancer Subtyping.** We used the computed *GIS* to rank the subset of 3,018 robust features selected by the LASSO regularized Logistic regression model with a *tolerance* of 1e-2; a total of 2,986 of these features had a non-zero *GIS*. Based on the distribution of these non-zero *GIS* scores, the median was chosen and used as a threshold to select the highest-ranked features (1,493), and additional LASSO regularized Logistic regression models were trained using this set of features and 10 permutations of all training samples. The default *tolerance* (1e-4) and a $\lambda = 0.25$ were adopted and returned a robust feature set of size 224, comparable to the one (194) obtained from the entire feature space (17,294 genes) with the same *tolerance* value and stricter regularization (tuned $\lambda = 2$). We then performed the enrichment analyses described in Sect. 2.4, to evaluate the biological relevance of each subtype-specific feature subset. Results from this classification model on the 10 training sample permutations are reported in Table 3. Findings from subsequent enrichment analyses of the robust features (from intersection across the 10 permutations) are presented in Tables 4 and 5. Without degrading the performance on the validation and test sets, our proposal identified a new predictive set of features with an increased number of significant GO terms for each BRCA subtype, compared not only to the default case with the same *tolerance* value of 1e-4 but also to the case with *tolerance* of 1e-2. Our proposed approach was not able to increase the degree of overlap between the selected genes and the Malacards BRCA gene signature with respect to the LASSO regularized Logistic regression model with a *tolerance* of 1e-4. However, the number of selected genes overlapping with BRCA- and cancer-related KEGG pathway signatures is higher, as well as the number of GO statistically significant BRCA-associated terms according to MalaCards (10 vs. 7). Since we are proposing a two-step approach, we also evaluated how the final selection would change when using a smaller feature subset as input for the second step, i.e., how a lower *tolerance* threshold (1e-3) for the first step would affect the final selection in terms of performance and biological validation. Out of the 711 features selected for a *tolerance* of 1e-3 (see Table 3), 705 genes had non-zero *GIS*. As before, we selected the highest-ranked features in terms of *GIS* (352) as input to the classification model. We used again the default *tolerance* (1e-4) and $\lambda = 0.10$, obtaining a comparable feature set of 214 genes across 10 permutations. Results from this experiment can be found in the Appendix A.

**Colorectal Cancer Subtyping.** Based on the computed GIS, we ranked the subset of 4,296 robust features selected by the standard LASSO model with a *tolerance* of 1e-2; a total of 4,238 of these features had a non-zero *GIS*. We used the median to extract the highest-ranked features (2,119) and trained additional LASSO regularized Logistic regression models using this set of features, performing 10 permutations of all training samples. As before, the default *tolerance* (1e-4) and a $\lambda = 0.25$ were used. The robust feature set returned had a size of 389 genes, which is on par with the one obtained from the entire feature space (17,559 genes) with the same *tolerance* value and stricter regularization (tuned $\lambda = 2$). We then performed the enrichment analyses described in Sect. 2.4. Results of this classification model on the 10 training sample permutations are reported in Table 6, whereas findings from enrichment analyses performed are presented in Tables 7 and 8. Our proposed approach reached a comparable performance in cross-validation; however, the classification accuracy on the final test set was slightly lower (4% less), both compared to the larger feature sets (higher *tolerance*) and smaller ones (lower *tolerance*). Despite a small performance degradation, we were able to increase the number of significant GO terms, as well as the number of matches with CRC-related GO terms retrieved from Malacards. Moreover, the degree of overlap with cancer- and CRC-associated gene signatures extracted from KEGG and Malacards also increased with respect to the feature set obtained for the LASSO regularized Logistic regression model with a *tolerance* of 1e-4. As for the BRCA dataset, we conducted additional experiments to test how the initial *tolerance* value would impact the performance and biological validation. From the 1,075 features selected with a *tolerance* of 1e-3 (see Table 6), 1,057 had non-zero GIS. We chose the top 352 features based on GIS for classification, using the median as threshold. We trained the classifier with the default *tolerance* (1e-4) and $\lambda = 0.125$, obtaining a similar set of 376 genes across 10 permutations. The detailed results are provided in the Appendix A.

## 5 Conclusions

Our exploratory analysis suggests that the default LASSO regularized Logistic regression is effective in identifying a small subset of features with high predictive power. Yet, the enrichment analyses conducted highlight that these subsets often retain little biological information. We proposed a first approach to guide the LASSO feature selection in identifying a pool of predictive features with the highest biological relevance, in terms of Gene Information Score, while retaining high prediction performance. Our goal is to find a trade-off between the predictive power of each gene, with respect to the prediction task under study, and its relevance in terms of biological interpretability. We evaluated our proposal on two different application use cases concerning the prediction of cancer-related subtypes of patients based on gene expression data. Our proposed approach achieved similar prediction performance on the final test sets, with just a slightly lower classification accuracy on the Colorectal Cancer dataset, compared to the model with standard LASSO regularization and default *tolerance* value (1e-4), while simultaneously selecting features that can provide improved biological interpretability. It should be noted that our approach does not alter the LASSO optimization algorithm, preserving the model's convexity. Instead, we stop it earlier to extract a broader set of predictive features, which are then refined using a biological knowledge-based score before being fed to a second standard LASSO model. We believe the accuracy of our predictions might be further improved by developing an integrated approach that directly incorporates each gene's biological relevance, in terms of *Gene Information Score*, in the optimization process of the LASSO feature selection. Furthermore, additional sources of biological knowledge (e.g., KEGG or Reactome pathways) could be considered in the calculation of the Gene Information Score to define a more comprehensive estimate of each gene's known biological relevance.

**Funding Information.** This work was supported by the MUSA – Multilayered Urban Sustainability Action – project, funded by the European Union – NextGenerationEU, under the National Recovery and Resilience Plan (NRRP) Mission 4 Component 2 Investment Line 1.5: Strengthening of research structures and creation of R&D "innovation ecosystems", set up of "territorial leaders in R&D".

**Conflict of interests.** The authors declare to have no conflicts of interest.

# A Additional Experiments

Given that we are proposing a two-step approach, we also assessed how the final selection is influenced by using a smaller feature subset as input for the second step. Specifically, we examined the impact of applying a lower *tolerance* threshold (1e-3) in the first step on the final selection, concerning both the classification performance and biological validation. The results from the additional experiments we conducted are reported in Tables 9, 10, and 11 for the BRCA dataset and in Tables 12, 13, and 14 for the CRC dataset. In terms of classification performance, we obtained slightly lower accuracy values when we reduced the size of the feature set used as input to the second step (*tolerance* of 1e-3), as reported in Tables 9 and 12. Moreover, using a lower *tolerance* appeared to make the final classifier model more prone to overfitting, as demonstrated by the higher CV accuracy and lower test accuracy values on both datasets. Results from the enrichment analyses (Sect. 2.4), instead, appeared to be comparable, returning similar degrees of overlap between the selected genes and the Malacards and KEGG signatures for both datasets. Overall, both *tolerance* thresholds returned similar results, but using a higher value appeared to improve generalization.

**Table 9.** Classification performances and number of features selected as **Union**, **Intersection**, or **Not robust** across 10 permutations on the BRCA dataset.

| Tolerance[a] | CV[b] accuracy | Test accuracy | Union | Intersection | Not robust[c] |
|---|---|---|---|---|---|
| 1e-2 | 0.90800 (± 0.0) | 0.93069 (± 0.0) | 225 | 224 | 1 |
| 1e-3 | 0.93039 (± 0.00017) | 0.92574 (± 0.0) | 217 | 214 | 3 |

[a] *Tolerance* value for the first step of the proposed method.
[b] Cross-validation (CV).
[c] Genes selected as features not in all 10 permutations.

**Table 10.** Enrichment analysis results for the BRCA dataset.

| Tolerance[a] Subtype | 1e-2 | | 1e-3 | |
|---|---|---|---|---|
| | N | GO terms | N | GO terms |
| Basal | 23 | 1,791 (365, 20.4%) | 28 | 2,017 (515, 25.5 %) |
| Her2 | 47 | 2,879 (488, 17.0%) | 55 | 3,181 (450, 14.1%) |
| LumA | 83 | 3,471 (365, 10.5%) | 104 | 3,823 (429, 11.2%) |
| LumB | 93 | 3,762 (162, 4.3%) | 90 | 3,974 (215, 5.4%) |

[a] *Tolerance* value for the first step of the proposed method.
N: size of the selected feature subset.
GO terms: number of GO annotation terms retrieved (and the corresponding number and percentage of significantly enriched terms, with FDR p-value $< 0.05$).

**Table 11.** Additional enrichment analysis results for the BRCA dataset.

| Tolerance[a] | Malacards BRCA genes | KEGG BRCA pathways | KEGG Cancer pathways | Malacards BRCA GO terms |
|---|---|---|---|---|
| 1e-2 | 34 | 9 | 27 | 10 |
| 1e-3 | 42 | 11 | 32 | 10 |

Number of matches between the robust selected features and the Malacards and KEGG gene signatures and the Malacards BRCA-associated GO terms.
[a] *Tolerance* value for the first step of the proposed method.

**Table 12.** Classification performances and number of features selected as **Union**, **Intersection**, or **Not robust** across 10 permutations on the CRC dataset.

| *Tolerance*[a] | *CV*[b] accuracy | Test accuracy | Union | Intersection | *Notrobust*[c] |
|---|---|---|---|---|---|
| 1e-2 | 0.75804 (± 0.0) | 0.79032 (± 0.0) | 390 | 389 | 1 |
| 1e-3 | 0.81066 (± 0.00098) | 0.76452 (± 0.00322) | 383 | 376 | 6 |

[a] *Tolerance* value for the first step of the proposed method.
[b] Cross-validation (CV).
[c] Genes selected as features not in all 10 permutations.

**Table 13.** Enrichment analysis results for the CRC dataset.

| Tolerance[a] Subtype | 1e-2 | | 1e-3 | |
|---|---|---|---|---|
| | N | GO terms | N | GO terms |
| CRIS A | 76 | 3,482 (170, 4.9%) | 88 | 3,489 (190, 5.4%) |
| CRIS B | 86 | 3,565 (132, 3.7%) | 105 | 3,829 (37, 0.97%) |
| CRIS C | 88 | 3,838 (108, 2.8%) | 110 | 4,170 (189, 4.5%) |
| CRIS D | 91 | 3,718 (112, 3.2%) | 110 | 4,094 (132, 3.2%) |
| CRIS E | 89 | 3,593 (93, 2.6%) | 101 | 3,808 (32, 0.84%) |

[a] *Tolerance* value for the first step of the proposed method.
N: size of the selected feature subset.
GO terms: number of GO annotation terms retrieved (and the corresponding number and percentage of significantly enriched terms, with FDR p-value $< 0.05$).

**Table 14.** Additional enrichment analysis results for the CRC dataset.

| Tolerance[a] | Malacards CRC genes | KEGG CRC pathways | KEGG Cancer pathways | Malacards CRC GO terms |
|---|---|---|---|---|
| 1e-2 | 41 | 4 | 47 | 2 |
| 1e-3 | 41 | 3 | 48 | 2 |

Number of matches between the robust selected features and the Malacards and KEGG gene signatures and the Malacards CRC-associated GO terms.
[a] *Tolerance* value for the first step of the proposed method.

## References


1. Ding, C., Peng, H.: Minimum redundancy feature selection from microarray gene expression data. J. Bioinform. Comput. Biol. **3**(2), 185–205 (2005)
2. Bommert, A., Welchowski, T., Schmid, M., Rahnenführer, J.: Benchmark of filter methods for feature selection in high-dimensional gene expression survival data. Briefings Bioinformatics 23(1), bbab354 (2022)
3. Cascianelli, S., Galzerano, A., Masseroli, M.: Supervised Relevance-Redundancy assessments for feature selection in omics-based classification scenarios. J. Biomed. Inform. **144**, 104457 (2023)
4. Tibshirani, R.: Regression shrinkage and selection via the Lasso. J. R. Stat. Soc. Ser. B-Methodol. **58**, 267–288 (1996)
5. Weinstein, J.N., Collisson, E.A., Mills, G.B., et al.: The cancer genome atlas pan-cancer analysis project. Nat. Genet. **45**(10), 1113–1120 (2013)
6. Ashburner, M., Ball, C.A., Blake, J.A., Botstein, D., Butler, H., Cherry, J.M., et al.: Gene ontology: tool for the unification of biology. Gene Ontol. Consort. Nat. Gen. **25**(1), 25–29 (2000)
7. Network, C., et al.: Comprehensive molecular portraits of human breast tumours. Nature **490**(7418), 61–70 (2012)
8. Pallotta, S., Cascianelli, S., Masseroli, M.: RGMQL: scalable and interoperable computing of heterogeneous omics big data and metadata in R/Bioconductor. BMC Bioinformatics **23**(1), 123 (2022)

9. O'Leary, N.A., Wright, M.W., Brister, J.R., Ciufo, S., Haddad, D., McVeigh, R., et al.: Reference sequence (RefSeq) database at NCBI: current status, taxonomic expansion, and functional annotation. Nucl. Acids Res. **44**(D1), D733–D745 (2016)
10. Parker, J.S., Mullins, M., Cheang, M.C., Leung, S., Voduc, D., Vickery, T., et al.: Supervised risk predictor of breast cancer based on intrinsic subtypes. J. Clin. Oncol. **27**(8), 1160–1167 (2009)
11. Cascianelli, S., Barbera, C., Ulla, A.A., Grassi, E., Lupo, B., Pasini, D., et al.: Multi-label transcriptional classification of colorectal cancer reflects tumor cell population heterogeneity. Genome Med. **15**(1), 37 (2023)
12. The Gene Ontology Consortium: Creating the gene ontology resource: design and implementation. Genome Res. **11**, 1425–1433 (2001)
13. Upton, G.: Fisher's exact test. J. R. Stat. Soc. Ser. A Stat. Soc. **155**(3), 395–402 (2018)
14. Rappaport, N., Twik, M., Plaschkes, I., Nudel, R., Iny Stein, T., Levitt, J., et al.: MalaCards: an amalgamated human disease compendium with diverse clinical and genetic annotation and structured search. Nucl. Acids Res. **45**(D1), D877–D887 (2016)
15. Kanehisa, M., Goto, S.: KEGG: kyoto encyclopedia of genes and genomes. Nucl. Acids Res. **28**(1), 27–30 (2000)
16. Acharya, S., Saha, S., Nandyala, N.: Unsupervised gene selection using biological knowledge: application in sample clustering. BMC Bioinformatics **18** (2017)
17. Teng, Z., Guo, M., Liu, X., Dai, Q., Wang, C., Xuan, P.: Measuring gene functional similarity based on group-wise comparison of GO terms. Bioinformatics (Oxford, England) **29**, 1424–1432 (2013)

# Extraction of Attributes from Electrodermal Activity Signals Applying Time Series Fuzzy Granulation for Classification of Academic Stress Perception in Different Scenarios

Sandra E. Barajas-Montiel(✉) and Carlos A. Reyes-García

Computer Science Department, Instituto Nacional de Astrofísica, Óptica y Electrónica, Luis Enrique Erro No. 1, Sta. María Tonantzintla, Puebla, Mexico
{sandybarajas,kargaxxi}@inaoep.mx

**Abstract.** Fuzzy granulation is a technique that allows us to represent complex data in a more interpretable and simple way. Granules provide a higher level of abstraction by capturing the essence of the data within specific intervals or ranges. In this work we propose Electrodermal activity (EDA) feature extraction through fuzzy granulation to classify four different scenarios of academic stress. EDA is a physiological signal controlled by the sympathetic nervous system, for this reason it is considered an important component of animical states as stress. According to the research carried out, the detection of stress using EDA is based on the use of morphological attributes or statistical characteristics obtained both in the frequency domain and in time domain. As a time series EDA data often exhibit complex patterns, traditional feature extraction methods may not fully capture these variations, and fuzzy granulation offers an alternative. We have experimented with four different techniques of fuzzy granulation, our results allow us to verify that fuzzy granulation is a useful approach to characterize EDA signals to adequately distinguish the effects of listening to music while performing a stressful task reaching classification accuracy up to 98.9%.

**Keywords:** Fuzzy granulation · Time series · Electrodermal activity · Academic Stress · Music

## 1 Introduction

Long periods of stress or being under this state continuously can cause serious damage to health, diminish academic performance and provoke maladaptive behaviors [1]. This motivates the development of a solution that allows detecting different levels of academic stress in order to provide tools that allow determining

This work was supported by Conahcyt (grant number: 49245).


M. Vettoretti et al. (Eds.): CIBB 2023, LNBI 14513, pp. 308–320, 2025.
https://doi.org/10.1007/978-3-031-90714-2_22

strategies to mitigate its effects. In this work we focus on the automatic classification of biosignals to identify stress, observing if listening to determined kind of music has effects on stress levels as a palliative for this noxious state. Different studies have observed the effects of music on human stress, sometimes with contradictory results, for example in [2] the authors point out an increase in cortisol levels, the stress hormone, while [3] mentions stress reduction in physiological and psychological outcomes as an effect of listening to music. Therefore, we propose the use of biosignals processing for the detection of stress and changes in this state induced by listening to music.

Among the different physiological signals used for automatic stress detection electrodermal activity (EDA) is one of the most widely used biosignals. EDA is directly related to the activation of the sudomotor system controlled by the sympathetic nervous system. In this way, the activation of EDA is associated with an increase in stress levels. Most of the research carried out to automatically detect and classify stress by analyzing EDA signals extracts statistical and morphological features in time and frequency domain [4,5,10,11].

We propose EDA time series fuzzy granulation methodology to extract features for classification of academic stress with the objective to observe and analyze changes in this state induced by listening to music. Listening to music while performing stress-inducing academic activities may act as a distractor and thus alter the perception of this unhealthy state, possibly reducing its effects. EDA is a signal that can be recorded using noninvasive and easy-to-set-up devices, it can detect subtle changes in the emotional state, the recording of these signals during a stressful academic task while listening to music and its automatic classification can provide a deeper and more accurate understanding of how different types of music affect stress levels and thus support the development of personalized interventions for academic stress management.

The use of fuzzy granules is proposed as a novel technique for the extraction of biosignal features that can facilitate their operation, interpretation and improve the accuracy of the classification of emotional states, in addition to establishing an initial approach to later apply this technique to other types of physiological data. In this work, our initial focus is to verify that fuzzy granules allow to effectively classify electrodermal activity signals corresponding to different academic stress scenarios and thus observe that listening to music induces changes in this state perceptible in a biosignal that is easy to record.

To collect the data base we instructed 16 subjects to perform a cognitive task to induce stress in four different scenarios: silence, listening to relaxing music, listening to the preferred music, and listening to hated music. Our results show that listening to music induces effects on stress levels, and that the application of fuzzy granulation methodology is effective to extract characteristics to discriminate between different stressful states with an accuracy classification up to 98.9%.

## 2 Related Work

Electrodermal activity is useful for detecting stress since it is a response of the body that cannot be controlled. The sympathetic nervous system, part of the autonomic nervous system, has among its functions the activation of the body's responses to stressful stimuli, such as sweating. The variation in the activity of the sweat glands causes a change in the conductivity of the skin that can be recorded by a noninvasive EDA sensor. In [4], the authors present a survey where we can find different research papers that use EDA as one of the biosignals measured for stress detection. In general, the papers listed in this survey make a binary classification between the state of rest and stress using statistical characteristics. Also, in [4] there are works with different sample sizes, relatively small, using biosignals from only two participants up to a maximum of 100.

Works such as [10–13] generally extract statistical features in time domain, frequency band characteristics, and morphological properties that quantify the shape of the signal from its two main components: the Skin Conduction Response (SCR), that measures the transient changes in skin conductance that occur in response to stimuli, also called phasic component; and the Skin Conduction Level (SCL), that represents the baseline level of skin conductance, also called tonic component. There are few works in which other methodologies are used to obtain attributes from EDA such as [14] where a convolutional network is used to extract new features using deep learning. Table 1 lists some of the works related to stress classification and the use of EDA that we have reviewed up to the time this work was carried out.

## 3 Data and Methods

To observe the effects of music in academic stress, we collected data from 16 subjects ranging from 16 to 18 years old using `NeXus-10` for measuring EDA, also referred as Galvanic Skin Response (GSR) or Skin Conductance (SC). These signals were processed through a fuzzy granulation algorithm for feature extraction, and then the resulting feature vectors were classified using Random Forest (RF) machine learning algorithm. The different stages of this project are described below.

### 3.1 Data Acquisition

The database collected consists of EDA signals from 16 healthy subjects, 8 females and 8 males, all of which signed an informed consent agreement. One male subject is left-handed and the rest are right-handed. To acquiring EDA signals a `NeXus-10` device was used. The skin conductance sensor consists of two electrodes used to read a single channel signal with a frequency sample rate of 32 Hz. The experiment to carry out the sampling begins with the placement of the EDA electrodes, one on the index finger and the other one on the middle finger of the non-dominant hand. The reading of the signal is verified and the participant is asked to sit in a comfortable position in front of the computer to perform a cognitive task.

Participants are asked to close their eyes when the recording of the signal begins, this stage lasts 30 s. Then the recording is divided into 3 recording blocks: Simple additions (180 s), simple multiplications (180 s), compound operations (90 s). In all the blocks several different operations are shown, each with a time limit of 5 s to be answered, if they are answered incorrectly the user is given a negative feedback.

The number of operations performed depends on the speed with which the participant solves the arithmetic operations presented in each block, on average, the participants answer 70 additions in the first block, 40 multiplications in the second block, and 10 operations in the third block. Between blocks participants has approximately 30 s to rest. At the end of the three blocks of operations, the participants are asked to close their eyes again and remain at rest for 30 s. Here the recording of the EDA signal ends, having an approximate total duration of 9.5 min.

Four sessions on different days were recorded for each subject: the participants carried out the task in silence, listening to relaxing music, listening to the preferred music, and listening to hated music. For the relaxing session Pachelbel's Canon in D major was used, which according to [3] is effective in reducing stress effects. In both the preferred music and the hated music sessions, each participant chose the music to listen to. As preferred music, the participants were asked to choose pieces of music that they considered enjoyable or attractive. As hated music, participants were asked to choose pieces of music that make them feel upset or uncomfortable.

In Fig. 1 we can see an illustration of each session of the recording protocol. We based the design of our sampling protocol in the work described in [15] where it is explained that protocols of this nature have proven to be effective stress triggers taking into account not causing serious problems in the participants.

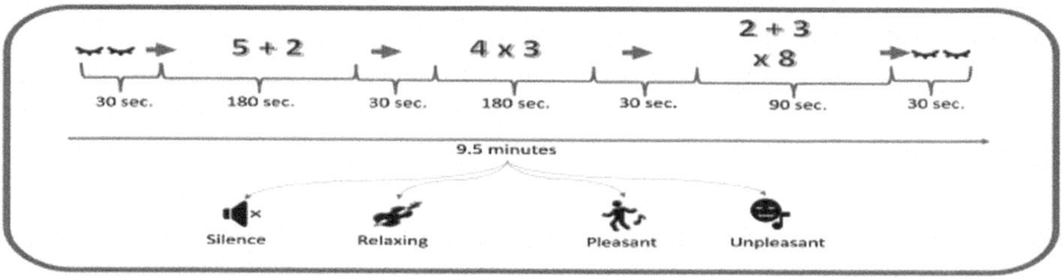


**Fig. 1.** Simple description of the designed protocol for EDA signal acquisition.

### 3.2 EDA Signal Preprocessing

In the preprocessing stage the electrodermal signal is broken down into its two main components: the tonic component that varies slowly and constantly changes

**Table 1.** Summary of related work.

| Reference | Classification Task | Number of subjects | Extracted Features | Model | Accuracy | F1 score |
|---|---|---|---|---|---|---|
| Ninh et al. [5] | Stress - No-Stress | 15 | SCL and SCR Peaks, SCR Onsets, SCR Amplitud | SVM | 90.5% | - |
| Ninh et al. [5] | Stress - No-Stress | 10 | SCL and SCR Peaks, SCR Onsets, SCR Amplitud | SVM | 81.1% | - |
| Ninh et al. [5] | Stress - No-Stress | 7 | SCL and SCR Peaks, SCR Onsets, SCR Amplitud | SVM | 92.8% | - |
| Hosseini et al. [6] | Stress - No-Stress | 15 | 87 Statistical and Morphological Characteristics. | Exhaustive and iterative evaluation of features + AdaBoost | 97.0% | - |
| Jaber et al. [7] | Stress - No Stress | 15 | EDA Mean, Median, Mode, Maximum, Minimum, and Standard Deviation SCR Mean duration, SCR Standard Deviation SCL Mean | Random Forest | - | 78.0% |
| Anusha et al. [8] | 3 stress levels | 41 | SCR Amplitud SCR Onset Latency Peak time Recovery time Derivatives of the EDA. Haar Wavelet Transform | Quadratic Discriminant Analysis | 85.0% | - |
| Campanella et al. [9] | Stressful and non-stressful tasks. | 29 | EDA Mean, Median, Mode, Maximum, Minimum, and Standard deviation, SCR Mean duration, SCR Mean Amplitude, SCR Average Amplitude, Mean SCR Rise Time, Number of Peaks | Random Forest | 76.5% | - |

within a subject, according to its hydration, dryness of the skin or autonomous regulation; and the phasic component, which exhibits noticeably more rapid changes, is responsive to particular emotionally stimulating events.

The phasic component is the one that is of interest for stress classification. A basic median filter removes the tonic component of the signal and gives us the phasic signal as a result Fig. 2.

### 3.3 Feature Extraction by Fuzzy Granulation

The extraction of characteristics in biosignals leads to the generation of high-dimensional vectors that make classification models very complex, requiring considerable computational resources for the practical use of these models. Unlike most of the works found in our research, we propose the extraction of features through fuzzy granulation as it can be an effective approach to reduce the dimensionality of signals for stress classification. Fuzzy granulation provides an analogy to how human reasoning tends to describe entities in terms without precise

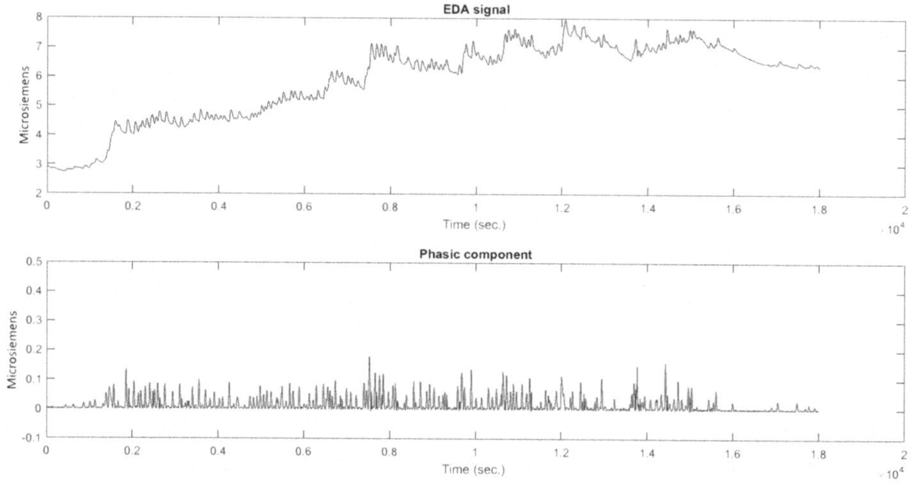


**Fig. 2.** EDA signal preprocessing. Extraction of phasic component.

boundaries like high, medium, low; therefore, it could be applied in order to discover a family of information granules to characterize time series.

Our feature extraction strategy is inspired in the methodology presented in [16]. We applied a fuzzy granulation process consisting of 3 stages: The time series is segmented into windows of the same size; then each segmented window is ordered and fuzzy granules are extracted; finally, the granules are associated with a linguistic term. To segment the signal, each sample is divided into 1 (1 s W), 3 (3 s W), and 5-s (5 s W) windows with no overlap (Fig. 3). This window size was chosen because the triggering of the phasic response occurs between 1 and 5 s after receiving an emotional stimulus.

From each EDA signal, 3 granules were extracted for each of the segmented windows: low, medium and high. Each segmented window $W = \{w_1, ..., w_n\}$ is sorted resulting in the ordered time series $W' = \{w'_1, ..., w'_n\}$ then it is divided in two subseries, the first half $W'_l = \{w'_1, ..., w'_{n/2}\}$ and the second half $W'_h = \{w'_{(n/2)+1}, ..., w'_n\}$.

First, we extracted the granules using the Triangular (Tri), Trapezoidal (Tpz) membership functions, and MinMax (mM) operator. Granules *low* are extracted form $W'_l$.

$$Tri \quad low = \frac{2\sum_{j=1}^{n/2} w'_j}{n/2} - median\{w'_1, ..., w'_n\} \tag{1}$$

$$Tpz \quad low = \frac{2\sum_{j=1}^{n/2} w'_j}{n/2} - w'_{n/2} \tag{2}$$

$$mM \quad low = w'_1 \tag{3}$$

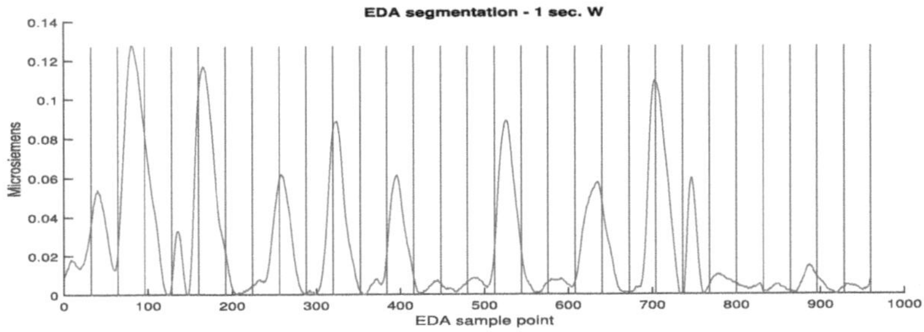


**Fig. 3.** Illustration of segmentation of one-second windows in one-minute EDA signal.

The median of the ordered time series $W'$ is calculated to represent the *mid* granule in all cases.

The granules *high* are extracted from $W'_h$.

$$Tri \quad high = \frac{2\sum_{j=n/2+1}^{n} w'_j}{n/2} - median\{w'_1, ..., w'_n\} \tag{4}$$

$$Tpz \quad high = \frac{2\sum_{j=n/2+1}^{n} w'_j}{n/2} - w'_{n/2+1} \tag{5}$$

$$mM \quad high = w'_n \tag{6}$$

The granulation process is illustrated in Fig. 4. Each of the granules granulated using Tri, Tpz membership functions and mM operator generate a different set of characteristics.

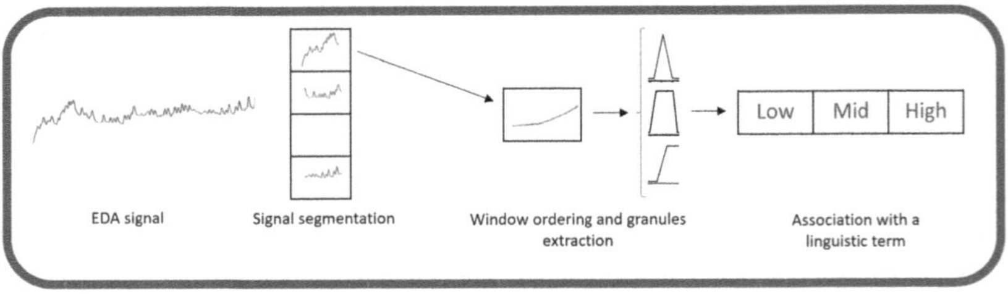


**Fig. 4.** Illustration of the three stages of fuzzy granulation applied to a time series..

Then, we applied fuzzy clustering to extract granules. Fuzzy clustering is an unsupervised learning method that assigns data points to groups with different membership degrees, each object having either high or low membership to a group. To extract granules by fuzzy clustering, we calculated three clusters in each window of the signal applying Fuzzy C-Means (FCM) algorithm [17]. As the relevant information of the clusters is represented by their centroids, we take them and order them, having the smallest granule as *low*, the middle granule as *mid* and the largest as *high*. This process is illustrated in Fig. 5.

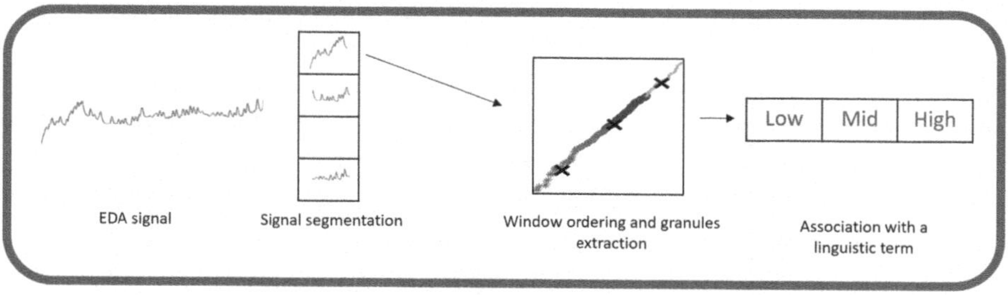


**Fig. 5.** Fuzzy C-Means granulation. Granule extraction using fuzzy clustering.

## 4 Experiments and Results

Granules extracted from recorded EDA signals were used to build a model to classify the four different stress scenarios described in the protocol. These four scenarios define the balanced classes: 1,2,3,4, for stress induced in silence, stress induced while listening to relaxing music, stress induced while listening to the preferred music and stress induced while listening to the hated music, respectively. From the recorded signals, each lasting approximately 9.5 min, we have only chosen to work with one minute of the signal, from minute 5.5 to minute 6.5, given that, according to our observations, it is the period of time in which the participants are more immersed in the arithmetic stressing task.

According to the proposed stress induction protocol and the sampling performed, 64 signals are obtained corresponding to the 16 subjects exposed to the 4 scenarios described. Each of these signals is used to extract different sets of features. Since GSR spikes occur between 1 and 5 s after the onset of a stimulus, the selected minute of signal is granulated into 1 s W, 3 s W and 5 s W. Each of the 4 classes has a total of 60 1-s window samples, 20 3-s window samples, and 12 5-s window samples. For each window of the signal, we compute granules low, mid and high, these granules are the attributes with which each object is characterized.

Experiments were performed modeling independent classifiers for each subject using Random Forest. As Physiological responses are intrinsically subject-dependent, separate models were generated for each participant in order to determine if our model manages to distinguish between the different stress scenarios, indicating that music induces changes in stress levels. Each model is trained with samples from four signals, one for each scenario. The experiments corresponding to the features extracted in one-second windows train each model with 240 samples, those corresponding to 3-s windows train each model with 80 samples and those corresponding to 5-s windows each model is trained with 48 samples. According to the reviewed literature and publicly available databases for

**Table 2.** Percentage of accuracy in the classification. Samples have three characteristics: granules low, medium, and high extracted using different window (W) lengths and the Triangular (Tri), Trapezoidal (Tpz) membership functions, and MinMax (mM) operator.

| Subj | 1 sec. W | | | 3 sec. W | | | 5 sec. W | | |
|---|---|---|---|---|---|---|---|---|---|
| | Tri | Trp | mM | Tri | Trp | mM | Tri | Trp | mM |
| 1 | 80.4% | 80.8% | 77.5% | 85.0% | 85.0% | 90.0% | 76.3% | 82.5% | 88.8% |
| 2 | 88.3% | 88.3% | 92.1% | 95.0% | 92.5% | 98.8% | 93.8% | 90.0% | 88.8% |
| 3 | 75.0% | 75.0% | 75.0% | 75.0% | 75.0% | 75.0% | 75.0% | 75.0% | 75.0% |
| 4 | 98.8% | 98.6% | 98.8% | 96.3% | 98.8% | 98.8% | 98.8% | 98.8% | 98.8% |
| 5 | 98.8% | 98.8% | 98.8% | 98.8% | 98.8% | 98.8% | 98.8% | 98.8% | 98.8% |
| 6 | 72.1% | 72.5% | 68.3% | 55.0% | 71.3% | 72.5% | 68.8% | 63.8% | 72.5% |
| 7 | 88.3% | 90.4% | 91.3% | 88.8% | 91.3% | 93.8% | 96.3% | 93.8% | 93.8% |
| 8 | 67.9% | 74.6% | 69.2% | 61.3% | 58.8% | 60.0% | 75.0% | 70.0% | 68.8% |
| 9 | 84.2% | 79.6% | 84.6% | 80.0% | 71.3% | 80.0% | 82.5% | 90.0% | 93.8% |
| 10 | 70.4% | 73.8% | 64.2% | 63.8% | 68.8% | 77.5% | 66.3% | 70.0% | 66.3% |
| 11 | 72.9% | 77.1% | 75.0% | 83.8% | 77.5% | 80.0% | 81.3% | 81.3% | 72.5% |
| 12 | 98.3% | 98.8% | 98.3% | 96.3% | 98.8% | 98.8% | 98.9% | 98.9% | 98.8% |
| 13 | 65.8% | 65.8% | 63.8% | 66.3% | 65.0% | 56.3% | 61.3% | 57.5% | 66.3% |
| 14 | 93.3% | 95.0% | 90.0% | 90.0% | 92.5% | 91.3% | 93.8% | 91.3% | 90.0% |
| 15 | **98.9%** | **98.9%** | **98.9%** | **98.8%** | **98.9%** | **98.9%** | **98.8%** | **98.9%** | **98.8%** |
| 16 | 77.9% | 76.7% | 80.8% | 85.0% | 88.8% | 81.3% | 72.5% | 73.8% | 83.8% |
| Avg | 83.2% | **84.1%** | 82.9% | 82.4% | 83.3% | **84.5%** | 83.6% | 83.4% | **84.7%** |

stress classification [4–15,18], one recording per participant/scenario with different trails is used to identify the presence of stress and for automatic stress classification, this is the approach we have used for our work.

To assess the performance of the generated model, leave-one-out cross-validation was used, as it provides a much less biased measure of a model's performance by fitting $n$ models with a set of $n$-$1$ samples and therefore tends not to overestimate the performance of the model. Experiments were made with 50, 100, 150 and 200 trees, there was no significant difference in the results when increasing the number of trees, so it was decided to model the classifier with only 50 trees. For the granules extracted in 1-s windows, the membership function that gave the best results for the classification of EDA signals into classes 1,2,3, and 4 described above, was the Trp function. Average accuracy is higher when extracting features in 5 s W. For 5 s W average accuracy is close to the one obtained with 3 s W, and the higher accuracy in classification is obtained when using mM operator to granulate features from EDA signal. The best results reached 98.9% accuracy in the classification and correspond to subject number 15. Table 2 shows classification accuracy percentage obtained using granules extracted by Triangular, Trapezoidal and MinMax operator.

**Table 3.** Percentage of accuracy in the classification using granules extracted by FCM.

| Subj | 1 sec. W | 3 sec. W | 5 sec. W | Subj | 1 sec. W | 3 sec. W | 5 sec. W |
|---|---|---|---|---|---|---|---|
| 1 | 83.5% | 88.1% | 88.5% | 9 | 80.8% | 82.8% | 81.5% |
| 2 | 89.3% | 92.6% | 93.4% | 10 | 68.4% | 72.0% | 68.2% |
| 3 | 61.4% | 71.5% | 69.7% | 11 | 75.4% | 74.2% | 78.6% |
| 4 | <u>**98.5%**</u> | <u>**98.7%**</u> | <u>**98.7%**</u> | 12 | 98.8% | 97.8% | 98.4% |
| 5 | 96.6% | 98.7% | 98.4% | 13 | 63.1% | 61.5% | 61.7% |
| 6 | 70.1% | 67.1% | 65.1% | 14 | 80.8% | 84.5% | 88.5% |
| 7 | 87.6% | 89.6% | 91.4% | 15 | 98.5% | 98.4% | 98.4% |
| 8 | 72.5% | 70.0% | 65.8% | 16 | 79.1% | 82.8% | 75.7% |
| | | | **1 s W** | **3 s W** | **5 s W** | | |
| | | **Avg** | 82.4% | **84.5%** | 83.7% | | |

The same experiments were performed with the granules extracted using FCM. As Table 3 shows, the best classification result was obtained with 3 and 5 s W for subject 4. The classification accuracy average percentage indicates that using 3 s W is the best way to distinguish between the four stress scenarios when extracting granules with FCM.

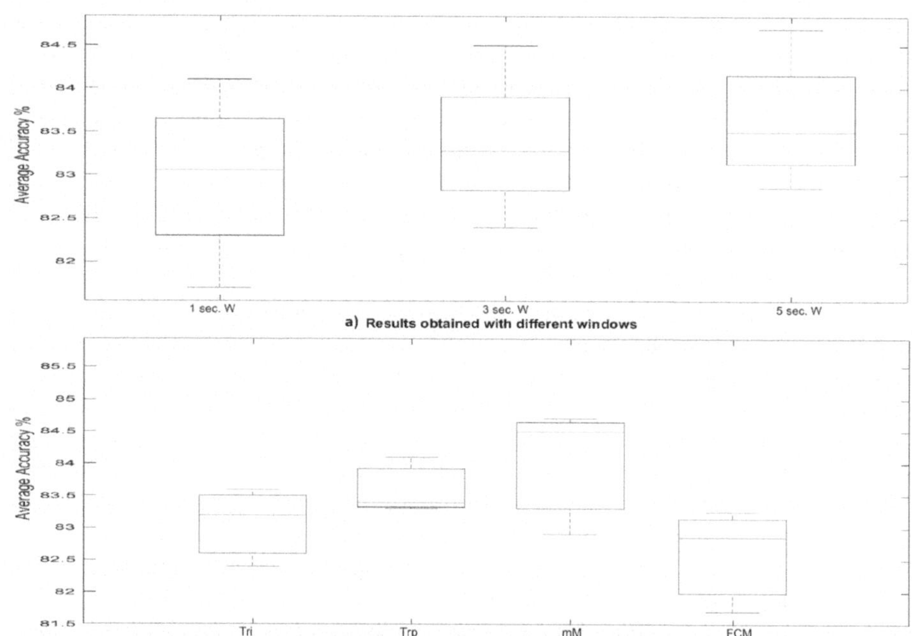


**Fig. 6.** Results comparison. Graph a) results with different window lengths. Graph b) results with different granulation techniques.

When comparing the results obtained in all the experiments performed, it is observed that in general the segmentation with 5 s W is the one that on average produces the best results, see Fig. 6 a). Regarding the granulation techniques, the mM operator generated the granules that allow to reach a better average in the classification accuracy of four stress scenarios, see Fig. 6 b).

## 5 Conclusions

To prevent the negative effects on health that stress causes, it is necessary to build tools that allow detecting its presence in early stages and propose some tools that provide palliative effects to this harmful state. The recording of biosignals and their automatic classification provide an approach for objective and opportune detection of stress. In our work we use the Electrodermal Activity signal to show its utility to classify four different stress scenarios in order to observe the effects of listening to music while performing a cognitive task.

According to our research, characteristics extraction techniques from EDA signals generally correspond to statistical measurements. In the present work we propose fuzzy granulation of time series to extract features from EDA signals. We experimented extracting fuzzy granules with four different techniques: applying membership Tri, Trp, mM operator and fuzzy clustering using FCM. We classified EDA signals from 16 subjects into four classes corresponding to performing a cognitive task in complete silence, while listening to relaxing music, while listening to pleasant music, and while listening to unpleasant music. In some subjects very similar classification results are observed despite using different window sizes, as well as different membership functions and operators for granulation. Based on our findings, these results are due to the fact that the extracted granules have very similar values among the different windows and membership functions. Fuzzy granulation is used to group data into fuzzy sets and therefore the choice of membership function for some subjects does not significantly affect the values of fuzzy granules.

The best average accuracy in classification was achieved with 5 s W segmentation and extracting granules by mM operator. Our results show that by using three fuzzy granules as characteristics of EDA signals recorded during a cognitive task, it is possible to clearly identify changes in the perception of stress when listening to music with an accuracy up to 98.9%. Our results show that using only three fuzzy granules as features of EDA signals allows us to identify changes in stress perception when performing a mathematical task while listening to music. The accuracy of the classification achieved allows us to say that the proposed method is effective and could be applied in various practical areas, such as the design of personalized interventions for stress management. This may open new lines of research and improve our understanding of how different stimuli, such as music, affect our emotional and cognitive responses. At this stage of our work our main contribution consists of feature extraction through fuzzy granulation, which is a novel approach for biosignal processing and analysis and stress classification to observe changes in its perception due to listening to music.

Furthermore, this approach can lead to establishing interpretable classification models, which is planned to be developed as future work.

## References

1. Vermunt, R., Steensma, H.: How can justice be used to manage stress in organizations? In: Handbook of Organizational Justice (pp. 383–410). Leiden University, The Netherlands (2005)
2. Suda, M., et al.: Emotional responses to music: towards scientific perspectives on music therapy. **19**, 1 (2008). https://doi.org/10.1097/WNR.0B013E3282F3476F
3. de Witte, M., et al.: Effects of music interventions on stress-related outcomes: a systematic review and two meta-analyses. **14**, 2 (2020). https://doi.org/10.1080/17437199.2019.1627897
4. Panicker, S.S., Gayathri, P.: A survey of machine learning techniques in physiology based mental stress detection systems. **39**, 2 (2019). https://doi.org/10.1016/J.BBE.2019.01.004
5. Ninh, V.-T., et al.: Analyzing the Performance of Stress Detection Models on Consumer-Grade Wearable Devices (2022). arXiv preprint arXiv:2203.09669
6. Hosseini, E. et al.: A low cost EDA-based stress detection using machine learning. In: IEEE International Conference on Bioinformatics and Biomedicine, pp. 2619–2623 (2022). https://doi.org/10.1109/BIBM55620.2022.9995093.
7. Jaber, D., et al.: Medically-oriented design for explainable AI for stress prediction from physiological measurements. BMC Med. Inform. Decis. Mak. **22**, 1 (2022). https://doi.org/10.1186/s12911-022-01772-2
8. Anusha, A.S., et al.: Electrodermal activity based pre-surgery stress detection using a wrist wearable. IEEE J. Biomed. Health Inform. **24**(1), 92–100 (2020). https://doi.org/10.1109/JBHI.2019.2893222
9. Campanella, S., Belli, A., Pierleoni, P., Palma, L.: A method for stress detection using empatica E4 bracelet and machine-learning techniques. Sensors **23**(7), 3565 (2023). https://doi.org/10.3390/s23073565
10. Tiwari, S., Agarwal, S.: A Shrewd Artificial Neural Network-Based Hybrid Model for Pervasive Stress Detection of Students Using Galvanic Skin Response and Electrocardiogram Signals **9**(6), 427–442 (2020). https://doi.org/10.1089/big.2020.0256
11. Sánchez-Reolid, R., et al.: Deep Support Vector Machines for the Identification of Stress Condition from Electrodermal Activity. **30**, 7 (2020). https://doi.org/10.1142/S0129065720500318
12. Viqueira Villarejo, M., García Zapirain, B., Méndez Zorrilla, A.: A stress sensor based on galvanic skin response (GSR) controlled by ZigBee. Sensors **12**, 6075–6101 (2012). https://doi.org/10.3390/S120506075
13. Acerbi, G., Rovini, E., Betti, S., et al.: A Wearable System for Stress Detection Through Physiological Data Analysis (2016). https://link.springer.com/chapter/10.1007/978-3-319-54283-6_3
14. Aqajari SAH, Naeini EK, Mehrabadi MA, et al: GSR Analysis for Stress: Development and Validation of an Open Source Tool for Noisy Naturalistic GSR Data. In: arXiv: Human-Computer Interaction (2020). https://arxiv.org/pdf/2005.01834.pdf. Accessed 20 Mar 2024
15. Reyes Galaviz, R.S.: Análisis con electroencefalografía (EEG) de la escucha de música para el estudio de estrés académico, http://inaoe.repositorioinstitucional.mx/jspui/handle/1009/2048. Accessed 20 Mar 2024

16. Shen, F., Chao, J., Zhao, J.: Forecasting exchange rate using deep belief networks and conjugate gradient method. Neurocomputing **167**, 243–253 (2015)
17. Bezdek, J.C., Ehrlich, R., Full, W.E.: FCM: the fuzzy c-means clustering algorithm. Comput. Geosci. **10**, 191–203 (1984). https://doi.org/10.1016/0098-3004(84)90020-7
18. Schmidt. P., Reiss. A., Duerichen. R., et al.: Introducing WESAD, a multimodal dataset for wearable stress and affect detection. In: International Conference on Multimodal Interfaces (2018). https://dl.acm.org/doi/pdf/10.1145/3242969.3242985

# Transfer Learning and AutoML as a Support for the Pneumonia Diagnosis Using Chest X-ray Scan

Antonio Agliata[1,2](✉), Francesco Bardozzo[1], Salvatore Bottiglieri[2], Giovanni Marco Di Vincenzo[1,2], Mariacarmen Sorrentino[2], and Roberto Tagliaferri[1]

[1] Dipartimento di Scienze Aziendali, Management and Innovation Systems, Università degli Studi di Salerno, 84084 Fisciano, Italy
{aagliata,fbardozzo,gdivincenzo,robtag}@unisa.it
[2] BC Soft, Centro Direzionale, Via Taddeo da Sessa Isola F10, 80143 Napoli, Italy
{salvatore.bottiglieri,mariacarmen.sorrentino}@bcsoft.net

**Abstract.** In the medical field, image classification is crucial for identifying respiratory diseases. Researchers propose combining the NIH Chest X-rays dataset with the Chest X-Ray Images dataset, creating a 14-class dataset including conditions like pneumonia, viral infections, and COVID-19. The goal is image classification.

Given that an image can have multiple labels, the problem is treated as multi-label classification, transformed into a binary classification where each sample can have one or more labels. In AutoML, this task is framed as multiclass classification.

The primary techniques employed are transfer learning and AutoML. Transfer learning fine-tunes pre-trained CNN models on the target dataset, while AutoML optimizes the entire process using automated pipelines. Existing studies differ in their focus, some employing transfer learning on all 14 classes, while others using AutoML for only three.

Evaluation metrics like AUROC, precision, F1-score, and recall are used to assess performance, providing insights into accuracy and prediction quality. By comparing results from transfer learning and AutoML, researchers aim to determine the most effective approach for accurately classifying respiratory diseases in medical images.

**Keywords:** AutoML · multilabel · multi classification · medical image · NIH Chest X-ray

## 1 Introduction

The primary objective of this article is to offer a comprehensive overview of the current state-of-the-art in utilizing Deep Learning for medical diagnostics using convolutional neural networks (CNNs). The focus is primarily on comparing two methodologies: the conventional approach, transfer learning, which has been extensively studied in the literature, and the innovative approach, AutoML. Two models and methodologies are compared: the first involves the use of a pre-trained network for feature extraction of images resulting in a binary relevance approach using Transfer Learning [11], the other uses AutoML.


M. Vettoretti et al. (Eds.): CIBB 2023, LNBI 14513, pp. 321–329, 2025.
https://doi.org/10.1007/978-3-031-90714-2_23

COronaVIrus Disease 19 (Covid-19), is a class defined in relation to the presence of Pneumonia in an advanced and therefore viral state, as reported by various medical studies [18].

The initial dataset employed in this study is a widely utilized dataset of chest radiographic images, NIH Chest X-rays, which has been extensively referenced in the literature on transfer learning articles [1–10, 14].

## 2 Data and Methods

### 2.1 Data

The dataset used in the experiments consists of chest X-ray images, NIH Chest X-rays, (https://www.kaggle.com/datasets/nih-chest-xrays/data), together with the Chest X-Ray Images dataset (https://www.kaggle.com/datasets/paultimothymooney/chest-xray-pneumonia) for identifying the patient with Covid and/or interstitial pneumonia [1–9].

Specifically, the National Institutes of Health (NIH) Chest X-rays database released by the National Institutes of Health Clinical Center includes 108,948 frontal view X-ray images of 32,717 unique patients, where the images are labelled into 14 classes of pulmonary diseases and a fifteenth, undefined class. Labels are extracted from text (where each image can have several labels), associated radiology reports using natural language processing, NLP with 90% accuracy (Table 1).

**Table 1.** Stats NIH X-Ray dataset.

| Labels | Count before pre-processing | Count after pre-processing |
|---|---|---|
| Atelectasis | 11559 | 4893 |
| Cardiomegaly | 2776 | 1397 |
| Effusion | 13317 | 5345 |
| Infiltration | 19894 | 8042 |
| Mass | 5782 | 3017 |
| Nodule | 6331 | 3602 |
| Pneumonia | 1431 | 608 |
| Pneumothorax | 5302 | 2158 |
| Consolidation | 4667 | 1446 |
| Edema | 2303 | 269 |
| Emphysema | 2516 | 1188 |
| Fibrosis | 1686 | 1300 |
| Pleural_Thickening | 3385 | 2200 |
| Hernia | 227 | 159 |
| No Finding | 60361 | 22452 |

The second COVID19 chest X-ray dataset collected by Dr. Joseph Paul Cohen of the University of Montreal is composed of anterior and posterior chest images of patients, comprises lung diseases including COVID-19, and X-rays of healthy patients.

It consists of 5,232 chest X-rays of which 3,883 images are bacterial (2,538) and viral (1,345) pneumonia specimens, the others, 1,349, specimens are healthy lung X-ray images.

The age range of COVID-19 cases in the is 18–75 years (Table 2).

**Table 2.** Stats Covid X-Ray dataset

| Labels | Count before pre-processing | Count after pre-processing |
|---|---|---|
| Covid-19 | *584* | 128 |

The NIH dataset is correlated with two csv files containing labels in relation to the radiographs, specifically the Data_Entry_2017.csv file and the BBox_List_2017.csv file, specifically the Data_Entry_2017.csv file has the following labels:

Image Index, Finding Labels, Follow-up #, Patient ID, Patient Age, Patient Gender, View Position, Original Image Size and Original Image Pixel Spacing.

The second dataset, too, is accompanied by the metadata.csv file, which contains the following labels: patientid, offset, sex, age, finding, RT_PCR_positive, survival, intubated, intubation_present, went_icu, in_icu, needed_supplemental_O2, extubated, temperature, pO2_saturation, leukocyte_count,neutrophil_count, lymphocyte_count, view, modality, date, location, folder, filename, doi, url, license, clinical_notes,other_notes.

Subsequently, the two datasets were combined by extracting the "pneumonia/viral/COVID19" positive examples and healthy patient images from the COVID dataset, and incorporating them into the NHI dataset, which originally consisted of 14 lung diseases and healthy patients. The result is a dataset of 45,235 images, made up of 16 labels: 14 from the NHI, 1 from the covid dataset, 1 representing patients not affected by any of the above.

This is made up of seventeen columns, one containing the path relating to the radiographic image, the others each corresponding to one of the different pathologies, as well as a column representing the patients not affected by any of the diseases examined. Each row of these columns is composed as a binary vector in which the presence of 1 or a 0 in a column respectively indicates whether or not that image is labeled with that given pathology. The Table 3 summarizes the cardinalities of the single classes. This dataset was partitioned into two subsets, train and test set, with a ratio of 80:20 in a stratified manner, i.e. maintaining for each class the percentage of positive examples in both.

The final dataset, the union of the two datasets described above, will be represented by a single table containing the path to the front view image, PA, in the folder and the image class.

More precisely, it will contain 16 classes: No Finding, Covid19, Atelectasis, Cardiomegaly, Effusion, Infiltration, Mass, Nodule, Pneumonia, Pneumothorax, Consolidation, Edema, Emphysema, Fibrosis, Pleural_Thickening, Hernia.

The final size of the images, that the neural network use, will have of 512 × 512 and will be normalized.

**Table 3.** Stats Covid X-Ray dataset

| Labels | Count |
|---|---|
| Covid-19 | 128 |
| Atelectasis | 4893 |
| Cardiomegaly | 1397 |
| Effusion | 5345 |
| Infiltration | 8042 |
| Mass | 3017 |
| Nodule | 3602 |
| Pneumonia | 608 |
| Pneumothorax | 2158 |
| Consolidation | 1446 |
| Edema | 269 |
| Emphysema | 1188 |
| Fibrosis | 1300 |
| Pleural_Thickening | 2200 |
| Hernia | 159 |
| No Finding | 22452 |

## 2.2 Method

Transfer learning (TL) is a specialized field within machine learning (ML) and deep learning (DL) that leverages pre-trained models and customizes them by incorporating new layers. This approach is widely employed in disease diagnostics, particularly in the context of medical image analysis, as it enables supervised classification with reasonably high accuracy. Indeed, in the existing literature, TL has found successful applications in various anatomical domains, including skin, ocular retina, heart, breast, brain, abdomen, and prostate. (see Fig. 1).

This means that a given image can belong to more than one class. To address this, we employ the binary relevance method, which is a well-established approach in the literature.

The binary relevance method follows a 'divide et impera' (divide and conquer) approach. For each class 'c' in the set of classes 'C' with a cardinality of $|C| = m$, a binary classifier 'Bc' is allocated. This classifier predicts the probability of the image belonging or not belonging to class 'c.' To determine the classes to which a new image belongs,

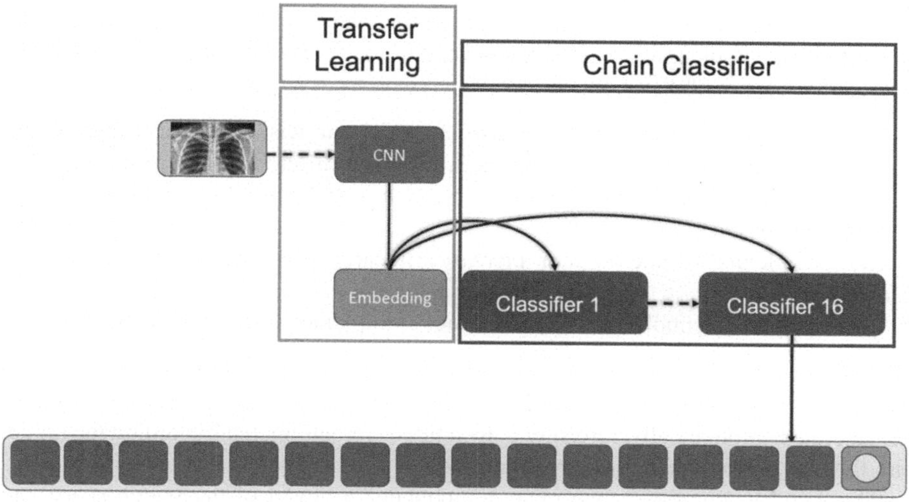


**Fig. 1.** Neural Network Architecture

we query all the classifiers and merge their predictions using a criterion selected from a wide range of criteria already utilized in the literature. The choice of this methodology is based on the following reasons:

The ability to handle multi-label classification problems. The ease of implementation and interpretation. The availability of established evaluation criteria in the literature, enabling a thorough performance evaluation (see Fig. 2).

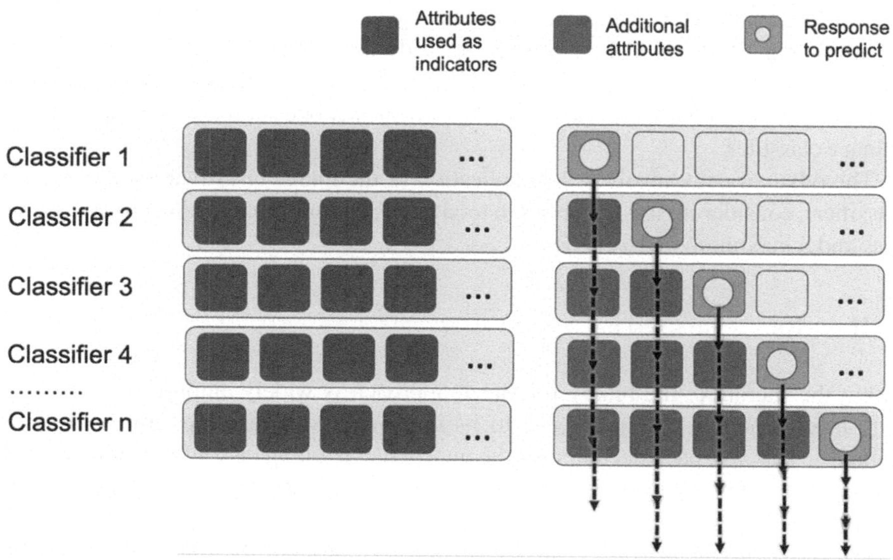


**Fig. 2.** Process Transfer Learning

By adopting this methodology, we aim to effectively address the challenges posed by multilabel classification in our study of lung diseases. The reasons for choosing this methodology are the following:

- It is agnostic with respect to the classifier model one wishes to use, which allows one to experiment both with neural networks of a different nature and with other models and, possibly, allows the use of different model ensembles, homogeneous or heterogeneous.
- The complexity of the system increases linearly with the number of classes considered.
- It is conceptually simple and therefore faster to implement than other methodologies.

Automatic Machine Learning (AutoML) attempts to use machine learning to automate itself. Data is passed to the AutoML application in an unprocessed form and models are generated automatically, referring to techniques for automatically finding the model with the best performance for a given data set.

The AutoML comprises two main components: a model of the search space and an optimiser. Initially, AutoML was only applied to standard prediction problems, i.e., single-label problems, but in recent years it is also being applied to muti-label problems.

AutoKeras will be used, which is an implementation of AutoML using the Keras API1, in particular the tf.keras API provided by TensorFlow 2, for deep learning.

We will use two distinct initialization strategies. First of all, the network will be initialized with pre-trained weights, using a transfer learning approach with multi-label, by comparing three pre-trained models of the CNN type (DenseNet, ConvNet and EfficientNet) [16].

Thanks to the sci-kit-multi-learn library it is possible to use binary relevance methods already implemented, in particular the classifier chain is used as it takes into account the relationships that exist between the variables.

In this regard, the model chain is composed of single multilayer Perceptrons. We have, as input layer, a number of nodes equal to the number of classes considered, an internal layer of 10 neurons and an output layer with only one neuron, since it implements a binary classifier.

The other strategy involves the application of the image classification API (Image Classifier), considering the 16 classes in total and each sample can belong to at least one class and a maximum of 16 classes.

## 3 Results

Within the literature, the binary relevance approach is widely employed for tackling multi-label problems, primarily due to its intuitiveness in learning from multi-label examples. This approach decomposes the multi-label learning task into several independent tasks with binary learning, with each task corresponding to a specific class label [11–14] (see Fig. 3).

Moreover, this methodology can be extended to incorporate the relationships between different labels. By considering these relationships, the binary relevance approach can

| | Model Name | AUROC | Accuracy | f1 Score |
|---|---|---|---|---|
| 0 | CustomNet | 0.740989 | 0.862637 | 0.259044 |
| 1 | DenseNet121 | 0.779881 | 0.866486 | 0.271429 |
| 2 | ResNet50 | 0.716902 | 0.854088 | 0.219912 |
| 3 | Inception | 0.650021 | 0.846562 | 0.151924 |
| 4 | Vgg16 | 0.671470 | 0.848543 | 0.164998 |

**Fig. 3.** Result metrics of paper [33].

capture dependencies and correlations among the labels, leading to improved performance in multilabel classification tasks. The extensions ensure that the binary classifiers not only make predictions for individual labels but also leverage the information from related labels to enhance the overall classification accuracy.

By utilizing the binary relevance approach and considering label relationships, our methodology aims to provide a comprehensive and effective solution for addressing the complexities of multi-label classification in the context of lung disease recognition. The preliminary results achieved, compared with those present in the literature [1, 2, 14, 15]:

The system used consists of two elements: a Convolutional Neural Network (CNN) among those selected. Its purpose is to take an input image and produce an embedding of the image.

This is achieved by removing the output layer and the remaining fully-connected layers in the original architecture, and retrieving the output emitted by the flattening layer. The obtained embedding has a cardinality that depends on the chosen CNN: 1024 for ConvNet, 2560 for EfficientNet, and 1920 for DenseNet. [14, 17, 18].

This embedding is then fed as input to a chain of binary classifiers, each implemented as a feed-forward fully connected neural network with a single hidden layer containing 100 nodes and ReLU activation function. The chain approach transforms a multi-class multi-label classification problem into a series of binary classification problems, capturing the potential relationships between the different classes. This is made possible because each classifier, except the first one, takes as input a vector containing the image embedding obtained from the CNN, along with the outcomes of the preceding classifiers in the chain. Specifically, for each i = 2,…,16, the i-th classifier in the chain takes as input a vector Vi with a cardinality equal to the sum of the embedding cardinality and (i-1), and outputs a value indicating whether the processed image belongs to the i-th class. The chain's output is a boolean vector, where each position corresponds to the outcome of the corresponding classifier.

In summary, the system employs a CNN to generate image embeddings, followed by a chain of binary classifiers that utilize the embeddings and the outcomes of preceding classifiers to classify the image into multiple classes. This approach effectively captures the interdependencies among the different classes.

The study presents comprehensive results, including detailed information on the neural network architecture, training, and validation processes. Table 4 displaying the results of various tests conducted is also provided.

**Table 4.** Result metrics

| Test Metrics | AUROC | Accuracy | Precision | Recall | F1Score |
|---|---|---|---|---|---|
| DenseNet | 0.7886 | 0.8376 | 0.8710 | 0.9360 | 0.9023 |
| AutoML | 0.9457 | 0.9654 | 0.9784 | 0.9745 | 0.9764 |

## 4 Conclusion

In this research, a custom dataset of radiographic chest images representing various lung diseases was created. A CNN architecture and AutoML techniques were utilized to develop a specialized model for automatic classification of lung images across 16 classes. The optimized architecture achieved state-of-the-art results for all 16 classes, although performance varied when analyzing different subsets of the dataset. Cross-validation was employed to ensure the model's robustness and generalizability.

The primary advantage of using deep neural networks is their ability to learn complex patterns directly from data without the need for explicit feature engineering. However, in certain cases, prior knowledge or a priori information played a critical role, particularly in accurately segmenting lung profiles. Integrating such information proved crucial for reliable discrimination in datasets with intricate image characteristics, while it was unnecessary or detrimental in other cases.

Future research will explore alternative models and methodologies to leverage label dependencies and incorporate segmentation information. These efforts aim to further improve the performance and applicability of the proposed approach in the classification of lung diseases based on medical images. [14, 18].

## References


1. A Survey: Identification of Different Thoracic Disease Using Convolutional Neural Network, 2021 IJRTI V6
2. Agliata – M.Sorrentino Deep Learning with TensorFlow2, 2021
3. Blog: https://lukeoakdenrayner.wordpress.com/2018/01/24/chexnet-an-in-depth-review/
4. ChestX-ray8: Hospital-scale Chest X-ray Database and Benchmarks on Weakly-Supervised Classification and Localization of Common Thorax Diseases
5. Oyelade, O.N., Ezugwu, A.E.: Deep Learning Model for Improving the Characterization of Coronavirus on Chest X-ray Images Using CNN di ORCID Profile
6. Gonzales, B., et al.: UMLS-ChestNet: a deep convolutional neural network for radiological-ndings, differential diagnoses and localizations of COVID-19 in chest x-rays. https://arxiv.org/pdf/2006.05274.pdf

7. Yue, H., et al.:
8. http://openaccess.thecvf.com/content_cvpr_2017/papers/Wang_ChestX-ray8_HospitalScale_Chest_CVPR_2017_paper.pdf
9. Lloyd-Jones, D.G.: Chest x-ray anatomy. Salisbury NHS Foundation Trust UK
10. Alipoor, S.D., Jamaati, H., Tabarsi, P., Mortaz, E.: Immunopathogenesis of Pneumonia in COVID-19. Tanaffos. 2020 Nov;19(2):79–82. PMID: 33262791; PMCID: PMC7680509
11. Tahir, M.A., Josef Kittler, J., Yan, F.: Pages 3738–3750, ISSN 00313203,https://doi.org/10.1016/j.patcog.2012.03.014
12. Tarekegn, A.N., Giacobini, M., Michalak, K.: A review of methods for imbalance d multi-lab el classification
13. Blais, M.-A., Akhloufi, M.A.: Deep learning and binary relevance classification of multiple diseases using chest X-Ray images. In: 2021 43rd Annual International Conference of the IEEE Engineering in Medicine & Biology Society (EMBC), Mexico, 2021, pp. 2794–2797 (2021)
14. Moradi, M., Madani, A., Karargyris, A., Syeda-Mahmood, T.F.: Chest X-Ray generation and data augmentation for cardiovascular abnormality classification. Proc. SPIE **10574** (2018). 105741M, https://doi.org/10.1117/12.2293971
15. Kermany, D.S., et al.: Identifying Medical Diagnoses and Treatable Diseases by Image-Based Deep Learning. https://doi.org/10.1016/j.cell.2018.02.010
16. https://keras.io/api/applications/
17. Tahir, M.A., Kittler, J., Bouridane, A.: Multilabel classification using heterogeneous ensemble of multi-label classifiers. Pattern Recogn. Lett. https://doi.org/10.1016/j.patrec.2011.10.019
18. Salehi, M., Mohammadi, R., Ghaffari, H., Sadighi, N., Reiazi, R.: Automated detection of pneumonia cases using deep transfer learning with paediatric chest X-ray images. Br. J. Radiol. **94**(1121), 20201263 (2021).https://doi.org/10.1259/bjr.20201263. Epub 2021 Apr 16. PMID: 33861150; PMCID: PMC8506182

# Author Index





M. Vettoretti et al. (Eds.): CIBB 2023, LNBI 14513, pp. 331–332, 2025.
https://doi.org/10.1007/978-3-031-90714-2

The manufacturer's authorised representative in the EU is Springer Nature Customer Service Centre GmbH, Europaplatz 3, 69115 Heidelberg, Germany. If you have any concerns regarding our products, please contact ProductSafety@springernature.com

Printed and bound by CPI Group (UK) Ltd, Croydon, CR0 4YY
30/09/2026
02225560-0010